Britton's Botanical Empire

The New York Botanical Garden and American Botany, 1888–1929

Britton's Botanical Empire

The New York Botanical Garden and American Botany, 1888–1929

PETER MICKULAS

Memoirs of The New York Botanical Garden, Volume 94
The New York Botanical Garden Press

Published by The New York Botanical Garden Press
200th Street & Kazimiroff Boulevard
Bronx, New York 10458-5126 USA

The paper used in this publication meets the requirements of the National Standards for Information Sciences-Permanence of Paper for Publications and Documents in Libraries and Archives, ANSI/NISO Z39.48-1992.

Printed in the United States of America.

MetLife Foundation is a leadership funder of The New York Botanical Garden Press.

Published on 26 October 2007.

International Standard Serial Number 0077-8931
ISBN-13: 978-0-89327-479-5
ISBN-10: 0-89327-479-8

CONTENTS

Acknowledgments and Dedication

I have incurred many debts while writing this history of the The New York Botanical Garden. I am especially grateful to Susan Schrepfer, my dissertation committee chair at Rutgers University. I could not have asked for a better intellectual mentor or a warmer or more generous advisor. Similarly, I am thankful to Philip Pauly, William Buck, Gerald Grob, and Kim Kleinman. Phil Pauly's patient counsel and expertise in the history of American science greatly enhanced my ability to say something useful about New York botany, whereas Bill Buck of the Botanical Garden made sure that what I tried to say was as accurate as possible. (Of course, the standard disclaimer applies here: any errors that might remain in this work are my own responsibility.) Gerry Grob, in active retirement, read my entire manuscript in various draft versions and provided invaluable encouragement and insight. Kim Kleinman's guidance from afar in St. Louis was tremendously encouraging. His deep knowledge of the history of botany enriched my own understanding of the subject; I am thankful also for our continuing friendship. Thanks also go to Jim Livingston for invaluable advice at the start of this project; Carla Yanni for key tutelage in the history of architecture; and Sharon Kingsland for sharing her knowledge of the history of the Garden and of American botany. I greatly appreciate the generosity of each of these scholars.

The entire staff of The New York Botanical Garden Library and Archives deserves recognition, especially LuEsther T. Mertz Library Director Susan Fraser, Archivist Stephen Sinon, and Librarian Marie Long. My many months of research in the Garden's Library's reading room were made enjoyable as well as productive by the warmth and graciousness of these dedicated professionals. Further thanks are due Sandi Frank, Nate Smith, and The New York Botanical Garden Press staff for helping shepherd this manuscript from dissertation to book form. Marlie Wasserman of Rutgers University Press deserves recognition for helping me navigate the business of academic publishing. I thank John Reed, former Mertz Library Director, for his helpful assessment of my manuscript, and Vassiliki Betty Smocovitis, whose reader's report and subsequent editorial direction as I revised the dissertation for publication were invaluable. Betty's expertise in the history of botany is unsurpassed, and I benefited immeasurably from her assistance as I attempted to place the founding of the Garden within the context of the intellectual history of American science.

I am grateful to the numerous library professionals who assisted me in my work outside The New York Botanical Garden, including the staffs of the following institutions: Rutgers University Libraries; Smithsonian Institution; Columbia University; Pierpont Morgan Library (now Morgan Library & Museum); the office of the Architect of the Capitol, Washington, D.C.; Carnegie Institution of Washington; Arnold Arboretum of Harvard University; New-York Historical Society; Municipal Archives of the City of New York; Rockefeller Archives Center; New York Academy of Medicine, and the public library of Floral Park, Long Island. I am grateful to both John Dixon Hunt of the University of Pennsylvania and the late Daniel Goodwin, editor of *New York History*, for editorial advice on published versions of these chapters (and for permission to use the previously published material in this book). Thanks for fellowship support go to Rutgers University and the United States Capitol Historical Society.

For friendship, encouragement, and hospitality as this project progressed in various forms, I thank Kristian Eshelman, Peter Tan, Eric Groepler, Alex Bain, Finis Dunaway, Sarah Buck, Dana Capell, Flip Herndon, Todd Uhlman, Anne Dobmeyer, Jason Rimmer, Rachel Campbell, Paul Benzon, Shamim Taherzadeh, and Wayne Wink.

I would like especially to acknowledge the support and love of my entire family, whose pride in my work inspired me more than I can adequately express. Most importantly, I thank my wife, Mary Katherine Gilbert, an astute reader and gentle critic, whose loving encouragement, patient advice, and, of course, financial support made this possible. I dedicate this work to her and our son, Benjamin, with love.

Britton's Botanical Empire

The New York Botanical Garden and American Botany, 1888–1929

Chapter 1
An Introduction to The New York Botanical Garden, 1888–1929

In a 1933 radio interview, Nathaniel Lord Britton, botanist and founding director of The New York Botanical Garden (NYBG), then retired, recounted the now often-quoted story of his wife's 1888 reaction to the beauty and prestige of the Royal Botanic Gardens, Kew, in England during their honeymoon in 1888. Elizabeth Gertrude Knight Britton reportedly asked, "Why couldn't we have something like this in New York?"[1] Although not officially documented at the time, this story serves as a fitting starting point, taking place as it did in an idealized garden, for consideration of the origin of NYBG's history. For the Brittons and their metropolitan New York contemporaries, Kew represented, metaphorically speaking, an Eden, although not in the biblical sense. Their New York garden would be a recreation of Kew built within their own diverse North American metropolis in order to serve an array of public and private needs.

By establishing NYBG when he did, Britton helped create a crucial venue for the pursuit of professional botanical research. Even as the Garden served as a key part of the park system of the City of New York, it remained, in conception and in practice, primarily a scientific institution, dedicated in the Britton years to continuing the unfinished taxonomic tasks of Western Hemisphere botany. At once admiring of the success of British botany and desirous of establishing scientific independence from Old World botanical knowledge, Nathaniel Britton acted upon his wife's idea and set about building a New York "Kew." He did so within the context of late-nineteenth-century New York City, a metropolis on the verge of consolidation, and whose civic elite remained wedded to environmentalist ideas regarding the social powers of landscaped parks, while harboring a patrician desire for impressive public spaces.

1 Original interview recorded October 27, 1933; see Nathaniel Lord Britton Vertical File, New York Botanical Garden Library. This origin account is chronicled, for example, in Ogden Tanner and Adele Auchincloss, *The New York Botanical Garden: An Illustrated Chronicle of Plants and People* (New York: Walker and Company, 1991): 48.

Botanical Gardens in Western Context

As landscapes of scientific and utilitarian activity, botanical gardens have a long history in Western culture. Although Britton's garden was entirely secular in character, the origins of western gardens can be traced to biblical ideas. During the fourth, fifth, and sixth centuries of the current era, Christian Latin writers began the process of fusing Greco-Roman creation myths with the biblical myths of Eden found in the Book of Genesis. Typical themes in these poetic descriptions became familiar motifs in Western conceptions of Edenic garden settings: lush foliage, brightly colored meadows, fragrant plants, and abundant water.[2] Though some writers early in the Christian era considered this Genesis account to be a figurative one, the overwhelming majority of westerners, led by the literalist examples of Saints Augustine and Thomas Aquinas, "chose realism when it came to the paradise of Adam and Eve."[3] In fact, through the medieval period, into the Renaissance and even the eighteenth century, historian Jean Delumeau notes, there was a strong strain of literalism in biblical interpretations of the creation myth, with a concomitant understanding of Eden as an actual, historical place. Eden, it was believed, still existed on earth, though in an inaccessible or unrecognizable location.[4]

Medieval commentators conceived of this original garden as "a place... in which trees of different kinds had been planted that met all possible needs."[5] Renaissance views of gardens underscored their physical beauty and resemblance to Eden, while considering them suitable refuges for those sensitive enough to appreciate such charms. Descended from the typical contours of the medieval garden, "an open area within a mass, generally organized around a center," Renaissance gardens maintained these essential elements up until the development in the eighteenth century of the English picturesque ideal. Such gardens, even in sacred settings such as the cloister plots from which they derived, served a very real function as sites for the provision of food and medicine.[6] Thus an earthly paradise, granted historical meaning and existence in the mainstream of Western thought at least until the Enlightenment, had unmistakably utilitarian aspects, which might be recreated in part through clever management and design of land, water, and plants. In this way, gardens in both sacred and secular settings were spaces for production, providing the necessities of human life. This was the role, for example, of "physick gardens," found throughout Europe by the seventeenth century, where plants from various corners of European empires could be gathered, grown, and studied. The cultivators of these plots managed to

2 Jean Delumeau, *History of Paradise: The Garden of Eden in Myth and Tradition* (Urbana: University of Illinois Press, 2000): 13.

3 Ibid., 18–19.

4 Ibid., 155.

5 Honorius of Autun is cited in Delumeau, *History of Paradise*, 20, see note 2.

6 Terry Comito, *The Idea of the Garden in the Renaissance* (New Brunswick: Rutgers University Press, 1978): 1–3, 25.

combine the earliest rudiments of an experimental method with a vestigial belief in magic in their efforts to alleviate various human sufferings. In the process, these gardeners heeded Francis Bacon's injunction regarding the creation of a utilitarian human dominion over nature.[7]

The earliest of these enclosed spaces that might be termed "botanic gardens" in a modern sense had been laid out in various European cities during the sixteenth and seventeenth centuries. Most were constructed near universities. Botanic gardens were built at Padua and at Pisa in 1543, at Florence in 1545, at Bologna in 1567, at Leiden in 1587, at Heidelberg and Montpellier in 1593, at Oxford in 1621, and at Paris in 1626. At these university locations, the gardens added another aspect to their traditional spiritual focus: they became utilitarian centers of study, teaching, and experimentation for the discovery of the "precise requirements of the cultivation of medicinal herbs." Often, common aesthetic devices such as statuary and fountains gave way to regular, rectangular beds designed to afford maximum efficiency in cultivation.[8]

A British Model: Creating a Botanical Empire

Even as a literal understanding of the biblical creation myth waned and innovative garden ideas developed, the essential utilitarian conception of space embodied in Western gardens endured. It found its most notable expression in the system of botanical transfer stations headquartered at Kew and run by Joseph Banks after 1772.[9] The British botanical empire, taking its cue from Continental models, involved the creation of a central repository of botanical specimens, both live and dried, thus making of the Royal Botanic Gardens a museum of imperial raw materials and a source of critical scientific advice for colonial administrators.[10] By the late 1830s, however, this botanic empire was threatened with irrelevance and disuse. In 1837, of the 10 British imperial botanic gardens in existence, only eight remained active and two were inactive; eight others, dating from the 1700s, had disappeared entirely. Historian Donal McCracken attributes this tenuous position for British botany to a growing "little England" movement. Spurred on by the Irish famine and free trade, the English Whig ruling class began to sour on the concept of empire, with some referring to

7 Richard Drayton, "Imperial Science and a Scientific Empire: Kew Gardens and the Uses of Nature, 1772–1903" (PhD dissertation, Yale University, 1993): 28–33.

8 Lucia Tongiorgi Tomasi, "Botanical Gardens of the Sixteenth and Seventeenth Centuries," in *The Architecture of Western Gardens: A Design History from the Renaissance to the Present Day*, Monique Mosser and Georges Teyssot, eds. (Cambridge, MA: MIT Press, 1991): 81–82.

9 See Drayton, "Imperial Science,"note 7. Useful overviews of the development of landscape naturalism, one of these innovations, can be found in Paul Shepard, *Man in the Landscape: A Historic View of the Esthetics of Nature* (New York: Knopf, 1967): 72–88.

10 Drayton, "Imperial Science," i–vi, see note 7. See also Ray Desmond, *Kew: The History of the Royal Botanic Gardens* (London: Harvill Press, 1995).

colonies as "millstones." In this political environment, 1830s British botanic gardens had been known to devolve into mere "governors' allotments," providing firewood for the hearths of colonial administrators.[11]

William Hooker's arrival at Kew in 1841, and his son Joseph Dalton Hooker's subsequent directorship from 1865–1885, saved British imperial botany. By the last decades of the nineteenth century, the Hookers had provided a clear, expansive model of nationalistic botany for Nathaniel Britton and his New York cohort to emulate.[12] Britton recognized Kew as a large network of colonial botanic stations involved in a coordinated transfer of plants from the tropics to Europe as well as to other colonial stations. Kew botanists thus coordinated efforts to cultivate species of economic value, including cinchona and new varieties of cane sugar.[13] By then, British botanic stations in India had been reorganized and were helping to administer a botanical survey of that subcontinent, while the colonial garden at Ceylon had become one of the largest of the entire Empire. Thus, by the time the Brittons arrived in London in 1888, Kew administrators, confident of their role at the center of imperial science and economy, had been training botanists and horticulturists, conducting basic taxonomic research, acclimatizing plants, publishing journals, and exchanging herbarium specimens for decades.[14]

The New York Botanical Garden: An American Space for Taxonomy

Britton attempted to establish NYBG on this British model of nationalistic botany even as British botanical colonialism was entering a new phase of development. Kew's research agenda began shifting its focus from broad floristic surveys toward laboratory studies of plant genetics and pathology. As a protégé of Columbia College geologist J. S. Newberry, Britton's career straddled the era of United States (US) natural history and the development of modern academic science. He came of age professionally at the height of the British colonial era and viewed British taxonomic research as a suitable institutional model to advance the scientific stature of his native city. Given his age and background, he understood a major portion of a public garden's mandate to include a large, well-planned set of taxonomic and economic museum displays, based upon Victorian notions of the pedagogic power of objects arranged in orderly glass cases.[15]

11 Donal McCracken, *Gardens of Empire: Botanical Institutions of the Victorian British Empire* (London: Leicester University Press, 1997): 1–18.

12 Ibid., 18–24.

13 D. R. Headrick, *The Tentacles of Progress: Technology Transfer in the Age of Imperialism, 1850–1940* (New York: Oxford University Press, 1988): 211; 216–217.

14 Drayton, "Imperial Science," 6, see note 7; McCracken, *Gardens of Empire*, 80, see note 11.

15 See Steven Conn, *Museums and American Intellectual Life, 1876–1926* (Chicago: University of Chicago Press, 1998).

Given his timing at the end of the nineteenth century, should it be argued that Britton's conception of the proper facets of a botanical garden and museum were rooted in a scientific system that was actually obsolete? According to one historiographic portrait of late nineteenth century American science, in the decades following the Civil War there occurred a dramatic decline of the "natural history tradition," a mode of inquiry characterized by the participation of avocational naturalists seeking to compile a descriptive compendium of knowledge about a wide variety of botanical, zoological, and geological specimens. Observation, description, and classification were the primary methodologies of this intellectual framework, with the ultimate, implicit end being, prior to Darwinism, the iteration of a coherent, defensible, natural theology (and post-Darwin, the delineation of phylogenetic relationships). This naturalist tradition, "the dominant view from ancient times through most of the nineteenth century," in general involved an interest "in whole organisms as opposed to their isolated parts," and "took its cue from field naturalists . . . concerned with organisms in their natural state, while [simultaneously] encompass[ing] museum studies of preserved specimens with an emphasis on taxonomy and classification."[16] These were the activities in which the Brittons were well-versed and that they pursued vigorously throughout their New York careers.

In contrast, historians of science have also identified a "new botany," which had an experimental emphasis and "found particular expression in the study of animal and plant physiology, was tied much more closely to the laboratory, and often viewed organisms wholly removed from their natural surroundings."[17] In his biography of nineteenth century botanist Charles Bessey, Richard Overfield characterized the University of Nebraska professor, administrator, and textbook author as a crucial American "new scientist," stimulated by the development of evolutionary theory to

16 Garland Allen, "The Transformation of a Science: T. H. Morgan and the Emergence of a New American Biology" in A. Oleson and J. Voss, eds., *The Organization of Knowledge in Modern America, 1860–1920* (Baltimore: Johns Hopkins University Press, 1979): 174–175.

17 Ibid. Allen characterized this late-nineteenth-century decline of the natural history tradition as precipitous, hastened by the arrival from Europe (specifically Germany) of an "experimental biology," a methodology marked by a concern with the functional structure of discrete parts of organisms, microscopic observation of cytological components, and compilation of quantitative data garnered from observation in artificially manipulated settings. The experimentalists assumed that "the principles of function determined in experiments with isolated parts (e.g., organs) could be applied to an understanding of the organ's function within the intact animal." They also asserted that information gathered under these conditions could be "extrapolated to the field situation," while consciously modeling their methodologies on those of their counterparts in the physical sciences. "Many experimentalists," Allen noted, "had as their stated aim to make biology as 'scientific' as physics and chemistry." Allen characterized natural history, both pre- and post-Darwin, as concerned largely with the cataloguing of organisms and the delineation of their place in either chains of being or, post-Darwin, with phylogenetic history.

re-evaluate the methodologies and assumptions of his discipline. Concerned with the international status of American science in general, and American botany in particular, Bessey and like-minded colleagues altered the focus of their labor from plant collection and description to evolutionary classification, structural physiology, plant pathology, and ecology.[18] Due in large part to new opportunities available at land grant colleges, experiment stations, botanical gardens, and the US Department of Agriculture, American botanists changed during this period from a group of collectors and identifiers to one of the "most energetic and expanding groups of American scientists." In addition, Overfield asserts, more people were attracted to botany by the controversies surrounding evolutionary theory as well as the increased impact of German educational practice on American schooling. The overall trend in US botany, Overfield writes, was away from taxonomy to a new morphology, physiology, embryology, and pathology. These new disciplines began siphoning young talent away from herbarium botany, as traditional taxonomic tasks became seen by some as the province of antiquarians.[19]

Leading figures in this intellectual trend included Bessey; Lucien Underwood (one the first botanists employed at NYBG); John Coulter, editor of the *Botanical Gazette*; and Missouri Botanical Garden director William Trelease. These men advocated a vision of botany as an experimental and laboratory science, and, Overfield asserts, sparked a wave of imitation of the German stress on investigating cellular structures and tracing evolutionary descent. Classic herbarium botany thus suffered a loss of intellectual prestige in the late nineteenth and early twentieth centuries.[20]

Though these changes were undeniably taking place within both US and European botany, this did not mean that taxonomic work of the sort Britton planned to coordinate from the Bronx was either outdated or inappropriate. First, it must be noted that the sharp disciplinary schism as described above is in some sense a literary interpretive device. In reality, various methods of investigation survived and complemented one another productively and do so up to the present day.[21] In addition, with the development of something called "the new systematics" by intellectually influential scientists, taxonomy by the 1930s and 1940s once again would be situated

18 Richard A. Overfield, *Science with Practice: Charles E. Bessey and the Maturing of American Botany* (Ames, IA: Iowa State University Press, 1993).

19 See Keith Vernon, "Desperately Seeking Status: Evolutionary Systematics and the Taxonomists' Search for Respectability 1940–60," *British Journal for the History of Science* 26 (1993): 207–27.

20 Ibid.

21 These themes are explored in greater detail in Chapter 7. In addition, herbaria remained important and respected centers for botanical teaching and research from Britton's era up to the present day. See R. C. Rollins, "The Role of the University Herbarium in Research and Teaching," *Taxon* 14 (1965): 115–120, cited in Barbara Briggs, "One Hundred Years of Plant Taxonomy, 1889–1989," *Annals of the Missouri Botanical Garden* 78 (1991): 23.

at the center of biologists' attempts to understand evolution.[22]

Of course, Britton's taxonomic work in North America and the US colonization of portions of the Western Hemisphere were only beginning in earnest decades after a more mature British botany had already begun the immense task of compiling floras and monographs of plant families throughout a long-established British empire.[23] Given the later rise of an American political empire, it is only natural that scientific surveys and their concomitant publishing projects took place after similar British efforts. (Britton's studies in the US Southwest, Latin America, and Puerto Rico, which are discussed in Chapters 5 and 6, all fall into this category.) Even though much of the utilitarian space in an American empire was occupied by the US Department of Agriculture, Britton deftly exploited his contacts in the federal scientific bureaucracy and at the Carnegie Institute of Washington to situate NYBG alongside experiment stations and agricultural colleges within the profitable context of practical science. The Garden's role in American overseas adventures should be understood as a specific Brittonian strategy to secure his place in US science. (Given taxonomy's relative decline in prestige during the Garden's early years, his success in institution-building is even more remarkable.)

This was a successful philanthropic strategy also deployed by Henry Fairfield Osborn at the American Museum of Natural History (AMNH). Osborn, Britton's colleague at Columbia, was the nephew of J. P. Morgan and a solid member of the city's wealthiest social class. As such, he was highly successful at salvaging a useful social function for the Museum's vertebrate paleontology program, even after the science itself, which did not lend itself to newer methods and the utilitarian goals of twentieth century American science, had become something of an expensive anachronism. Osborn successfully placed the AMNH at the heart of New York's elite philanthropic project; for the Museum's trustees, his fossils "were important because they were rare, valuable, and visible facts."[24] Osborn thus helped AMNH science retain relevance by underscoring its political and social objectives. Britton, while doing the same, was not faced with the same disciplinary irrelevance. Taxonomic botany as practiced at NYBG remained an important field of inquiry, open to new methods of investigation. To the extent that resources allowed, Britton encouraged these developments at the Garden, even as he continued the classic herbarium work

22 Vernon in "Desperately Seeking Status" portrays the efforts of Ernst Mayr and G. G. Simpson, two architects of what came to be known as the "Evolutionary Synthesis," to rehabilitate the reputation of taxonomy. See Vernon pp. 210–15 and 224–26, note 19. This effort will be discussed in further detail in Chapter 7.

23 Kew, of course, continued directing taxonomic work into the twentieth century and does so at present.

24 R. Rainger, *An Agenda for Antiquity: Henry Fairfield Osborn and Vertebrate Paleontology at the American Museum of Natural History, 1890–1935* (Tuscaloosa: University of Alabama Press, 1991): 20–23.

that characterized his career. As Britton's efforts in both the US Southwest and in the Caribbean colonies attests, botanical study was readily adapted to the utilitarian cast of American science. Chapters 5, 6, and 7 treat Britton's goal of securing a place for Garden botany within this overall context of pragmatic scientific endeavors in the US.[25]

As important as the timing of a US botanical empire in preserving a relevant space for the botanical garden in the life of the city was NYBG's additional role as an urban park. As landscaped public space, the Botanical Garden exhibited elements of a patrician "reform" park of the mid-nineteenth century, designed in accordance with Olmstedian ideals that retained relevance for a significant portion of New York's elites. In light of this element of the Garden's pedigree, it is useful to situate the development of NYBG within the historiographic contexts of the American Gilded Age and Progressive Era. As historian Paul Boyer noted, American Protestantism of the era "intensified its long-standing effort to develop institutional modes that would enable it to exert moral influence . . . over . . . increasingly non-Protestant urban masses," and did so partly through a belief in environmentalist reforms such as park-building, a belief that retained currency into the 1890s.[26] The Brittons, upper-class Protestants themselves, were prone to view the Garden as a moral influence over the city's growing immigrant population. As Chapter 4 suggests, both Nathaniel and Elizabeth were dismayed by any signs, such as vandalism or plant theft, that the objects of their moralizing influence failed to heed the salutary example and refining influence of the Garden.

The Garden as a Professional Research Institution

It is crucial to understand that NYBG was designed primarily as a research institution. Of paramount importance in its development was Britton's own vision of US botanical science, taxonomic research, and the continuation of a century-long tradition of learning within New York City. As a site of intellectual work, Britton's NYBG would have much in common with the natural history museums that had provided him with a model of institutional creators of knowledge.[27] The scientific

25 To underscore Britton's willingness to explore all possible avenues of advancement for the Garden is not to deny that, given his age and his personal background, he was capable of missteps and the occasional anachronistic endeavor. His efforts around the turn of the twentieth century to organize a "Scientific Alliance" and to secure a single building for New York science at that time were clearly outdated and suitable to antebellum New York. See D. Sloan, "Science in New York City, 1867–1907," *Isis* 71: 73 (1980).

26 Paul Boyer, *Urban Masses and Moral Order in American, 1820–1920* (Cambridge, UK: Cambridge University Press, 1978): 121–125.

27 See, e.g., J. M. Kennedy, "Philanthropy and Science in New York City: The American Museum of Natural History, 1868–1968" (PhD Dissertation, Yale University, 1968). Kennedy characterized the American Museum as primarily the product of its trustees' desire to educate a potentially subversive

apparatus within the Garden's public park setting were of primary importance to the institution; the landscape around the Museum and Conservatory itself were critical components of that investigative equipment. The very nationalism Britton advocated in his displays of a quintessentially "American" plant, the cactus, and his concerted efforts to establish US preeminence in New World taxonomy, grew out of his desire to continue a New York (and American) scientific tradition inaugurated decades earlier by botanists David Hosack, John Torrey, Asa Gray, and John Strong Newberry. More than the desire to inculcate Protestant values in an increasingly heterogeneous city populace, Britton's garden-building efforts were driven by a native son's urge to elevate the scientific facilities of his city to world-class status.[28]

Britton's development of NYBG needs to be understood not solely as the product of empire-building (although he looked to an imperial British model for inspiration). Rather, his conception of scientific nationalism, though hospitable to utilitarian motives, centered on his effort to establish US scientific independence from Old World knowledge producers. In Britton's vision, the products of this work, i.e., the specimens of American plants collected by Garden explorers, would be placed in exhibits designed to impress upon visitors both the scope of American botany and the civic grandeur of New York City.

Finally, it is important to note that by developing the Garden when he did, Britton became part of a general transition from an "amateur" to a "professional" New York science. He served as a crucial transitional figure. Though he and Elizabeth Britton were not nearly as wealthy as the Garden's major benefactors, they possessed enough personal wealth to underwrite various Garden functions. A botanist, Britton was also a philanthropist, a position that enhanced his prestige while linking him to a nineteenth century tradition of gentleman practitioners. Mrs. Britton worked in her capacity as taxonomist and curator without salary (although as director-in-chief her husband drew the highest pay of any Garden employee). Britton in part earned his living from science and, thus, was among the city's earliest group of professionals. Yet his significance was primarily as an entrepreneurial institution-builder who made it possible for others (primarily men) to earn middle-class incomes from practicing botany. Britton's collaboration with amateur enthusiast Judge Addison Brown on a

---

public, while bringing civic prestige to New York. The Museum's priorities, Kennedy asserted, were set by philanthropists' physiocratic notions of nature: displays of habitat dioramas were designed to convey an underlying natural order that, in turn, would provide the touchstone for morality of public life.

28 Also, as Chapter 4 demonstrates, such desires for social control would have been hubristic at best anyway. An enthusiastic, multiclass city populace embraced the available new public space and, even after it was developed with clearly demarcated purposes in mind, attempted with considerable success to implement its own multifaceted vision for its use.

monumental flora of the Northeastern US (a text meant to challenge Asa Gray's standard in the field) was followed by his work with professional botanist Joseph Rose of the Smithsonian on a definitive monograph of the cactus family. In the 1890s, Britton sought practical support from a wealthy amateur. In the first decades of the twentieth century, he helped establish the contours of a modern system of science that enabled men (mostly) to earn a living doing botanical work. In this sense, then, Britton's career fostered the professionalization of American botany.

In his lifetime, Nathaniel Britton did not fulfill his stated ambition of displacing Kew as the world's most significant botanical research institution. Nevertheless, his efforts enabled New York's public garden to rank, by the end of his career, as one of the New World's most productive taxonomic research institutions and enabled the city to burnish its image as a cultural capital and center of learning. Yet despite its significance in the history of the city's development as a scientific and cultural center, NYBG has received surprisingly little scholarly scrutiny. Simon Baatz's history of the New York Academy of Sciences contains a brief and valuable portrait of Britton's work, and Douglas Sloan long ago considered the establishment of the Garden within the context of the history of Columbia University.[29] To date, Sharon Kingsland's history of American ecology has been the most thorough examination of the Garden's role in American science; her discussion of Britton's role in the late-nineteenth and early-twentieth-century debates over botanical nomenclature is definitive.[30] This narrative of the Garden's first four decades is partly the story of a concerted effort by a legendarily energetic scientific entrepreneur to build a premier botanical research center. The following chapters examine the major events of the Britton years, while placing Britton's personal ambition and institutional priorities within the context of the development of New York and botanical science from the 1890s up until the Great Depression. They attempt to begin elucidating the history of an important New York and American institution.

29 Simon Baatz, *Knowledge, Culture and Science in the Metropolis: the New York Academy of Sciences, 1817–1970* (New York: New York Academy of Sciences, 1990); Sloan, "Science in New York," 1967, see note 25.

30 Sharon Kingsland, *The Evolution of American Ecology, 1890–2000* (Baltimore: Johns Hopkins University Press, 2005). Overfield's *Science with Practice* (see note 18) contains a similarly excellent account of the nomenclature debates.

Chapter 2
Early Botanical and Horticultural Efforts in New York

In a bitter 1837 letter to Charles Hovey, an influential Boston nurseryman who edited the *Magazine of Horticulture and Botany*, a member of the New-York Horticultural Society placed blame for his organization's recent decline on his neighbors, asserting angrily that while societies in other cities prospered, "New Yorkers [were] so eat up with avarice and immorality that they [could] not afford to patronize a horticultural exhibition."[1] The Society's attempts at displaying horticultural treasures, the irate member noted, had met with singular failure for the preceding five years, as "people [had been] so engaged in the engrossing of fictitious wealth, that they had not time to admire the beauties of nature, except so far as building lots . . . were concerned."[2] With this jeremiad, an old accusation was leveled against New York: its citizens were too consumed with a passion for wealth to cultivate the nobler pursuit of knowledge. In 1837, as Hovey's correspondent witnessed the decline of a formerly thriving organization, this charge seemed to have merit. While the New-York Horticultural Society could barely muster a quorum, a flourishing Massachusetts Horticultural Society was enjoying the return on its fortuitous investment in a landscaped cemetery that served an enthusiastic urban populace as a park. In Philadelphia, the Pennsylvania Horticultural Society boasted a membership of several hundred and had begun sponsoring a series of successful floral exhibitions.[3] Did the absence of similar success in New York in the 1830s lend credibility to the city's critics?

Although the thriving city on Manhattan's southern tip had been saddled with a reputation for commercial mania since the colonial era, in fact nineteenth-century New Yorkers nurtured a significant interest in horticultural science, botany, and landscape design. Early in the century, the city's men of wealth and education allied themselves with experienced businessmen, western explorers, and horticulturists to

1 "This is not on account of hard times," he asserted, "for we now have six [crowded] theaters open every evening." *Magazine of Horticulture, Botany, and All Useful Discoveries and Improvements in Rural Affairs* (December 1837): 389. Hard times, however, undoubtedly did contribute to the society's downfall; see below.

2 Ibid.

3 Andrew Jackson Downing to John Torrey, October 11, 1836, Papers of John Torrey, New York Botanical Garden Archives.

create institutions to generate knowledge in a host of areas, from natural history to the development of steam engines. At century's end, a similar alliance resulted in the founding of a scientific public garden. This institution, the New York Botanical Garden (NYBG), represented the culmination of a century of institutional and scientific trends in the city. With the forging of the last in a series of nineteenth century alliances among residents of means and those who commanded certain experience of the botanical and horticultural world, New York succeeded in attaining the stature as a center for the production of scientific knowledge commensurate with its rank as *the* American metropolis.[4]

Nathaniel Lord Britton, the man largely responsible for this union of Gilded Age wealth and expertise, was a skilled scientific entrepreneur.[5] Acutely aware of his role in New York's scientific history, Britton sought to continue the scholarly and institutional tasks of his nineteenth century predecessors. Upon obtaining a state charter for the new botanic garden in 1891, he approved for its corporate seal a telling image. Over the Latin motto "pro scientia populique salute" ("on behalf of knowledge and the welfare of the people"), stood a shield patterned with books and flowers, flanked on one side by a scholar in full academic regalia and on the other by a working gardener, clad in the rugged clothing of a competent horticulturist.[6] Britton's intention for the new institution was thus apparent from the outset: it would serve the city and the nation in an expert capacity that was twofold. From the standpoint of "pure" science, Britton's staff of itinerant botanists would pursue the completion of the great desideratum of nineteenth-century American botany: the compilation of a descriptive and taxonomic account of North American plant life. Meanwhile, his staff of trained horticulturists would strive to create a public garden suitable for the health, recreation, and civic prestige of the recently consolidated city. They would do so in accordance with a half-century's precedent of American landscape architecture, a republican pursuit enthusiastically adopted by the region's rapidly expanding middle class. Together, professional botany and elite horticulture would combine to form a new experimental garden of the first rank, ending in the process a century's worth of

4 For New York's growth as a metropolitan intellectual center see Thomas Bender, *New York Intellect: A History of Intellectual Life in New York City, from 1750 to the Beginnings of Our Own Time* (Baltimore, MD: Johns Hopkins University Press, 1988). The description of New York as "the" American metropolis is Bender's.

5 Horticultural and botanical knowledge was not among the priorities of organized educational institutions; Columbia College maintained a classical curriculum. In addition, city and state authorities failed to support such endeavors. Finally, between the two branches of natural science, botany was given more attention by those in the position to advance knowledge, including Torrey.

6 Britton and his collaborator in the *Northeastern Flora*, Addison Brown, had been responsible for choosing the design of the Garden's seal. Minutes of the Scientific Directors of the New York Botanical Garden, January 9, 1896, New York Botanical Garden Archives.

false starts and compromised institutions. The history of NYBG, then, must begin with a description of what came before: the development of organized horticulture and botany in the city of New York.

New York's First Botanical Garden

The most prominent of New York's early nineteenth century horticultural enthusiasts was David Hosack, a respected physician and professor of materia medica at Columbia College.[7] Born in 1769, Hosack had been educated at both Columbia and Princeton and had taken medical degrees from the Universities of Pennsylvania and Edinburgh. After studying botany in London, he returned to New York where he taught at Columbia for over a decade before resigning to take a chair at one of the city's medical schools, the College of Physicians and Surgeons. Admired for his learning, he was at the same time a colorful and controversial figure in the city's public life. A successful diagnostician, Hosack was an early advocate of vaccination against recurrent smallpox epidemics, then a hotly contested issue among medical practitioners. In addition, he courted the wrath of fellow physicians by his opposition to the use of mercury in treatment, while also espousing the use of quarantines to combat contagion. An ardent standardizer of the practice of medicine, Hosack served as president of the New York County Medical Society, and used this platform to compile and publish one of the first codes of professional medical ethics and to urge the municipality to adopt stricter public health measures.[8] Hosack was also a legendary bon vivant and hosted lavish affairs at his home for the city's cultural elite and its smaller circle of scientific men. A college classmate and close friend of the equally controversial politician and patron of learning DeWitt Clinton, he created the nearest thing New York had to a salon at that time.[9] Until his death in 1835, Hosack was one of the city's leading physicians.[10]

7 In her classic account of American horticulture, Ann Leighton discusses a "chain" of individuals responsible for the development of an American landscape aesthetic and correctly identifies Hosack as the first in a series of crucial precedent-setters for Britton. See *American Gardens of the Nineteenth Century: 'For Comfort and Affluence'* (Amherst: University of Massachusetts Press, 1987): 119–122.

8 Chauncey D. Leake, "A Gentleman Physician of New York: Dr. David Hosack, the Great Practitioner of the Early Nineteenth Century," paper presented at the Medical History Seminar, Madison, Wisconsin, February 4, 1920. (Typescript in New York Botanical Garden Library.) Hosack was not, however, entirely free from contemporary medical convention; he did not hesitate to prescribe bloodletting for patients to relieve pain. Undated letter, David Hosack to Montgomery Livingston, New York Academy of Medicine Archives.

9 Christine Chapman Robbins, "David Hosack: Citizen of New York," *Memoirs of the American Philosophical Society* 62 (1964). Simon Baatz, *Knowledge, Culture, and Science in the Metropolis: The New York Academy of Sciences, 1817–1970* (New York: The Academy, 1990). Baatz estimates that "only a few dozen men had any active interest in science" in early-nineteenth-century New York (11).

10 Addison Brown, *The "Elgin Botanic Garden," Its Later History and Relation to Columbia College, the New Hampshire Grants and the Treaty with Vermont in 1790* (Lancaster: New Era, 1908).

Upon returning from his European education with a set of botanical volumes, herbarium specimens, and other natural history collections unmatched in the city, Hosack was invited by a group of gentlemen farmers and professionals to join the New York State Society for the Promotion of Agriculture, Arts, and Manufactures. Organized in 1792, this Society served as a forum for the discussion of innovative agricultural methods, horticulture, and husbandry. The group's founders included several illustrious New York names, including George Clinton; John Jay; Robert Livingston; and Dr. Samuel Latham Mitchill, a colleague of Hosack's on the medical faculty at Physicians and Surgeons and later one of the founders (in 1817) of the New York Lyceum, the precursor to the city's Academy of Sciences.[11] Hosack thus became a prominent member of the city's "natural history circle," a group that in the early decades of the nineteenth century grew increasingly rooted in academic training and away from its origins among amateur naturalists, gentlemen gardeners, and nurserymen.[12]

An accomplished student of distinguished European scientists, Hosack deplored the inadequate botanical education he had received in American colleges and lamented the lack of a publicly funded botanical garden that could be used as a teaching aid for his own lectures.[13] In 1800, after being rebuffed by the Columbia College trustees, Hosack petitioned the New York State Legislature for an appropriation to develop his ideal garden. He planned to acquire an accessible site near the city limits in which he could arrange specimens of rare and ornamental plants, both native and exotic, in discrete beds representing different botanical families. In addition, he intended to construct a conservatory and a large hall containing classrooms, a botany library, and a herbarium of dried plant specimens. Despite (or perhaps *because of*) these elaborate plans, the legislators also turned Hosack away without funding.[14]

11 Robbins, "David Hosack," 47–49, see note 9. Mitchill's polymathy (he was appointed "Professor of Natural History, Chemistry, Agriculture, and other Arts Depending Thereon") was an asset to the natural history enterprise of New York in the late 1780s and 1790s. By 1820, however, Mitchill's wide interests were ridiculed as those of an antiquated "savant" by the new and growing class of self-styled "scientists." For men such as Yale's Benjamin Silliman, for whom the ability to remain up-to-date on all new publications in mineralogy and chemistry was nearly impossible, Mitchill's work was hollow; it did not demonstrate true mastery of a multitude of subjects, but instead typified an eighteenth-century dilettantism. See Brooke Hindle, *The Pursuit of Science in Revolutionary America, 1735–1789* (Chapel Hill: University of North Carolina Press, 1956): 315 and Chandos Brown, *Benjamin Silliman: A Life in the Young Republic* (Princeton, NJ: Princeton University Press, 1989): 297–298.

12 Hindle, *The Pursuit of Science*, 308–311, see note 11

13 These lectures included discussions of plant taxonomy, anatomy, and the history of botany. "Syllabus of the course of Lectures on Botany, delivered in Columbia College, by David Hosack, M.D., Fellow of the Linnaean Society of London, and of the Royal Medical and Physical Societies of Edinburgh," reprinted in *The American Medical and Philosophical Register* 4 (April 1814).

14 Robbins, "David Hosack," 61–63, see note 9.

FIGURE 1 David Hosack's Elgin Botanic Garden in early 1800s on the site now occupied by Rockefeller Center.

In the wake of these unsuccessful appeals, in 1801 Hosack purchased 20 acres of central Manhattan land, then about three miles north of the heart of the city. Confident that the public utility of such a garden would be readily apparent,[15] Hosack spent his own money to create, by 1806, a botanical garden in which he successfully cultivated over 2,000 species of plants (FIGURE 1). Dubbed the "Elgin Garden," in commemoration of his father's Scottish birthplace, this new institution became the setting for Hosack's efforts to ascertain which plants were "most useful in agriculture, in medicine, and the arts," and which were "capable of being naturalized to New York's climate."[16] Anticipating the work Nathaniel Britton would initiate as director-in-chief of NYBG

15 Ibid., 62, 70–73.

16 Brown, *The Elgin Botanic Garden*. See note 10. Robbins, "David Hosack," 71, see note 9. David Hosack, *A Statement of Facts Relative to the Establishment and Progress of the Elgin Botanic Garden, and the Disposal of the Same to the State of New-York"* (New York: C. S. Van Winkle, 1811). This site is occupied currently by Rockefeller Center, where an inconspicuous plaque dedicated to Hosack and his garden is obscured by a park bench. Hosack's public garden had precedent in the colonial period, of course. Leaving aside John Bartram's garden outside Philadelphia (which was private) and such institutions as the Prince family's "Linnaean Botanic Garden" and Andre Parmentier's "New York botanical garden" (both on Long Island) because they were actually elaborate commercial nursery operations, it should be noted that the trustees of the Colony of Georgia had established an experimental garden for the promotion of agriculture, which they placed under the charge of a medical botanist, Dr. William Houston, in the 1730s. See Hindle, *The Pursuit of Science*, 31, note 11.

almost a century later, Hosack recruited a set of competent assistants from among his many medical pupils and began work on the publication of a North American flora, a comprehensive catalog of the continent's plant life.[17] Hosack's ambition, however, exceeded his means. After several unsuccessful attempts to secure public funding for his horticultural and botanical enterprise, he conceded that maintaining the garden had grown too burdensome and sold it to the State of New York.[18] Within several years of taking title to the land, the state granted it outright to Columbia College. With no professor of botany, no medical faculty, and extremely limited operating funds, the Columbia trustees were not inclined to maintain an experimental garden. Consequently by 1817, after three years of marginal upkeep, only Hosack's greenhouse remained standing. The city's first botanical garden was defunct.[19]

Throughout this process of seeking public funds, Hosack nurtured a sharp sense of resentment. Having built the garden with his own fortune in the expectation that he would ultimately win public financial support, he was quick to anger when state money was not forthcoming. He published a lengthy pamphlet recounting the chain of events surrounding the garden's development and underscoring its utility to the public welfare. He then circulated it, along with letters highlighting his perceived mistreatment by state authorities, among correspondents from Philadelphia to Boston.[20] Hosack implored his friend Stephen Higginson to read the pamphlet and "use [his] influence [in Boston] to make public the ingratitude with which you will perceive I have been treated." To his distinguished medical colleague Benjamin Smith Barton of Philadelphia, Hosack condemned the State of New York's "parsimony and meanness," asserting that the state's purchase price was based on an assessment of the land that was too low, thus causing him to suffer a considerable financial loss.[21] Other prominent recipients of Hosack's public broadside against New York State

17 Robbins, "David Hosack," 66–67, see note 9.

18 Brown, *"The Elgin Botanic Garden,"* 8. See note 10. Hosack's liberal spending habits (the Elgin Garden was just one example) left him deeply in debt, though his annual income in 1811 was a robust $11,600. (See the Hosack Memorandum book, New York Academy of Medicine). In addition, the state delayed paying him for the garden land, leading him to borrow money from other wealthy friends. His income also suffered a significant setback with the state-mandated reduction of fees for his courses at the College of Physicians and Surgeons at a time when individual instructors relied heavily upon income from course fees charged to each student. See Robbins, "David Hosack," 109, note 9; Hosack, *"A Statement of Facts,"* note 16; David Hosack to Stephen Higginson, Jr., March 16, 1811 and Hosack to Benjamin Smith Barton, March 16, 1811, New York Academy of Medicine Archives.

19 Robbins, "David Hosack," 98, see note 9.

20 Hosack, *"A Statement of Facts,"* see note 16.

21 Hosack to Higginson, Jr. and Hosack to Barton, see note 18.

included the American Philosophical Society; the Royal Society of London; the Liverpool Athenaeum; and even Joseph Banks, the Royal Society's president.[22]

The New-York Horticultural Society

Just as the Elgin conservatory was becoming the only visible vestige of Hosack's botanical ambitions, a group of New York's professional horticulturists, including commercial nurserymen, seedsmen and estate gardeners, began gathering regularly in a Broadway inn whose sympathetic owner offered a free meeting space. In 1818, several of these men banded together and established the nation's first organization devoted exclusively to horticulture, the New-York Horticultural Society (N-YHS). Their meetings attracted the notice of the gardeners' wealthy employers, many of whom subsequently joined as active members. In 1822, the Horticultural Society incorporated, and within the next two years, its members began to formalize and expand the organization. At the suggestion of a new member, William P. Van Ness, the US Judge for the Southern District of New York,[23] a board of councilors of the society was appointed, including, in an ex officio capacity, the mayor of the city and the president of New York Hospital. Unlike the city's genteel and ineffective Literary and Philosophical Society, New York's horticultural organization would boast a diverse membership whose ranks were not limited to a patrician elite but instead welcomed the participation of "practitioners," who, after all, had founded the organization.[24] Thus, it resembled the city's newly established and relatively egalitarian scientific club, the Lyceum of Natural History. Despite modest beginnings, the city's cross-class alliance of horticultural enthusiasts built an organization that by the late 1820s achieved a measure of prominence in New York's cultural life.

The N-YHS appeared at a propitious time. Interest in both agricultural methods and horticulture had been increasing in the city since the 1790s, and a multitude of associations seeking to advance theoretical, practical, and specialized knowledge were

22 David Hosack letter book, New York Academy of Medicine Archives.

23 This, coincidentally, was the position held 77 years later by New York Botanical Garden incorporator Judge Addison Brown.

24 Baatz, *Knowledge, Culture, and Science*, 11, 18–19, see note 9. The relatively diverse nature of the Horticultural Society is unsurprising in an organization founded by the city's practical horticulturists, most of whom were foreign-born. The overwhelming majority of vocational gardeners in 1820s New York were either Irish or Scottish; even three decades later, Andrew Jackson Downing estimated that only three percent of professional gardeners in the U S were native-born citizens. See Joel E. Springarn, *Henry Winthrop Sargent and the Early History of Landscape Gardening and Ornamental Horticulture in Dutchess County* (New York: Dutchess County Historical Society, 1937). Of course, the ranks of horticultural workers in the early decades of the nineteenth century also included slaves.

being successfully organized along the eastern seaboard.[25] Within five years of the group's founding, its ranks had swelled to 78 local members, and each monthly meeting brought the nomination and admission of several more.[26] To the company of commercial nurseryman Thomas Hogg and florist Michael Floy, the Society added to its roster in the 1820s several illustrious knickerbocker names, including members of the Schermerhorn, Livingston, Stuyvesant, Lenox, Lorillard, and Van Rensellaer [sic] families.[27]

Much of the activity of the young society remained dedicated to what was dubbed "practical" horticulture. The allure of raising healthy, profitable, and showy fruits and vegetables united the gentleman amateur with both his estate gardener and the commercial nurseryman who supplied him with seeds and stock plants. In 1822, all of the plants presented at the monthly meetings were vegetables, and only four types of regular premiums were awarded for flowers at exhibitions the group sponsored. During the growing season, a standing committee of members met weekly as an inspecting committee in order to award premiums of several dollars each to members presenting the best examples of various vegetables (corn, carrots, lettuce, beets, cabbages, potatoes); occasionally fruits (melons, strawberries); and rarely, flowers. Members earning premiums were expected to provide a written account of their successful cultivation effort for deposit with the Society.[28] These horticultural priorities were underscored by the man elected president of the N-YHS in 1824: David Hosack.

Hosack attributed the increasing success and prominent public profile of the society during the 1820s to the efforts of the organization's "practical members," without whose aid and "spirited exertions" the Society "could not claim the patronage it receive[d] from [its] fellow citizens." From "humble and unostentatious beginnings," Hosack noted, the Society became, during the 1820s, the principal

> means of increasing the variety, and of improving the quality of vegetables of [the] table; of totally changing the face of [the city's] markets; of introducing a great number of valuable fruits; of augmenting the number and variety of ornamental

25 Robbins, "David Hosack," 49, see note 9. For a discussion of the emergence of learned societies in the early Republic, see Alexandra Oleson, "To Build a New Intellectual Order," introduction to Alexandra Oleson and Sanborn C. Brown, eds. *The Pursuit of Knowledge in the Early American Republic: American Scientific and Learned Societies from Colonial Times to the Civil War* (Baltimore, MD: Johns Hopkins University Press, 1976): xv–xvi.

26 New-York Horticultural Society Records, 1818–1844, 1872, Minutes of the Society, New York Botanical Garden Archives.

27 *The American Gardener's Magazine* (July, 1835).

28 Minutes of the New-York Horticultural Society, April and May, 1822, New York Botanical Garden Archives.

plants, both indigenous and exotic, and thereby of spreading a taste for this innocent, yet instructive and delightful source of enjoyment.[29]

Each of these major aims, Hosack proclaimed, was being met admirably by the Society's own practical men.

Hosack's approbation was genuine, and not merely a rhetorical recognition of the group's vocational roots. Yet credit for the Horticultural Society's increased public stature belonged as well to its more socially prominent members, who encouraged the concrete goals of the Society even as they transformed the group into something beyond an association of like-minded vocational horticulturists.[30] Under the influence of these high-profile members, the organization became a hybrid: for a brief, successful period in the 1820s, it enjoyed the social status of a learned society that, while encouraging such practical pursuits as the development of marketable produce, also helped sponsor the publication of a journal and kept a library; conducted correspondence and carried on an exchange of books, seeds, and botanical specimens with similar groups abroad; elected honorary and corresponding members that included the leading names of European and American art, science, and politics;[31] and even hosted lavish banquets whose proceedings were extensively chronicled by the city's popular press.[32] With this organizational structure, the city's horticultural society provides an early nineteenth-century illustration of Nathan Reingold's

29 Ibid., 17.

30 Seventy years later, Nathaniel Britton acknowledged a similar dynamic at work among Garden incorporators that would benefit the newly established institution. Recognizing the presence (and practical necessity) of including the city's richest families among the Garden's patrons, he privately characterized these wealthiest contributors as "ornamental members" (of the Torrey Club) from whom basic quotidian necessities from books to laboratory instruments could be procured and political influence derived. N. L. Britton to E. Britton, Records of Elizabeth Gertrude Knight Britton, Series 1: Correspondence files, 1877–1934, New York Botanical Garden Archives.

31 Minutes of the New-York Horticultural Society, New York Botanical Garden Archives. In 1824 alone, the Society elected to corresponding or honorary membership status: DeWitt Clinton, Thomas Nuttall, Benjamin Silliman, Asher B. Durand, Thomas Jefferson, John Adams, James Madison, the Marquis de Lafayette, William Hooker, Andre Michaux, and Albert Gallatin.

32 Minutes of the New-York Horticultural Society, New York Botanical Garden Archives. I use the term "learned society" here as it was defined by John Voss: the N-YHS served the city as an institutional "promoter and disseminator of knowledge" that "organized research, provided opportunities for the exchange and diffusion of knowledge through publications and meetings, and advanced the professional interests of their members." See John Voss, "Foreword: The Learned Society in American Intellectual Life," in Oleson and Brown, eds., *The Pursuit of Knowledge in the Early American Republic*, vii, note 25. This status as a learned society was given further credence by the Horticultural Society's inclusion in the ranks of associations that were granted space in the "New York Institution of Learned and Scientific Establishments," the excess municipal space used

descriptive paradigm in which he delineates a network of "cultivators," members of a learned culture whose livelihood did not depend upon their participation in that group or production of scientific knowledge but who supported it financially and socially, and "practitioners," whose employment required competence in the techniques of horticulture but no underlying theoretical scientific knowledge. In this situation, then, the Horticultural Society's wealthy, estate-owning members served their practical comrades in a horticulturally apt role as "cultivators."[33]

Seeking Trans-Atlantic Legitimacy: British Precedents

Just as Nathaniel and Elizabeth Britton would find inspiration for their garden vision in England, so too had their predecessors in New York's botanical and horticultural circles. In late 1822, the N-YHS initiated an intellectually and materially profitable series of exchanges with its counterpart in London, sending a letter to that group's corresponding secretary that read in part:

> We . . . have not hitherto cultivated this fair field of science with sufficient diligence or success; but we are surrounded with a luxuriant nature and she does abundantly offer to our exertions her hidden treasures.The best interests of our respective countries are involved in the improvement of the soil, and we are persuaded that in point of civil oeconomy and domestic comfort Horticulture stands preeminent as a means of improving the conditions and comforts of the poor, and increasing the enjoyments of all classes of the community; it recommends itself to those whose wisdom governs the State, as well as to the Colleges. . . .[34]

Here New York City's allied horticultural enthusiasts explicitly evaluated their efforts in terms of civic utility, a refrain Hosack had used when seeking support for his garden, and a tactic that would be employed with great success by Britton as he organized NYBG seven decades later. Their endeavor was linked in the minds of members both

---

by various groups with governmental sanction in the 1820s. See Bender, *New York Intellect*, 62–65, note 4. In an 1823 report on the N-YHS's process of incorporation, William Wilson, one of the founding "practical members" of the society, explicitly referred to the group as a "learned body." See letter, Robert Manning to unspecified recipient, February 13, 1872, New-York Horticultural Society Records, New York Botanical Garden Archives. Manning was the Secretary and Historian of the Massachusetts Horticultural Society at that time.

33 See Reingold's "Definitions and Speculations: The Professionalization of Science in America in the Nineteenth Century," in Oleson and Brown eds. *The Pursuit of Knowledge* (1976), note 25 and Elizabeth Keeney, *The Botanizers: Amateur Scientists in Nineteenth Century America* (Chapel Hill: University of North Carolina Press, 1992): 4–5. Reingold's third group, the "researchers," (leading scientists conducting studies and aiming to advance and diffuse science) corresponds in this portrait with the growing ranks of the city's botanists, discussed below.

34 Minutes of the New-York Horticultural Society, December 24, 1822, New York Botanical Society Archives.

practical and patrician with that of "science," national interest, community, progress, and even charity. Although overtly apologetic, this solicitation of recognition was a confident one, cloaked in a westward-facing republican argot of cultivation not limited to the New York region, but which was given special resonance in the minds of city residents by the contemporary development of the Erie Canal.

This appeal to the London Horticultural Society (LHS, founded in 1804) represented an obvious desire for legitimacy and international recognition. As such, it must be considered within the context of contemporary British attitudes toward botany, horticulture, and an emerging class-driven set of distinctions between high and low culture. First, it should be noted that at the time, British disdain for (what was perceived to be) the squalid nature of American domestic settings was pronounced. Such attitudes had led to Englishman William Cobbett's condescending 1818 account of American urban and rural landscapes, in which he speculated that Americans had such a surfeit of land that they treated their small home plots contemptuously.[35]

More significant than English horticultural snobbery, however, was the emergence in early nineteenth century Britain of a distinction between a working-class, "artisan" botany and a more literate, upper-class science.[36] While New Yorkers of varied economic strata (estate gardeners, nurserymen, and landed professionals) were gathering under one roof to indulge horticultural and botanical pursuits, their English counterparts were steadily diverging along class lines. Although there were tensions within these rough divisions, nevertheless two basic groups could be identified at the time that the New-York Horticultural Society sought English approval and cooperation. The men who received that overture were members of a literate elite, dedicated to ornamental horticulture. Simultaneously, however, interest in botany and horticultural innovation was fostered in what had become an entirely working-class setting: the public houses of the English countryside. As Hosack and his own cross-class cohort met and began holding commemorative dinners in New York restaurants, English working-class gardeners, nurserymen, and self-taught botanists had been relegated to a separate domain from their higher-status countrymen and had forged their own space for the

35 William Cobbett, A Year's Residence in the United States of America (London: Sherwood, Neely, and Jones, 1818), cited in Christie H. White, "Reform and the Promotion of Ornamental Gardening" in Peter Benes, ed. *Plants and People* (Boston: Boston University Press, 1996): 104.

36 Anne Secord, "Artisan Botany," in N. Jardine, J. A. Secord, and E. C. Spary, eds. *Cultures of Natural History* (Cambridge, UK: Cambridge University Press, 1996): 378–393. Secord is careful to describe the emergence of working-class, "artisan" botany as an autonomous outgrowth of the development of craft pride and values, as well as communal tastes for conviviality. She highlights the extent to which these attitudes represented an implicit challenge to the attempt by upper-class Britons to appropriate the cultural power of science for themselves. This challenge, however, was only partly successful, because the literate productions of science aimed at dissemination of such knowledge eventually *did* become the exclusive province of middle- and upper-class British horticulturists.

pursuit of knowledge within their local pubs. These were convenient settings for the acquiring of botanical knowledge for workingmen pressed for time, where they could pool their resources of knowledge, literacy, and books.[37]

Despite this apparent divergence between the American and British models of cooperative botanical endeavor, in which the Americans were initially unified across class lines and the British divided, the trans-Atlantic efforts shared a common theme. Both British and American horticultural societies understood the necessity of recognizing crucial work in horticulture accomplished by working-class men. Even in class-segregated London, the genteel members of the LHS realized the extent to which they were dependent upon "artisan" botanists as collectors and were generally quick to acknowledge the contributions of such men in the written forums provided by elite gardening periodicals.[38] In New York, Hosack had been eager to recognize his own debt to the working gardeners who had, after all, founded the city's (and the nation's) first horticultural organization.

In part, these divergences between an American and an English model of organized horticulture can be traced to the more advanced stage of industrialization at which the British found themselves. The atomization along class lines of a British gardening public was a relatively unsurprising development in a nation undergoing rapid industrialization, with its concomitant creation of a large middle class. This class was at the forefront of the innovative amalgamation of botany and gardening that would become the "science" of horticulture practiced on both sides of the Atlantic. Members of this growing segment of British society became the beneficiaries of increased leisure time, which allowed them, as the proprietors of modest suburban plots, to test and cultivate an array of horticultural skills. British middle-class reformers such as J. C. Loudon believed such efforts would help the nation to address the social problems attendant upon industrialization, improve public health, and cultivate the intellect and morals of citizens. As New York's nascent horticultural alliance of practical gardener and genteel cultivator looked to London for guidance, British horticulture became a widespread middle-class pursuit, perfected in middle- and upper-class learned societies but also established in the casual public settings occupied by artisans interested in botanizing. Each of these discrete segments of British society, in pursuing horticulture, appropriated botanical science for its own economic and personal ends.[39]

It was to the elite London Society, however, that the Americans turned for recognition and advice. By August of 1823, the same year that the LHS broke ground

37 Ibid., 380.

38 Ibid., 385.

39 Abigail Lustig, "The Creation and Uses of Horticulture in Britain and France in the Nineteenth Century," (PhD Dissertation, University of California, Berkeley, 1997): 1–10, 18–19.

for its own garden at Chiswick, the N-YHS members had an answer to their overture. Their English counterparts had responded cordially, offering reciprocal cooperation, a list of their members, a printed report of their garden committee, and, most significantly, news of the impending arrival of their collector of North American plants, David Douglas.[40] Hosack and nurseryman Thomas Hogg, Sr. were part of a committee immediately designated to greet the London Society's horticultural envoy.[41]

Celebrating Horticulture

The civic concerns of New York's horticulturists were publicly articulated by enthusiastic attendees of the Society's festive anniversary celebrations throughout the 1820s. After electing and installing new officers for the ensuing year, N-YHS members listened to presidential inaugural addresses, entertained distinguished guest speakers, and then engaged in a spirited round of toasts in which they addressed those concerns. In 1828, for example, at the height of the Society's prominence, Hosack's presidential paean to local horticultural skill was delivered to a throng of 200. Assembled in Broadway's Niblo's Garden theatre to hear Hosack were, in addition to Society members, the Governor of Ohio, various military officers, and members of the city's clergy. The inaugural was followed by the Reverend Schroeder's discussion of the history of horticultural arts. As he traced this narrative, with emphasis on the greatest names in horticulture from antiquity forward, Reverend Schroeder asserted that such endeavors had been in part responsible for all of humanity's "physical and moral enjoyments." Following the speeches, the assembled guests were led to an adjacent banquet hall and were seated at long tables laden with "rich and various" products of local growers. A stage at one end of the hall had been decorated with paintings and the walls hung with flags, while cuttings of flowering shrubs and other plants were displayed within vases throughout the hall.[42] The accumulated fruit of the members' harvest was memorialized by an attending newspaper reporter fully in tune with the spirit of the gathering:

> The display of flowers and green-house plants . . . contained several specimens of extraordinary beauty. The peaches, pears, apples, and plums were in general exceedingly fine; and then, the nectarines! . . . We never saw any look so rich and beautiful before. . . . [U]pon the center table there was a real substantial peach tree, its branches bending under the weight of their fruit, handsomely illuminated. This illumination of a fruit tree was a novelty introduced with fine effect . . . [and] was so

40 Ibid., 163.

41 Minutes of the New-York Horticultural Society, August 26, 1823, New York Botanical Garden Archives.

42 *New York Commercial Advertiser* (August 28, 1828) and *Proceedings of the New-York Horticultural Society, at the Celebration of Its Tenth Anniversary, August 26, 1828* (New York: E. Conrad, 1828).

> attractive and pleasing that we presume decorations of a similar kind will be more extensively and carefully introduced hereafter.[43]

Then, in the banquet room, with the fermented products of members' grape harvests flowing freely, the oratory extended into the evening. Twelve regular toasts were made at the start of the dinner, including tributes to the President of the United States, to the late DeWitt Clinton, and to the Horticultural Societies of London, Edinburgh, and Paris. Other celebrants acknowledged the role of the nation's armed forces in scientific and horticultural reconnaissance and, echoing Hosack's enumeration of the Society's priorities, saluted horticulture itself as "one of the most useful and elegant arts of peace." Subsequent speakers, in between renditions of patriotic music played by a band from West Point, exalted the "national" influence of the host society. These well-wishers boldly predicted a future in which representatives of every state in the Union would attend the annual celebrations in order to pay homage to the work of an organization that would transform the city from a well-kept garden into a "Paradise in which there would be no forbidden fruit, and where science might be cultivated to the full extent of the capacity of the human mind." Still other men rose to toast both the "art" as well as the "science" of horticulture, and to make a more quotidian call for the cultivation of the mulberry tree as a means of promoting the domestic silk industry.[44]

The evening finally closed after hours of bombast and tortured botanical metaphors with a series of songs sung by popular performers, some of which paid tribute to the heritage of "the industrious Scots gardeners of [the] neighborhood, [who] constitute a respectable number of the members of the [Horticultural] Society." Thus appeared Hosack's familiar refrain: despite the high proportion of the city's elite in attendance, due recognition was consistently granted the organization's practical horticulturists. In this way, all members could present themselves, in the words of G. W. Prentiss at the 1827 anniversary dinner, as "cultivators of the earth, the mind, and the good feelings of the heart."[45]

Seeking Hosack's Lost Eden

Despite the high spirits and the cross-class institutional camaraderie, President Hosack had injected a somber note into the anniversary proceedings. At the conclusion of his 1828 inaugural, he admonished his listeners that the inability of the organization to secure access to a suitable garden plot was a great disappointment. Hosack still felt

43 *New York Commercial Advertiser* (August 28, 1828).

44 *Proceedings of the New-York Horticultural Society*, see note 42.

45 Ibid. Society members' acknowledgement of their organization's roots had been even more explicit at the previous year's festivities: William Neilson, Esquire, had toasted "the operative gardeners of

keenly the loss of his Elgin Garden and, in fact, had been engaged throughout the 1820s in a series of negotiations with Columbia College authorities in an effort to reclaim it for use by the Society.[46] Efforts to persuade the Columbia Board of Trustees to lease the old Elgin site had begun almost as soon as Hosack had joined the Horticultural Society and had been appointed the association's "professor of botany and vegetable physiology." In late September of 1824, Hosack conferred with Columbia's president and then confidently reported that the outlook was favorable.[47]

The self-assurance of the early 1820s gradually withered over the course of the decade, however. By the time Hosack addressed the festive Society gathering from his presidential pulpit in 1828, that confidence had turned to bitter disappointment. Repeatedly rebuffed by Columbia's trustees in his attempts to regain access to his old land on behalf of the Society, Hosack had turned to a subscription fund in order to raise money for the purchase of a garden site. Shares in the future garden were available in November of 1824 for $10 each; the Society's working gardeners could pay in plants.[48] And yet these efforts, too, failed to secure either a new site or access to Hosack's old land, whose value as soon-to-be-prime Manhattan real estate was not lost upon Columbia authorities. If, as one celebrant at the anniversary dinner had predicted, the Society was destined to recreate a "paradise" of knowledge within the City of New York, by the end of the 1820s, that paradise was merely a distant goal. Almost a decade and a half after ceding title to the city's botanical garden and being summarily rejected by the legislature in his quest to retain at least curatorial control over the land, David Hosack's plans to recapture his lost personal Eden, the Elgin Garden, remained unfulfilled.

Hosack's zeal in promoting the garden cause, thus frustrated for years, reached a fever pitch as the 1820s ended. Turned away by Columbia, he availed himself of the platform that his position in the Horticultural Society provided to seek not only a garden for New York, but a system of federally supported botanical gardens that would serve the nation as a "means of diffusing a knowledge of the indigenous vegetable productions of the Western Hemisphere, as well as of those plants, the growth of foreign climes, that are the most useful in medicine, in agriculture, and in the arts." In 1828, in a letter to US Treasury Secretary Richard Rush, he lamented on behalf of New York's horticultural savants that

---

New-York, the founders of this society."

46 David Hosack, Anniversary Address to the New-York Horticultural Society, August 31, 1824 (New York: J. Seymour, 1824): 19 and *Proceedings of the New-York Horticultural Society*: 9, see note 42.

47 Minutes of the New-York Horticultural Society, September 28, 1824, New York Botanical Garden Archives.

48 Minutes of the New-York Horticultural Society, November 2 and November 18, 1824. New York Botanical Garden Archives.

> we fall very far short of every European nation in scientific gardening. We have no public botanical gardens of any consequence, nor experimental gardens, which are deemed of so much importance that every part of Europe, even places of minor notice, can display one of these necessary and beautiful deposites [sic], where all the vegetable kingdom is brought to the notice of every visitor . . . If three or four establishments were provided in different parts of the Union under the care of experienced gardeners to nurse and cherish the plants and seeds . . . to acclimate them to their proper place, the most important results may be anticipated.[49]

Hosack's remedy for American inadequacies was, in the context, predictable. "In order to effect this purpose," he counseled Rush, it was necessary to establish a set of three federally subsidized botanical gardens, one in the North, one in the South, and one centrally located along the Atlantic coast. And, just as predictably, he suggested that Rush consider the City of New York, with its "open access to all parts of the world . . . at all seasons of the year," as the preferred site of the northern garden.[50] Though Congress would eventually establish a greenhouse at the foot of Capitol Hill for the care of a portion of the botanical spoils of the Wilkes Expedition of 1838–42, Hosack's grand self-serving scheme was never realized.[51]

Failing in its efforts to secure federal financial support, the Society closed the 1820s with one more concerted attempt to acquire governmental sanction. In 1829, a committee of members petitioned the City itself for a grant of land with which to construct a "Horticultural or Botanic Garden." The petitioners based their claim to municipal munificence upon the planned garden's utility, describing its anticipated benefits in precisely the same language with which Hosack had attempted to secure assistance from Washington. (The arguments were also strikingly similar to those to be used six decades later by the Torrey Botanical Club.) "In such a garden," the Society contended, "experiments would be made to meliorate the quality of our fruit and vegetables." In addition, the garden would serve as a suitable repository of "remarkable or useful plants" gathered by the nation's expeditions, where efforts

49 David Hosack to Richard Rush, printed in *New York Farmer and Horticultural Repository*, 1 (2), (February, 1828).

50 Ibid.

51 The US Exploring Expedition, led by US Navy Lieutenant Charles Wilkes, was "one of the most consuming enterprises of the mid-nineteenth century," according to botanical historian Joseph Ewan. The success of Wilkes's collectors enabled the US to take its place with leading European powers as a scientific partner, Ewan contended, even though the published results of the various Wilkes investigations were comparatively inaccessible to the international scientific community. From Wilkes onward, US herbaria have only grown in importance, and it was in this context that Britton planned to compile the most comprehensive plant collections in the Western Hemisphere. See Joseph Ewan, *A Short History of Botany in the United States* (New York: Hafner, 1969): 42.

could be made to acclimatize and propagate them for economic, medicinal, or ornamental purposes. Finally, the committee wrote, the garden would "form a school for information respecting the theory and the practice of the science of Horticulture."[52] However, the city did not grant the Society any land. In the heated political climate that surrounded the city's educational institutions, it is unsurprising that repeated negotiations failed. Ironically (in light of the ultimate location of NYBG on Lorillard property along the Bronx River) in 1825, Society Vice-President Jacob Lorillard offered the organization land of his own in the Manhattan neighborhood of Bloomingdale. It is unclear why this offer was not accepted. Hosack probably persisted in his hope for success in negotiations with Columbia, which was unlikely because of his controversial relationship to the college and Columbia's need to maximize rental income from the property.

The N-YHS of the early 1820s, corresponding with an elite London counterpart and entertaining European botanists, had had reason for its confident outlook. Yet subsequent years of disappointment in its quest for a garden, coupled with mounting financial trouble meant that, despite the rhetorical flourishes of the partygoers of 1828, the Society entered its second decade precariously.[53] This series of disappointments, along with Hosack's sudden death in 1835 and the economic contraction of the late 1830s, resulted in a precipitous decline in membership and public prominence. Without Hosack, the Society lost much of its prestige, ceasing to function as a full-fledged learned society and reverting to the informal trade association it had been before incorporation in 1822. With the withdrawal of many wealthy amateur gardeners from the membership rolls, the Society's treasury, never robust, was further diminished.[54] In addition to social and financial pressures, the Society's fortunes were hindered by nature. The mid-1830s in New York were marked by periods of prolonged drought. Such unfavorable weather diminished the available number of entries for the Society's annual premium competitions. This forced inactivity of the inspecting committee was

52 "Petition of the New-York Horticultural Society to the Corporation of the City of New-York," March 10, 1829. New-York Horticultural Society Records, New York Botanical Garden Archives.

53 Even in the heady days of the 1820s, a considerable number of Society members were in arrears in dues. By 1826, over $1,000 in membership fees remained uncollected. Minutes of the Society, May 30, 1826, New York Botanical Garden Archives.

54 The effect of the Panic of 1837 upon gardening was noted by C. M. Hovey in 1840: "Commercial affairs, upon the prosperity of which horticulture is, in a degree, dependent, have been again plunged into embarrassment, and attention of a portion of the community drawn away from the peaceful and quiet pursuit of horticulture, to the din and bustle of the commercial world." *Magazine of Horticulture* 7 (January 1840): 1. In addition, the city had suffered debilitating damages (contemporary estimates put the figure at $20 million) from a fire that consumed over 600 structures on December 16, 1835. See *Journal of the American Institute of the City of New York* 1 (November, 1835).

followed, in the fall of 1835, by cancellation of a series of meetings for lack of a quorum.[55]

In 1836, in an attempt to recapture the success of the 1820s, the remaining members of the Horticultural Society elected John Torrey, a former medical student of Hosack's with a growing reputation as a taxonomic botanist, to the president's chair.[56] Though qualified to practice medicine, Torrey's primary interest was floristic botany. Along with his student Asa Gray, Torrey spent the 1840s, '50s, and '60s working to identify and categorize the wealth of botanical specimens that were collected by a series of US government surveys of the trans-Mississippi West.[57] This work, never completed, was designed to be published as a comprehensive North American flora, the task necessarily abandoned by Hosack in 1810. With a host of pedagogical responsibilities and a primary concern with taxonomic research rather than the applied scientific work of horticulture, Torrey was in no position to lead the floundering Horticultural Society back to the prominence it had enjoyed a decade earlier.[58] Within a year of Torrey's assumption of the presidency, C. M. Hovey reluctantly informed his readers of the travails of New York's horticulturists: [the society], "owing to the apathy of its members, [was] not able to get up an exhibition this year. . . . This we were sorry to hear and we regret to learn . . . that some of the most active members have already withdrawn their connection from it."[59] By September 1844, after several years of sparsely attended meetings and postponed exhibitions, seven remaining members of the Horticultural Society gathered to discuss the propriety of disbanding

55 *Magazine of Horticulture* (1835): 463–464 and Minutes of the New-York Horticultural Society, 1835–1836, New York Botanical Garden Archives.

56 Torrey's relationship with Hosack had been strained; wary of Hosack's flair for self-promotion, Torrey had been reluctant to accept the senior physician's patronage, going so far as to accept the vice-presidency of a short-lived rival horticultural society formed by a group of prominent New Yorkers in 1824. See Christine Robbins, "John Torrey: His Life and Times" *Bulletin of the Torrey Botanical Club* 95 (1968) and Nathaniel L. Britton, "Dr. Torrey as a Botanist: Memorial of Dr. John Torrey, 1796–1873: Exercises at Torrey Day, June 27, 1900" (New York: Torrey Botanical Club, 1900).

57 Gray later became nationally famous as a Harvard professor of natural history, Charles Darwin's primary American ally, and the leading US botanist until his death in 1888. He was largely responsible for the introduction of the natural system of taxonomy to US students of botany, who, prior to the publication of his *Elements of Botany* in 1836, had been consistently taught the artificial sexual system of Linnaeus, in which plants were categorized according to the number of reproductive parts of their flowers. See Emmanuel Rudolph, "The Introduction of the Natural System of Classification of Plants to Nineteenth Century American Students," *Archives of Natural History* 10 (1982): 464.

58 Torrey also served as botanist to the New York State Geologic Survey and edited an American edition of John Lindley's 1831 Introduction to the Natural System of Botany, all while holding (at times simultaneously) faculty positions in chemistry at West Point, Princeton, and Columbia. See Robbins, "John Torrey," note 56 and Britton, "Dr. Torrey as a Botanist," note 56.

59 *Magazine of Horticulture, Botany, and All Useful Discoveries and Improvements in Rural Affairs* (December 1837): 449.

the group. With $2.04 in the treasury, $2.00 of which would be needed to pay for the use of their meeting space, a final entry was logged into the Society's ledger: "Finis, requiascet in pace."[60] With the Society went contemporary hopes for the construction of a botanic garden.

In a manner similar to that of the horticultural reformers of Great Britain, New York's first alliance of men of prominence with men of ("practical") experience had begun the 1820s confident of their organization's proper role in public life. Horticultural Society members perceived themselves to be acting upon a mandate: the improvement, through a multifaceted horticulture, of the educational standards of their schools; the living standards of their fellow citizens; the economic prospects of their maritime-based business pursuits; and the physical environment of their bustling, commercial city. As a successful learned society, the New York horticulturists' organization of the 1820s represented an effective, if ultimately short-lived, union of the practical outlook of working horticulturists and the more specialized orientation fostered by those botanists and natural historians motivated by a basic quest for knowledge.[61] That their efforts and goals were not confined to a newly defined, middle-class sphere of private life did not mean that they received public support in the form of governmental subsidy, however. Rather, in another parallel with their British counterparts and correspondents, New York's antebellum scientific horticulturalists never secured direct financial support or land from either the city of New York or the federal government. Thus, ironically, though members of the city's horticultural society entertained an ambitious sense of the possibilities of their efforts to help create a humane and cultivated metropolis, it would be left, five decades later, to a decidedly middle- and upper-class set of men and, crucially, women, to secure public funds and land to build a public garden in New York.

Northeastern Horticulture at Mid-Century

The demise of organized horticulture in mid-nineteenth century New York was particularly ironic in light of contemporary developments elsewhere in the Northeast. Similar societies in Boston and Philadelphia prospered. In many ways, the New Yorkers' contemporaries in New England shared similar aims and had established the Massachusetts Horticultural Society as an institutional outlet for their concerns. Boston's

60 Minutes of the New-York Horticultural Society, September 24, 1844, New York Botanical Garden Archives.

61 In "The Underside of the Learned Society in New York, 1754–1854," Brooke Hindle notes that though most premium societies of the colonial and early republican eras had been motivated primarily by a desire to disseminate practical knowledge, these types of organizations were not antiscientific. Wherever 'pure' scientific research (botany in the case of the Horticultural Society) was useful, it was encouraged. In Oleson and Brown, eds. *The Pursuit of Knowledge*, 84–116, see note 25. London reformer J. C. Loudon, himself a nursery-trained gardener, encouraged the use

wealthy merchants, conscious of their elite position within a post-Revolutionary city wedded to republican ideals, actively embraced in the 1820s and 1830s the rhetoric of a yeoman-run agriculture in order to legitimize their own social standing. Within this context there was an "inextricable link between horticultural developments and the emergence of landscape design in America."Interest in ornamental horticulture by prominent plantsmen, "country gentlemen . . . nurserymen, practicing agriculturists, and farmers" served as "a fulcrum in the implementation and acceptance of landscape design as both a legitimate and desirable profession in the United States."[62]

It was at this time that Andrew Jackson Downing became "the medium through which both the aesthetics and the more practical aspects of 'designing' the land were funneled and synthesized."[63] Even as the N-YHS waned, Downing was establishing a specifically American landscape architecture through the astute modification of English gardening ideals. In the late 1840s, he approached the height of his career as an estate designer and editor of an influential horticultural journal in which he elaborated upon his vision of a republican rural art suitable for a nation of freeholders.[64]

In the late 1830s, while the city's horticultural society struggled to stay afloat, Downing secured Torrey's nominal participation in yet another organization dedicated to horticulture and "rural pursuits." This new group would welcome members from a wider geographic area: the "river counties" between Albany and New York City. With encouraging words from Hovey's influential journal, Downing convened the Horticultural Society of the Valley of the Hudson in 1838 in an attempt to revive the region's sagging horticultural fortunes. Downing was elected corresponding secretary and Torrey Vice President from Manhattan. Although Downing optimistically predicted that this new organization, embracing as it did a somewhat disparate array of climates and soils, would escape the fate of its city-based predecessor, it too lapsed

---

of basic botanical science in all classes of society. See Lustig, "The Creation and Uses of Horticulture," 19, note 39.

62 See Tamara Plakins Thornton, *Cultivating Gentlemen: The Meaning of Country Life among the Boston Elite, 1785–1890* (New Haven: Yale University Press, 1989) and Leighton, American Gardens of the Nineteenth Century, 104, note 7.

63 Brenda Bullion, "Hawthorns and Hemlocks: The Return of the Sacred Grove." *Landscape Journal* 2 (1983): 114.

64 Judith K. Major, *To Live in the New World: A. J. Downing and American Landscape Gardening*, (Cambridge, MA: MIT Press, 1997): 4–5, 17. The literature on Downing and his influence on US landscape architecture and the subsequent work of Frederick Law Olmsted is substantial. Major's work is particularly adept at conveying the principles behind eighteenth-century debates over the "beautiful," the "sublime," and the "picturesque" and on the evolution of Downing's attitude toward the use of native species in his designs. In addition to Major's excellent overview, see David Schuyler, *Apostle of Taste: Andrew Jackson Downing, 1815–1852* (Baltimore, MD: Johns Hopkins Press, 1996).

into inactivity after mounting only a handful of exhibitions.[65] Thus, even Downing, the most influential and talented of horticulturally interested New Yorkers, failed to revive the region's moribund organizations, and he abandoned the later attempt to form a similar union that encompassed the entire Hudson Valley. Before his accidental death in 1852, however, he took up Hosack's earlier call for a botanical garden, urging his readers in dispatches from London to support the establishment of an institution modeled after the Royal garden at Kew.

Alternative Horticultural and Agricultural Venues at Mid-Century: The American Institute of the City of New York

Fortunately for the city's vocational gardeners, the disappearance of New York's Horticultural Society did not leave them entirely bereft of an outlet for their horticultural prowess. Since 1829, New York had been home to the American Institute, an organization charged by the State Legislature with the responsibility for promoting a rather catholic array of agricultural and manufacturing interests in the region. Modeled after the Franklin Institute, Philadelphia's successful sponsor of mechanical and agricultural innovation, the American Institute of the City of New York mounted a series of increasingly elaborate annual fairs throughout the nineteenth century. It also maintained a library of several thousand volumes and eventually maintained its own headquarters in a four-story brownstone at 351 Broadway.[66] Even after a devastating 1858 fire that consumed an entire set of horticultural and mechanical exhibits, these fairs continued to serve as popular annual venues for horticultural displays.[67] Dozens of Institute premiums (usually copies of horticultural

65 See the *Magazine of Horticulture* (1838): 353; and *The Horticulturist* 5 (October 1850): 153–156.

66 Regarding the Franklin Institute, see Bruce Sinclair, "Science, Technology, and the Franklin Institute," in Oleson and Brown, eds. *The Pursuit of Knowledge*, 194–205, note 25. Though founders of the Institute paid homage to their Philadelphia forebears, toasting "the Franklin Institute . . . we follow where they have led [and] may they recognize our Fairs as offsprings of theirs," the two organizations did differ significantly. The Franklin Institute gradually became a conductor and publisher of original scientific research, while its New York counterpart remained steadfast in the belief of the efficacy for technological progress of its premium system. See Hindle, "The Underside of the Learned Society," in Oleson and Brown, 108–109, note 25 and *Journal of the American Institute* 1 (November 1836). On the Institute's headquarters, see Annual Report of the American Institute of the City of New York, 1851.

67 This was the 1858 destruction of the city's "Crystal Palace," site of New York's first World Exposition in 1853–54. See Ivan D. Steen, "America's First World's Fair" *New-York Historical Society Quarterly* 47 (1963): 257–287 and Horace Greeley, *Art and Industry as Represented in the Exhibition at the Crystal Palace, New York, 1853–54, Showing the Progress and State of the Various Useful and Esthetic Pursuits. From the NY Tribune* (New York: Redfield, 1853). On the American Institute, see Hindle, "The Underside of the Learned Society," in Oleson and Brown, eds. *The Pursuit of Knowledge*, 105–110, note 25. The Institute's 1835 fair had been attended by 22,000; its 1850 fair by 88,000; and its 1852 fair by 110,000. See Annual Reports of the American Institute of the City of New York, 1851–52.

journals, medals, or certificates) were awarded, for example, for flower arrangements, fruits, vegetables, and preserves at the 1847 fair. Recipients included estate gardeners and nurserymen, as well as amateur men and women from throughout the city, the Hudson Valley, Long Island, and New Jersey. In addition, a few premium winners had traveled from farther upstate, New England, Philadelphia, and the Midwest.[68] Even A. J. Downing found the American Institute fairs to be congenial outlets for exhibiting the products of his Hudson Valley orchard.[69]

These activities, and the emphasis on a shared intellectual heritage with the Franklin Institute, were indicative of a characteristic mid-nineteenth-century mindset regarding farmers and cultivation. Seventeenth- and eighteenth-century commentators had highlighted a gendered role for the husbandman to civilize and fertilize a feminine earth, thus cultivating a bountiful garden within a former wilderness. By the early decades of the nineteenth century, *husbandmen* had become, simply, "farmers," a more moderate epithet designating a "maturing technical expertise" and the ability to master the land. As New York's various vocational cultivators (nurserymen, horticulturists) sought another public outlet to display products and share specialized knowledge, they found that their skills made them natural partners with members of the region's "Farmers' Club," who had begun to envision themselves "engineers whose work . . . demanded new skills and rewarded modest innovation."[70]

In the 1840s, this new Farmers' Club (a section of the American Institute) was designated by the New York State Assembly to serve as the Agricultural Society for New York County (Manhattan Island). As an official component of a highly successful statewide agricultural fair movement, the American Institute, then, was among the first sets of New York cultivators to secure public funds and official sanction for their efforts. This group's conception of cultivation, however, differed somewhat from that of their neighbors in Hosack's fold, and their aim was not a public botanical garden, but a model experimental farm. The Farmers' Club members, as the name indicated, pursued innovation in *agriculture*, rather than ornamental horticulture or taxonomic botany, although members did discuss viticulture and the cultural requirements of cherry, apple, peach, and pear orchards.

In 1847, led by politically influential members including one-time New York Congressman James Tallmadge, members of the Livingston family, and sitting judges,

68 Thomas Bridgeman, "Horticultural Report: To the Managers of the American Institute, Presiding at the Twentieth Annual Fair, Held at Castle Garden, October, 1847," *Transactions of the American Institute of the City of New York* (1847): 47–64.

69 See C. M. Hovey on the American Institute's fourteenth annual fair, in *The Magazine of Horticulture, Botany, and All Useful Discoveries and Improvements in Rural Affairs* 8 (1842).

70 John R. Stilgoe, *Common Landscape of America, 1580–1845* (New Haven, CT: Yale University Press, 1982): 137–138.

the Farmers' Club of the American Institute lobbied Albany legislators for a grant of 200 acres of land and a $10,000 appropriation with which to establish an experimental farm and agricultural college near the city, and presented a petition at the State Capitol with over 19,000 signatures in support of their cause.[71] Reiterating the republican agrarian rhetoric of New York's horticulturists of the 1820s, a delegation of American Institute members presented the Assembly in 1847 with a list of "reasons for the establishment of an Agricultural School and Experimental Farm near the City of New York," contending that "agriculture is the most important and extensive employment of the people of this country . . . sustaining all other occupations, and forming the basis of public and private wealth and prosperity."[72] "To complete the plan of the Institute," the legislators were informed, an "establishment for teaching, both practically and scientifically, by means of an agricultural college or school and an experimental farm, is deemed essential."[73]

Thaddeus Wakeman, an American Institute founder, joined the drive for an experimental farm in the last year of his life, depicting the proposed institution as "a charity [that] would save many a worthy lad [by setting] him to a noble work immediately" and the American Institute as a worthy farm sponsor whose annual exhibitions of agricultural machinery would serve indispensably as an educational adjunct. At these fairs, Wakeman contended, "more could be learned at once . . . than an individual could [acquire] by traveling about the country . . . in a year." Throughout the late 1840s, the American Institute pressed its case in language that underscored its institutional agenda of a union of science and the "arts," in this case, the art of agriculture. "The cultivation of the earth, and all the processes of agriculture," members contended, "requires a knowledge of natural science in all its departments," as well as a "practical application of [these] principles." Farmers in general, they noted, lacked the time and means to conduct experiments. Instead, the faculty and students of the proposed farm school would engage in such pursuits and would do so in the vicinity of "the great markets," vast population, and established educational and scientific resources of "the metropolis."[74] Among the tasks of the farm's faculty

71 General James Tallmadge, 1778–1853; lawyer, militia officer, congressman. While in Congress, he introduced the antislavery Tallmadge Amendment to the Missouri statehood bill and helped set in motion the legislative debates leading to the Missouri Compromise. He also served as Lt. Governor of New York in the 1820s and helped to found New York University. *Dictionary of American Biography* (New York: Scribner's, 1946–58).

72 "Sixth Annual Report of the American Institute of the City of New York, Made to the Legislature March 25, 1848," *Transactions of the American Institute* 1847 (Albany, NY: Charles Van Benthuysen, 1848).

73 *Annual Report of the American Institute* (1847).

74 *Transactions of the American Institute* (1847): 19–20, 766.

would be to "test fairly" the various food seeds and plants that arrived in the city before distributing them to the farmers of the region.[75]

The American Institute's vision of a public agricultural college would not be realized until 1862 and the passage of the Morrill Act by Congress. Their efforts were significant, however, as yet another precedent upon which Nathaniel Britton could draw in the 1890s. Though there were clear differences between the push for a botanical garden by the N-YHS of the 1820s and the 1847 petition of the Farmers' Club to set up a model farm, both actions underscored the desirability of maintaining a scientific establishment in which horticultural and botanical research would be underwritten by public funds. In addition, both efforts involved an alliance of New Yorkers across social lines whose official, republican-inflected rhetoric involved the recognition of the role of patriotic, ingenious "practical" men, or "mechanics."[76] After the defeat of the experimental farm bill, some members, in the hope of reviving the stalled project, even advocated a scaled down version of the farm that could operate solely as a botanic garden. This latter effort also failed.[77]

Although it espoused experimental farming, the American Institute remained wedded *primarily* to the principles of its Philadelphia predecessor, which entailed the elevation of artisanal dignity through the practical application of scientific knowledge and principles. It failed to translate this promise fully to its agricultural and horticultural members, however. American Institute authorities, including its president Horace Greeley, shared the Franklin Institute's faith in the "intelligent mechanic," whose understanding of the scientific principles behind his craft would enable him to foster the progress of technology in a republican society.[78] "Mechanics and their interests," the Institute's *Journal* proclaimed in 1836, "were the favored objects of the Institute."[79] Horticulture, though granted display space at the annual fairs and fostered in the 1840s by the Institute's "Farmer's Club," was never the organization's main priority. Professional gardeners and seedsmen were not "mechanics" in the sense of the word employed by American Institute leaders. The Institute's prominent benefactors considered the organization and its system of awards to be a stimulus to an "American System" (in Henry Clay's famous phrase) of domestic manufactures that, in turn,

75 "Resolutions to Establish an Experimental Farm and College" Ibid., 774–775.

76 For a typical example of such rhetoric see "To the Public," *Journal of the American Institute* 2 (1837): 169–170.

77 Peter Mead, nurseryman Thomas Bridgeman's successor as head of the American Institute's horticultural department, thought the garden project so important in light of the farm bill's defeat that he maintained it should be founded without governmental subsidy if necessary. Interestingly, the Institute also failed in its stated goal to establish a school of mines. Both projects would ultimately appear in New York only under the institutional auspices of Columbia University.

78 Sinclair, "Science, Technology, and the Franklin Institute," 194–197, see note 66.

79 *Journal of the American Institute* (1836): 89.

would bolster personal and national economic independence.[80] Peter Mead, the Institute's horticultural editor for the report on the 1849 Fair doubted whether horticulture's "real importance [was] fully understood" by all members, and complained that his area of expertise "seem[ed] to be generally regarded as a thing of secondary importance." "[The] general prevalence of [such sentiment] among the members," Mead added, "is much to be deprecated."[81] The American Institute was thus not a natural home to citizens interested primarily in natural history or horticulture. Despite the annual opportunity to exhibit the products of their nurseries, by the immediate post-Civil War years, New York's professional plant growers continued to search for an organization that would meet their specific needs.

In fact, notwithstanding the ominous 1844 epitaph penned by the Horticultural Society's secretary, the city's vocational horticulturists had begun meeting again outside the auspices of the Farmers' Club by the 1850s and had continued to publicize themselves, if less prominently, as the "New-York Horticultural Society." Gathering early in the autumn, weeks before the opening of the American Institute's fairs, the city's professional growers displayed their floral and pomological products eagerly. Apparently no N-YHS had sent delegates to a Congress of Fruit Growers convened by the American Institute in 1849.[82] Yet by 1855, C. Reagles, editor and publisher of the extremely short-lived *New York Horticultural Review*, could assert that the Society, operating in effect as an informal trade association of professional horticulturists,

80 Hindle, "The Underside of the Learned Society," in Oleson and Brown, eds. *The Pursuit of Knowledge,* 105–110, see note 25. In an address to American Institute fairgoers in 1853, William Seward highlighted this theme, asserting that "the increase of personal power and skill diminishes individual dependence, and individual independence, when it pervades the whole state, is national independence." *The True Basis of American Independence: An Address before the American Institute, New York, October 20, 1853, in the Broadway Tabernacle.* (New York: Fowler and Wells, 1854). A Whig Party leader, Speaker of the House, and member of the US Senate, Henry Clay (1777–1852) advocated an economic system based upon extensive federally financed, transportation projects; a protective national tariff; and a national bank. Each was intended to develop an overall US economy that transcended geographic regions. See Daniel Walker Howe, "Henry Clay" in Eric Foner and John A. Garraty, eds. *The Reader's Companion to American History* (Boston: Houghton-Mifflin, 1991): 190–191.

81 *Transactions of the American Institute* (1849): 99–100. During the 1840s, the Institute expended approximately $4,000 annually on agricultural matters, or one-third of its annual budgets, not an inconsiderable amount. This would seem to give lie to Hindle's assertion that the Institute was decidedly *not* an agricultural society. Yet Hindle's overall portrait of the American Institute as first and foremost an organization for the promotion of the technical arts (McCormick's reaper, Morse's telegraph, and Colt's repeating firearms all made their debuts at AI fairs) remains valid. See *Annual Report*, 1849.

82 See *Transactions of the American Institute* (1849): 191. This Congress, at which S. B. Parsons, proprietor of a prosperous Flushing nursery, played a prominent role, resulted in the compilation of an approved "American Pomological Congress General Fruit List" six pages long of fruits deemed "worthy of general cultivation."

was holding regular meetings and had sponsored a recent exhibition that was "in every respect the best ever given in the city."[83]

Though less prominent than its 1820s predecessor, the Horticultural Society of the 1850s captured the attention of the *New York Tribune*, which published a laudatory account of at least one of its exhibitions.[84] This version of the horticultural society does not seem to have thrived during the city's tumultuous Civil War years, however, and once again its constituents, the practicing horticulturists of the New York region, were devoid until the end of the Reconstruction Era of a well-organized effort to promote their interests.[85]

The Torrey Botanical Club

Held in late autumn, the American Institute fairs were imperfect outlets for the public efforts of the region's 20-odd commercial nurseries, whose products ripened well before then. Yet another reorganization of the dormant horticultural society in the 1870s thus represented an opportunity not only to reclaim social status through an alliance with the city's Gilded Age plutocracy, but to capitalize upon the advantages of association once again to spearhead a drive for a botanical garden.[86] The Society's rejuvenation had begun in the late 1870s among, once again, the city's vocational gardeners and nurserymen. Within four years of this latest attempt to resuscitate the nation's oldest horticultural society, many of New York's wealthy had joined, lending prominence to the group by serving as officers, and, crucially, bringing their socially active wives into the fold. Once more, in a departure from British precedent, the city's horticultural organization, begun and sustained by practitioners, was advanced successfully by the inclusion of wealthy amateur patrons. The ideal of accessible, practical, scientific knowledge, fostered through a cross-class institution and bestowing status on intelligent practitioners of horticultural arts, remained potent even into the 1880s.[87]

83 C. Reagles, "Regular Exhibition of the New-York Horticultural Society," *New York Horticultural Review* 1 (1855).

84 *New York Tribune*, ca. 1855. Descriptive superlatives, however, seem to have been used with some regularity. The Society's 1857 fair was described by Hovey in precisely the same manner.

85 In compiling a list of horticultural societies active in 1866, Andrew Fuller, editor of Woodward's Record of Horticulture (New York: Woodward, 1867) does not include an entry for the New-York Horticultural Society. Significantly, Charles Downing, brother of the late Andrew and prominent Hudson Valley pomologist, was listed as a member of the Western New-York Horticultural Society.

86 "List of Principal Nurserymen, Florists, and Seedsmen," *American Horticultural Annual* (1867): 148; Thomas Bridgeman, *Horticultural Report of the Nineteenth Annual Fair of the American Institute, Held at Castle Garden, October, 1846* (New York: Van Gelder and Riley, 1846).

87 Barbara Gutman Rosenkrantz, "Early American Learned Societies as Informants of Our Past: Some Conclusions and Suggestions for Further Research," in Oleson and Brown, eds. *The Pursuit of*

Horticultural journals took notice of New Yorkers' efforts. Garden writers asserted by 1880 that the revived society, operating under the same corporate charter Hosack's group used in 1822,[88] "promises . . . to become the most popular society in the United States."[89] With a roster of officers that included J. P. Morgan and a membership comprised of Roosevelts, Auchinclosses, Hewitts, Lows, Tildens, and Vanderbilts, this reincarnation of the old New York alliance of prestige and experience seemed prepared finally to match the success of sister organizations in neighboring cities. By 1883, the Society numbered over 500 members, once again held its own exhibitions, and began calling for the gathering of the horticultural and botanical "products of distant lands . . . under one roof [to] afford thousands a knowledge of their attributes, uses, and culture. . . ." With these examples, not only would science and industry benefit, members contended, but, through a healthy "spirit of emulation," an "elevated taste" for the good in landscape design would be diffused among "private gardeners, [their wealthy] employers, florists, and nurserymen.[90] Sounding a refrain that gained particular relevance in the late Victorian era, speakers at events sponsored by the reconstituted Society urged women to lend their salutary influence to the Horticultural Society.[91]

Despite grand intentions and undeniable cachet, however, the 1880s N-YHS ultimately required the assistance of the Torrey Botanical Club (TBC) in order to realize David Hosack's decades-old garden ideal. The Torrey Club, a society of amateur and professional botanists, began as an informal gathering of natural historians and plant collectors who met during the 1860s in John Torrey's Columbia College office. There, they read papers, examined books, and exchanged herbarium specimens. After Torrey's death in 1873, the group continued to meet, eventually incorporating and publishing the *Bulletin of the Torrey Botanical Club*, a respected journal of taxonomic

---

Knowledge, 349, see note 25.

88 "Charter of New-York Horticultural Society," (New York: J. Milton Ferry, 1877) in New-York Historical Society Archives; New-York Horticultural Society, Monthly Reports, 1883–1886, NYBG Archives; and William Bennett, "Essay on Horticultural Societies, Their Uses, Objects, and Advantages, Read before the N-Y Horticultural Society, April 6 1880," (New York: Sears and Cole, 1880), Library of the New-York Historical Society. Note that this is *not* the Horticultural Society of New York, which survives today. This latter organization was formed with the assistance of N. L. Britton in the early 1900s.

89 *The Gardener's Monthly* (June 1880): 192.

90 Bennett, "Essay on Horticultural Societies" (1880): 15, see note 88. The idea of a diffusion of elevated taste in landscape gardening through "emulation" as enumerated by Bennett was an explicit reiteration of a theme advanced by Downing and others decades earlier. *The New York Times* reported in 1876 that New Yorkers expended approximately $5 million annually on flowers and plants. Cited in *The Gardener's Monthly and Horticulturist* (March, 1876): 92.

91 Samuel Parsons, "Women in Horticulture, Read before the N-Y Horticultural Society, May 4, 1880" (New York: Sears and Cole, 1880). Archives of the New-York Horticultural Society.

botany. The founding editor of this journal, classics scholar William Henry Leggett, took an active interest in the promotion of botany within New York City. A frequent correspondent of Asa Gray and a source of botanical information for Charles Darwin, Leggett used the *Bulletin*, which grew from its original four-page format to a monthly exceeding 50 pages by 1900, to advance the cause of taxonomic botany.[92] In the issue of November 1873, Leggett published a directory of botanists and plant collectors, a "crucial contribution," that fostered the correspondence among plant scientists that was the lifeblood of their taxonomic efforts.[93]

In addition to Leggett, the Club was led by George Thurber, botanist to the 1850–53 US Survey of the Mexican Boundary, and John Strong Newberry, Nathaniel Britton's mentor. Early in the 1870s, these men, along with Torrey (who held a position in the college's chemistry department) had petitioned Columbia College to dedicate a sizable set of resources to the study of botany by granting the discipline full departmental status. Botany, Leggett asserted, was a branch of biology about to "claim its share of attention" from the entire scientific community, and he admonished the trustees that "somewhere there must be a school where capable men are thoroughly instructed in the science and its methods." New York, as a metropolis situated strategically between North and South, would be the logical setting for such an educational endeavor, and Columbia, already in possession of a "noble herbarium [Torrey's] perhaps unsurpassed in collections of American plants," should meet this challenge by building up its library and providing scholarships for interested students.[94] Once again, a New York learned society had begun agitating for the organized advancement of botanical and horticultural interests. Though these 1870s efforts were not immediately successful, this petition foreshadowed a mature Torrey Club's efforts with Columbia's assistance almost two decades later to establish NYBG.

By the late 1880s, after a decade in which membership had doubled and members had begun their efforts to secure city parkland for the establishment of the botanical garden, the TBC had become a significant presence on the New York scientific scene. In addition to the original *Bulletin*, the TBC in 1888 had begun publishing a series of *Memoirs*, which contained monographic treatments of various plant genera, and by 1891, had joined with other local learned societies to form the Scientific Alliance, a federation of organized scientific groups that jointly publicized meetings, lectures,

92 E. Burgess, "The Work of the Torrey Botanical Club," *Bulletin of the Torrey Botanical Club* 27 (1900): 552–558.

93 John H. Barnhart, "Historical Sketch of the Torrey Botanical Club" *Memoirs of the Torrey Botanical Club* 17: 16.

94 Draft proposal to Trustees of Columbia College, ca. 1870. William Henry Leggett Papers, New York Botanical Garden Archives.

and outings.[95] By 1889, the TBC counted 109 active members, primarily from New York City, New Jersey, and Pennsylvania. Several of these members, including professional horticulturist Samuel Henshaw (later first head gardener of NYBG), and members of the Hogg family (prominent city florists), enjoyed concurrent membership in the revitalized Horticultural Society. In addition, 138 corresponding members from remote locales in Hawaii, South America, Italy, and Syria kept the New Yorkers supplied with exotic specimens for its herbarium.[96]

Not all of the Torrey stalwarts were pleased with the changes that transformed the Club as it grew into a formal learned society. In 1887, an early Club participant asserted acerbically that the days prior to Torrey's death "were much happier and more profitable than when the Club found it necessary to fetter itself by parliamentary proceedings . . . which may be necessary and more productive in a financial way, but not so good for botany pure and simple." After Torrey's death, he claimed, "the greatest charm was gone to the oldest members."[97] In all likelihood, however, "parliamentary proceedings" would have been welcomed by the Club's sizable female contingent. Few of New York's learned organizations could boast of a constitution that enshrined women's right to election "in the same manner as [male] members," and guaranteed them the opportunity to serve as Club officers. While Horticultural Society secretary Samuel B. Parsons implored the wives of his fellow horticulturists merely to "direct [their husbands'] attention to the checkbook at [their] elbows" in order to solicit donations for a Horticultural Hall, Torrey Club women had begun attending meetings and weekly field excursions, reading papers, and even editing contributions to the Club's publications.[98]

In fact, by 1889, fully 43 percent of the Torrey Club's active roster was female. Included among this number were Jeannette Greene, a physician, and Emily L. Gregory, one of only 17 American women to earn a doctorate in botany during the nineteenth century. At those Club gatherings, Gregory, with a confidence perhaps born of her unusual credentials, did not hesitate to challenge the assertions of male members when she could present evidence to the contrary.[99] In addition, Elizabeth Britton,

95 See Barnhart, "Historical Sketch," 16–17, note 93 and Baatz, *Knowledge, Culture, and Science*, 139–142, note 9.

96 Records of the Torrey Botanical Club, New York Botanical Garden Archives, 1889.

97 Isaac Hall to Torrey Botanical Club, January 10, 1887, Records of the Torrey Botanical Club, New York Botanical Garden Archives.

98 Parsons, "Women in Horticulture," see note 91.

99 Her degree was granted not by an American college, but by the University of Zürich,rich in 1886. See Emanuel D. Rudolph, "Women in Nineteenth Century American Botany: A Generally Unrecognized Constituency"*American Journal of Botany* 69 (1982): 1351. Minutes of the Torrey Botanical Club, January 23, 1889, New York Botanical Garden Archives.

who met and married Nathaniel while both were Club members, served for a time as an editor of the *Bulletin*, whereas Maria Steele served as the group's Recording Secretary and Louise Merrit Stables as a curator.[100]

As the Torrey Club, stirred by Elizabeth Britton's report of her journey to Kew in 1888, began its botanic garden campaign in earnest, its members noted in a public appeal that the Club had "many lady members, [who] would address [themselves] to their fellow citizens of both sexes to take steps to make among us in this great city a home for the flowers, and trees and products of all nations." The three women on the Club's Botanic Garden Committee appealed for support explicitly to those "wealthy public spirited women whose private collections, and whose residences show forth their appreciation of God's lavish gift of beauty to the world."[101]

Torrey Club Women and Middle-Class Botany: Making the Public Case for a Garden

The social space that allowed women such a prominent role in these scientific and cultural endeavors was a function of contemporary Victorian assumptions regarding nature in general and botany in particular. Because plants, and specifically flowers, seemed the most graceful and delicate of providential creations, they were considered suitable for the attention of women and girls, whose characteristic feminine traits of perception and sensitivity would enable them to appreciate the morally uplifting benefits of such close study. That women would display these characteristics was considered a given. Such traits were attributed to women by late-nineteenth-century feminists and suffragists, who at times cleverly adapted the logic of social and biological Darwinism to undermine contemporary essentialist arguments that attempted to find a normative "natural history" for the social inferiority of women.[102] Thus ladies and their daughters were given license to "botanize" on rambles through the countryside. Excursions such as those sponsored each Saturday by the Torrey Club to rural locales including the New Jersey Pine Barrens and the North Shore of Long Island provided "healthy and agreeable exercise," well-suited to "gentle minds" and feminine constitutions. Indeed Elmira H. Lincoln, in her popular *Familiar Lectures on Botany*, a text aimed at both women and children, had even predicted that such pursuits would diminish the occurrence of scandal, which, she wrote, "oftener proceeds

100 Records of the Torrey Botanical Club, TBC Membership folder, 1875–78, New York Botanical Garden Archives.

101 Torrey Botanical Club, "The Need of a Botanical Garden in New York," 1889, New York Botanical Garden Archives.

102 See Rosemary Jann, "Revising the Descent of Woman: Eliza Burt Gamble," in *Natural Eloquence: Women Reinscribe Science*, Barbara T. Gates and Ann B. Shteir, eds. (Madison: University of Wisconsin Press, 1997): 147–163.

from a want of better subjects, than from malevolence of disposition."[103]

Meanwhile, popular authors developed elaborate lexicons of floral "language" through which messages could be conveyed in a socially sanctioned and chaste manner, and urged women that while healthfully engaged in the "gentler tasks" of the garden, they might "make the acquaintance of Flora," and thus ensure domestic happiness.[104] Though considered appropriate for women's attention, both botany and horticulture, it was felt, needed to be conveyed in a form palatable to feminine sensibilities. To this end, texts aimed at women and children were written in euphemized verse and illustrated by reassuringly domestic, allegorical scenes of familial harmony with a benign, fruitful nature. The Linnaean system of classification, thought simple enough to be taught to women and girls, was based upon the tallying of reproductive anatomical features of flowers. It was subsequently modified in a way deemed appropriate for feminine consumption and use: the intimate details of floral reproduction were considered far too lascivious for female minds. To be conveyed appropriately, writers of botany texts aimed at female and juvenile audiences resorted to poetic metaphors in which flowers, personified as romantic actors, participated in a fanciful set of maneuvers that conformed to the Victorian ideal of courtship.[105]

Once such "safeguards" had been effected, widespread public sanction of women's participation in both botanical and horticultural pursuits was a logical, if temporary, result. Popular authors, male and female, noted approvingly that feminine interest in the flora of the fields as well as the cultivation of products of the garden could claim biblical origins in the Edenic story of the Book of Genesis. "We are told in sacred history that the first gardener had a woman given him for a . . . partner," wrote Andrew Fuller, editor of *Woodward's 1867 Record of Horticulture*. "Why then should we

103 Cited in Benes, ed., *Plants and People*, 108, see note 35. See also William Darlington's preface to Alphonso Wood's 1851 *First Lessons in Botany*, ibid., 107; Minutes of the Torrey Club, 1883–1889, Records of the Torrey Botanical Club, New York Botanical Gerden Archives; and Keeney, *The Botanizers*, see note 33. The coeducational botanizing excursions of the Torrey Club represented a marked change from the rural rambles of a century earlier. The sexualized nature of Linnaean botanical taxonomy led eighteenth-century English moralists to decry such joint excursions as immoral. See Ann B. Shteir, *Cultivating Women, Cultivating Science: Flora's Daughters and Botany in England, 1760–1860* (Baltimore, MD: Johns Hopkins University Press, 1996): 16–28.

104 Leighton, *American Gardens of the Nineteenth Century*, 85–94, see note 7. See also Jack Goody, *The Culture of Flowers* (Cambridge, UK: Cambridge University Press, 1993). The prominence of female participation in botany by the end of the nineteenth century was such that it even became the fodder for newspaper cartoonists. When the American Association for the Advancement of Science held its annual convention in Denver in the summer of 1901, the front page of the newspaper was covered with humorous sketches depicting the meetings of the various sections of the Association. Under a drawing of a large congregation of formally dressed women stood the caption, "the botanical section, of course." *The Rocky Mountain News*, Tuesday, August 27, 1901.

105 Leighton, *American Gardens of the Nineteenth Century*, 85–94, see note 7.

not only admit, but encourage women to assist in producing those blessings that our Creator in his beneficence has given to mankind?"[106] Such reasoning did not extend to an endorsement of female labor in field agriculture; even the "coarse drudgery" of milder horticultural efforts, Fuller asserted, needed to be left in the rough hands of "such men as seem to have been created to be hewers of wood and drawers of water, and whose tastes and aspirations are but little above those of animals."[107] The xenophobic nature of Fuller's diatribe notwithstanding (it was clearly aimed at the overwhelming proportion of foreign-born professional gardeners employed on American estates, many of whom were Irish), his depiction of feminine participation in the "gentle" pastime of horticulture captures perfectly the prevailing attitude toward a feminine relationship with nature during the last half of the nineteenth century.[108]

There was, of course, another dimension to Fuller's plea for women's entry into horticultural circles (and societies). Implicit in his efforts to stimulate female interest in an appropriate relationship to the products of garden and field was a strain of anti-urbanism common to his contemporaries. His passion stemmed from a belief that "thousands of women, old and young [were] now crowded into cities [and] would gladly seek employment in the fruitful fields [though *not* farm fields] of the country if they could be assured that the finger of scorn would not be pointed at them." Would it be "less noble," he wondered, if more women were to emulate those few that had established their own commercial greenhouses, than to "wait upon customers in some close, half-stifling shop in a city," or to "ply a needle for sixteen hours a day . . . with no time . . . to enjoy the exhilarating pleasure of breathing the free air of the country?" His fear was such that he advocated entrance of women into the newly established agricultural colleges recently endowed with federal support by the Morrill Act. (Presumably, his ideal female students would study only the "smaller" fruits grown on campuses.)[109]

By the time prominent landscape designer and nurseryman Samuel Parsons exhorted women to lend their energies to further organized horticulture, he could draw upon a widespread cultural idiom that, by encompassing ideals of domesticity and presumptions of feminine compatibility with flowering plants, justified his appeal. In fact, these attitudes stemmed not only from late-century ideals of domesticity, femininity, and scriptural interpretation, but were rooted in Romantic European precedents that had associated woman with nature since at least the eighteenth

106 Fuller, "Women in Horticulture," *Woodward's Record of Horticulture*, 28, see note 85.

107 Ibid.

108 Downing estimated that only three percent of paid gardeners in the US in 1850 were native-born. See note 24.

109 Fuller, "Women in Horticulture," 29–30, see note 85.

century.[110] Both European and American plant enthusiasts knew that wives and daughters had served as botanical helpmates for decades as the exotic New World flora had been collected, catalogued, and transferred to Europe beginning in the eighteenth century.[111] These women, from the domestic spheres of their homes, had corresponded with leading (male) scientific figures; contributed reports to the growing number of learned journals; collected, mounted and exchanged specimens; and served as illustrators for published botanical works. At the century's start in England, women's role in botany was even considered an essential adjunct to motherhood, because basic scientific education took place within the home and women's familiarity with plant life often derived from their role as family healers conversant with herbal remedies.[112]

And yet, even this long tradition of participation in botany and horticulture in various guises did not guarantee women equal status within learned societies or, as the century progressed and science became the province of paid, credentialed professionals, in university settings. If scriptural allusions could provide an entryway for women into the larger public world of these particular branches of learned culture, they might also provide justifications for excluding them altogether. In the first decades of its existence, the Massachusetts Horticultural Society denied honorary membership to prominent women, since, as some members noted, "a woman made trouble in Eden."[113] By the early 1880s, as the Torrey Club prospered, many learned bodies, including the venerable American Philosophical Society, continued to deny admittance to women.[114] Even as cultural authority was invested in the proposition

110 See, e.g., Lisbet Koerner, "Goethe's Botany: Lessons of a Feminine Science" *Isis* 84 (1993): 470–495. In 1800, as the model German Romantic, Goethe practiced a dilettante's natural history, the only kind eventually deemed appropriate for both European and American women. Goethe's "feminine" science entailed substituting a personally experienced immediacy with nature for the quantified, technologically assisted observations of modern scientific botany. The latter was considered the province of male botanists.

111 For a fuller discussion of women in eighteenth and nineteenth-century trans-Atlantic botanical history, see Shteir, *Cultivating Women, Cultivating Science*, note 103; Nancy Slack, "Nineteenth-Century American Women Botanists: Wives, Widows, Work," in Pnina Abir-Am and Dorinda Outram, eds. *Uneasy Careers and Intimate Lives: Women in Science, 1789–1979* (New Brunswick, NJ: Rutgers University Press, 1987); Debra Lindsay, "Intimate Inmates: Wives, Households, and Science in Nineteenth-Century America" *Isis* 89 (1998): 631–652; and Helena Pycior, Nancy Slack, and Pnina G. Abir-Am, eds. (New Brunswick, NJ: Rutgers University Press, 1996).

112 Shteir, "Botany in the Breakfast Room: Women and Early Nineteenth-Century British Plant Study," in Abir-Am and Outram, eds. *Uneasy Careers and Intimate Lives*, 33–42, see note 111 and Shteir, *Cultivating Women*, see note 103.

113 Thomas Meehan, "Ladies as Members of Learned Societies," *The Gardener's Monthly and Horticulturist* 28 (June, 1886): 188.

114 *The Gardener's Monthly and Horticulturist* (1887): 251.

that the gentler pursuits of the garden suited feminine sensibilities, the scientific tasks of botany began to be defined as beyond the proper purview of women. In Britain, so often the model for American cultural pursuits, this process had begun as early as the 1830s. Women's careers in botany began to be problematic, culturally, as botany "came to have a modern profile as a science," leaving behind the private, domestic sphere in which women operated to enter the public realm of modern institutionalized botany.[115]

English, and subsequently American, authorities wished to defeminize botany and were determined to forge distinctions between a "polite botany" (which would continue to serve as a suitable diversion for ladies) and a rigorous, masculine, botanical science. Thus, in England early in the century and in the US several decades later, a deepening divide appeared between popular and academic, generalist and specialist, and the "high science" of professionalizing disciplines (including botany) and the "low science" of practitioners who diffused the knowledge for practical use, along with those dilettantes and autodidacts whose passion for natural history became increasingly unlike the dispassionate investigations of university-trained scholars.[116] This differentiation of recognizable types of botanical and horticultural practice, of course, occurred in Britain as part of the pronounced divergence, described above, of upper-class learned-society botany from artisan botany practiced by workingmen gathered in public houses. This gradual set of changes in nineteenth-century Anglo-American science involved a concomitant secularization of the realm of serious botanical knowledge. Whereas polite feminine "botanizing" had been infused with the tenets of evangelical morality, expert masculine botany, while not abandoning the idea of a "natural theology" altogether, became the exclusive province of academic professional men.[117] By century's end, then, male-dominated scientific botany was no longer considered an analogue to the domestic domain of the garden. With the notable (and anomalous) exception of a few women members of the TBC, it had been transformed into the domain of male experts, who, though still dogged in some quarters by botany's reputation as the pastime of women, children, and the sickly, began building a new institutional apparatus with which they might further advance their disciplinary aims.[118]

115 Shteir, *Cultivating Women*, 148–149, see note 103.

116 Ibid., 148–149.

117 Ibid., 154–157 and Lustig, "The Creation and Uses of Horticulture," 5–6, see note 39.

118 Victorian moralists and commentators had portrayed both horticulture and botany as feminine, but the marginalization of this realm of natural history had not ended there. By the 1880s, male botanical professionals feared a widespread (and potentially emasculating) public perception that no one *except* women, the very young, and invalids would find plant life, either cultivated or wild, interesting. See E. F. Andrews, "Botany as Recreation for Invalids" *Popular Science Monthly* 28 (1885–86): 779–781 and "Is Botany a Suitable Study for Young Men?" Science 9 (1887): 116–117.

Britton's Nineteenth-Century Botanical and Horticultural Precedents

By the 1890s, a less heterogeneous organization, one constructed along the lines of an upper-class British model and dedicated to erecting a clear homage to British botany, was able to secure land and money from New York City's government. In addition, it is crucial to note that with the culmination of a late-nineteenth-century Victorian cult of femininity, wedded to a gendered conception of botany, a significant aspect of this success cannot be overlooked: it was the result of a "coeducational" effort by the Torrey Club. Whereas New York's prior nineteenth-century efforts had been the province of cross-class organizations, by the late 1880s those efforts could be best prosecuted by a set of actors that was not only more homogenous in social status, but that also incorporated the "salutary" influence of women within a clearly gendered science. The editors of the *New York Tribune, The New York Times*, and *The Evening Post* all noted between January and May, 1891, that Maria Daly, the wife of Judge C. P. Daly, had convened a committee of the Torrey Club, along with a set of 50 prominent women of wealth in her parlor to plan to secure a legislative charter for a new botanical garden. These women had met with Judge Addison Brown; Nathaniel Britton; and Henry Cadman Potter, the Assistant Episcopal Bishop of New York, to discuss incorporation and selecting a site for the garden.[119]

The various societies producing and disseminating botanical knowledge that preceded the Torrey Club must be considered somewhat exceptional to the models of learned societies depicted in previous studies.[120] Whereas in most such organizations prestige was a concomitant of professional (or professionalizing) status, in nineteenth-century horticultural clubs, social status was contingent upon the participation of "amateur," gentlemen horticulturists, who in turn employed or patronized "professional" nurserymen and gardeners. Even as Hosack and other elite members of the New-York Horticultural Society continually acknowledged the origins of their organization among "practical men," these gentlemen gardeners and estate owners provided what little financial backing and social prestige the Society commanded during its heyday in the 1820s.[121] This idiosyncratic characteristic of organized New York horticulture served to differentiate it during the nineteenth century from the later incarnations of its sister science botany, a field of research that did conform to the traditional form of increasing prestige on the part of "professionals," and the marginalization of the formerly valuable contributions of autodidactic amateurs. As the century progressed, the associations with, and direct contributions to, practical

119 *New York Tribune*, January 9, 1891; *The Evening Post*, May 19, 1891, and *The New York Times*, May 20, 1891.

120 e.g., Hindle, "The Underside of the Learned Society" (note 61) in Oleson and Brown, eds. *The Pursuit of Knowledge*, see note 25.

121 *The New-York Farmer and Horticultural Repository* 1 (January, 1828).

organized horticulture by botanists John Torrey and Asa Gray diminished in direct proportion to the degree to which they each became immersed in the research tasks of floristic botany, which, in turn, would become a subject deemed suitable for elite graduate education within university settings.[122] Thus were the best botanical minds lost to organized horticulture. In nineteenth-century New York botany, the union of theory and practice characteristic of other early American learned or premium-granting societies never really materialized, ensuring in the process that although the city's botanical scientists professionalized and eventually thrived, its numerous professional horticulturists (gardeners, nurserymen, and seedsmen) failed in their efforts to create a prominent organized outlet for their own concerns.[123] The result of this severing of botany from horticulture, coupled with the geographic dispersion of the region's horticultural enthusiasts, was New York's marked inability at mid-century to support a horticultural organization. The city enjoyed only two periods of significant success in promulgating its own horticultural knowledge during the nineteenth century: the 1820s and the 1880s–1890s. In both periods, sometimes disparate groups aligned themselves around common perceptions of civic pride; personal pride (in lavish family estates); craft pride (an earnest interest in horticultural techniques on the part of the nurserymen); and financial self-interest. Until the latter era, a Gilded Age marked by vast accumulations of wealth and enthusiasm for the construction of institutions to serve as signifiers of civic progress, the ties that bound estate owners and commercial growers were too tenuous to support more than ephemeral organized horticultural efforts. With the maturing of institutionalized botany by the end of the century, however, a determined scientific entrepreneur was able to harness the wealth of New York's philanthropists, link it (through the horticultural and landscape needs of the new botanical garden) to the practical/applied knowledge of its horticulturists, and join them to the institutional resources and scientific expertise of its botanists to unite in a new and permanent institution those formerly disparate elements.

By the end of the nineteenth century, neither the Gilded Age plutocrats nor the professional botanists securing positions within permanent university settings neglected

122 Notable exceptions to this portrait of horticultural exceptionalism, of course, are Downing and, in Boston, C. M. Hovey. Both men managed to avoid the loss of prestige attendant with horticultural professionalism by editing influential journals in which they promulgated standards of taste in landscape design. Even Downing, however, sometimes met with patrician disdain: one genteel Philadelphian, after meeting Downing, remarked, "He is a Yankee and not a thoroughbred. Landscape gardening is with him a profession and not a liberal taste, and he talks with a professional air. I dislike 'bread studies and artisanship and the smell of the shop destroys my pleasure in any subject." Cited in Major, *To Live in the New World*, 33, see note 64.

123 Rosenkrantz, "Early American Learned Societies," 349 (note 87) and Hindle, "The Underside of the Learned Society in New York," 110 (note 61) in Oleson and Brown, eds. *The Pursuit of Knowledge*, see note 25.

the need to make their case for publicly supported New York botany by an appeal to the general public. (Britton's choice of a seal that touted the Garden's ability to promote the "welfare of the people" remained a perfect example of the ways in which this new professional/plutocratic elite manipulated public discussions in order to secure governmental support.)

Finally, the 1890s alliance of professional botany and plutocratic horticulture was contingent upon first, the participation of women in the public sphere (in an appropriate and socially sanctioned manner aimed at increasing civic prestige) and second, the effective termination of what had traditionally been a productive partnership across class boundaries between an upper-class learned elite and the city's set of working horticulturists, seedsmen, and nurserymen. The latter group had found an outlet for its concerns in the various organizations that had incorporated and held exhibitions in the city throughout the nineteenth century. Yet each of these groups had attained status and success primarily as a result of the presence in the membership ranks of wealthy New Yorkers, who, nevertheless, had been quick to acknowledge the contributions of their "practical"-minded social inferiors.

Elected to the ranks of the Torrey Club in 1877, Nathaniel Britton began building his career as a leading botanist within a context of male-dominated scientific specialization. As he organized the specimens left behind by John Torrey, Britton astutely promoted his own career by seeking to complete the labors of his local scientific predecessors. Britton's pronounced entrepreneurial skills were needed to bring to a conclusion the decades-old efforts of the city's botanists and horticulturists and forge an institution that made New York into a true producer, organizer, and disseminator of scientific knowledge. Britton considered his work to be the logical culmination of the efforts of Hosack, Torrey, and the various learned organizations that had labored throughout the nineteenth century to bring the city's scientific reputation up to a suitably commanding position in American letters.

Compared to contemporary British examples, the American botanical organization appears exceptional. The cross-class patterns of antebellum New York City horticulture and botany represented a decided departure from English models, although the scientific and cultural Anglophilia of the key Americans involved in such efforts was pronounced. New York's botanists and horticultural enthusiasts' ultimate success in securing a degree of public support involved, at first, making botany an adjunct of geologic surveys being conducted in the West as well as in New York State and New Jersey. At century's end, however, the practice of botanical science and public horticulture was the province of men and women of the Brittons' social class, including their compatriots in the TBC, as well as the city's unrivaled set of philanthropists. Even this set of powerful actors, however, found it necessary to make recourse to the middle- and upper-class women in their ranks in order to articulate a rationale for a great public garden to the New York State Legislature.

In light of Britton's ultimate success, little significance should be ascribed to the relative obscurity of New York's mid-nineteenth-century horticultural endeavors. In reality, the successes of other Northeastern horticultural societies in Boston and Philadelphia were exceptional. Learned societies had the potential to bring together wealthy sponsors with skilled investigators and researchers, but such collaborations were more effective for singular, short-term efforts than the sustained activity of long-lived institutional structures.[124] Thus, many more groups organized for the advancement of specialized forms of knowledge faltered than prospered.[125]

124 Sally Gregory Kohlstedt, "Institutional History," *Osiris* 2nd Series 1 (1985): 31.

125 Note, e.g., that even the Pennsylvania Horticultural Society had had difficulties: in the early 1880s it had not convened for regular monthly meetings "for some years," according to Thomas Meehan's *Gardener's Monthly and Horticulturist* (May 1884): 158.

Chapter 3
Creating New York's Second Botanical Garden

In November 1888, the editors of *The New York Herald* published a list of noteworthy European botanic gardens, ranking Russia's St. Petersburg Garden as "the best in Europe," ahead of such notable Old World landscapes as the Jardin des Plantes in Paris and the Royal Botanic Gardens at Kew, England. In sharp contrast to these venerable centers of horticultural, scientific, and educational expertise, the *Herald* reported, stood a representative American botanical institution: the US government's botanic garden in Washington, DC, "a swindle on the public simply useful in supplying Congressmen with flowers at the expense of taxpayers." "Shaw's Garden" (today's Missouri Botanical Garden), the sole American garden that deserved to be named in the same category as Europe's horticultural masterpieces, the writer noted, was in St. Louis. Clearly, the *Herald* maintained, a great new *American* garden was needed, and New York City should be the setting for an "artistic floral study ground" that would foster the "softening, civilizing education of both rich and poor."[1]

Calls for improving the city's public landscape were common in the late 1880s. In the years following the Civil War, as the city's population surpassed 900,000, New Yorkers had been eager to create public spaces and imposing structures of beauty. Other than the justly celebrated City Hall, situated in an area no longer at the heart of the expanding metropolis, there were very few buildings noteworthy for beauty or elegance. Gradually, however, in the 1870s, '80s, and '90s, private wealth began transforming New York's urban landscape, with stately hotels, grand residences, and, eventually, imposing museums erected along major thoroughfares.[2]

Members of New York's Torrey Botanical Club (TBC) joined the voices lamenting their own metropolis's "laggard" status in both botany and civic grandeur. From 1889

1 In their very brief survey of American gardens, *Herald* writers managed, of course, to omit mention of many notable ones, including those in Boston and Pennsylvania. No doubt they were aware of the existence of such established settings as Harvard's Arnold Arboretum, so their argument represented a form of regional civic boosterism, and their point was clearly understandable to New York readers. *The New York Herald*, November 26, 1888. Note that all nineteenth-century periodical articles cited in this and subsequent chapters can be found in the clippings files of the New York Botanical Garden Archives.

2 J. B. Jackson, *American Space: The Centennial Years, 1865–1876* (New York: Norton, 1972): 202–203.

through early 1891, the TBC called upon New York to establish a world-class botanic garden "in the interest of American Science; as a nucleus and a center for the collection, the cultivation, and the study, in their living form, of the Flora of this vast continent, and for the gathering, exhibition and study of the floral treasures of other lands." Such an institution, proponents contended, "would exert a great educational influence, and become a source of perennial instruction and delight to all classes of our people."[3] Citing the shining example at Kew outside of London, the Torrey Club asserted that the garden would be "a national delight" within the metropolis, providing the refreshing beauty of natural scenery and the mental refinement of the cultivation of flowers as well as the practical benefits of science. New York's scientific garden, Club members wrote, would be able to "preserve, perpetuate, acclimatize and distribute medicinal herbs, useful trees, seeds, [and] everything that distant lands contribute towards the sustenance of man, or that can alleviate and cure disease."[4]

Since the TBC, at Elizabeth Britton's urging, had first gone public with its plans in the late 1880s, the city's press and national botanical and horticultural periodicals had seemed unanimously supportive of the project. The *Herald*, citing the city's "botanical weakness," had explicitly expressed its approval early in the campaign, asserting that a botanic garden would help increase the level of education in the city.[5] Throughout the winter and early spring of 1889, other journals followed suit, chronicling with interest the efforts of the Torrey committee to bring botanical respectability to New York. In February, *The American Druggist*, after recounting the fate of New York's (David Hosack's) "lost garden," placed responsibility for reestablishing a botanic institution with Columbia College, already a key player in the effort to found the New York Botanical Garden (NYBG). *The Medical Record* suggested that physicians lobby their wealthiest patients to support the Garden cause. *The Commercial Advertiser, The Sun, The Post*, and *The Home Journal* all echoed these sentiments in their pages, describing in sometimes adulatory prose the composition of the Torrey botanic garden committee and anticipating the "wholesome influence" that a first-rate botanical garden would have upon the city. "A frequent communion with budding and growing things," *The Home Journal* intoned, would help awaken in the city's population "a sense of renewed life, of ever-increasing life-force." The new botanic garden would serve "not only as a treasury of facts, but of fancies, stimulating perception of form and color and training the visitor to distinguish in matters of taste."[6]

3 Torrey Botanical Club, "The Need of a Botanical Garden in New York," 1891. Pamphlet in New York Botanical Garden Archives.

4 Ibid. Although the actual *cultivation* of the proposed garden's flowers would be accomplished by hired staff members, instruction in home horticulture would presumably serve to diffuse these "elevating" intellectual benefits throughout the wider public.

5 *The New York Herald*, November 27, 1888.

6 The Medical Record, February 2, 1889; *The Commercial Advertiser*, February 2, 1889; *The New York*, Sun, February 10, 1889; *The Home Journal*, March 13, 1889. The *Post* and the *Sun* did, however, note

The American Art Journal was even more enthusiastic: in its February, 1889 report on the work of the Torrey botanic garden committee entitled "A Million Dollar Garden," the *Journal* favorably compared the project to New York's extant museums of Art and Natural History and justified the proposed expenditure of public monies on the basis of the Garden's predicted popularity. The *Journal* confidently asserted that this "people's garden" would attract thousands more visitors than libraries and colleges and would grow to be the "greatest popular scientific institution on this side of the Atlantic."[7]

A garden of the scale and stature proposed by the TBC would serve the city as an indispensable scientific and educational institution "quite as much so as a library or a college," garden partisans asserted, and would also serve horticultural, pharmaceutical, and recreational needs. By 1891, the Torrey Club could inform New Yorkers that, through the efforts of its members, the city Parks Department had been persuaded to reserve a portion of Bronx Park,[8] a recently annexed tract of city parkland north of the Harlem River, for the establishment of a botanic garden. "In natural beauty," Torrey members gushed, the tract was "the very paradise of artists," and its landscape replete with soil that exceeded Kew's in fertility, a profuse natural supply of fresh water, and ample protecting timber was "the very best that could be found for the purposes of a great botanical garden and arboretum."[9]

During the 1890s, a state charter and city parkland were granted, money was raised, and plans were laid for the physical layout of NYBG. These events were closely watched by contemporaries interested in the development of New York City's civic and scientific life and were contested by those, both within and outside of the city, who felt that things should proceed differently. NYBG was almost universally hailed as

---

that the first chairman of the Club's botanic garden committee, E.E. Stearns, began behaving erratically in mid-March, 1889. He feuded with other committee members regarding his use of printed circulars at a flower show to solicit subscriptions to the Garden fund and attempted to resign from the committee. Within days, he was admitted to Bellevue Hospital's "insane ward" presumably at the request of his wife, who reported that "he [had] religious mania" and had attempted to choke her. Stearns had begun declaring himself a "second Christ" destined to destroy New York within 44 years. Mrs. Stearns noted that he had been upset with worries regarding the botanic garden project, and that he had been in an asylum a decade earlier. See *The New York Evening Post*, March 11–14, 1889 and *The New York Sun*, March 16, 1889.

7 *The American Art Journal*, February 23, 1889. The Torrey committee indeed sought to establish a "people's garden," at least initially. From its earliest efforts in the late 1880s through the early 1890s, in addition to seeking large donations from wealthy benefactors, three-year, $30 subscriptions (payable in $10 annual installments) were available to city residents of more modest means. With the panic of 1893, however, and the subsequent dearth of available potential donors, the Garden incorporators would seek funds from the wealthiest of New Yorkers *first*, and later would make provisions for smaller annual membership fees.

8 This was initially the southern end of the park, though the parcel assigned would ultimately comprise the northern section.

9 Torrey Botanical Club, "The Need of a Botanical Garden in New York," see note 3.

a scientific resource, but access to it was hotly contested among the city's growing universities. The battle was joined: for the better part of the 1890s, various actors struggled to establish and define New York's final grand institution of the nineteenth century.

Nathaniel Britton, possessed of a modest personal fortune, recognized that the needs of New York's elite class to display affluence to rival that of any city could be met in part by the construction of the Botanical Garden. In a deliberate imitation of his British contemporaries, who had couched their own institutional needs in a rhetoric of public interest, Britton took care to tie the fortunes of this Garden to those of the growing metropolis.[10] He identified a natural set of allies to unite behind the cause of New York botany. Among these men and women were professional botanists as well as middle- and upper-class botany enthusiasts; philanthropists; and public officials eager either for the potential "reforming" influence of another landscaped city park or the opportunity for the patronage attendant with large public works projects (or both). Britton observed with envy the ways in which his British counterparts at Kew had articulated a concept of, in Richard Drayton's characterization, "national efficiency," in order to gain the respect and direct support of the British government and the wider public. Kew's successful administrators had gained financial support, Drayton notes, with arguments intended to underscore a link between the development of the Empire's human and material resources and their "rational administration" by scientific adepts willing to harness their expertise to advance the overall interests of the nation. This centralized "Continental" model of administration, Drayton writes, ultimately entailed a "marriage between knowledge and established power."[11] Such a productive alliance, Britton recognized, could only profit the city's botanical garden as a scientific institution, and throughout his career in New York, he utilized similar arguments to contend that the city and the nation needed to underwrite his personal vision for the Garden in order to realize its own potential in the Western Hemisphere. Within the context of American Progressivism at the turn of the twentieth century, with its own belief in rational administration for local and national ends, Britton's strategic borrowing of British models found eager listeners.

Britton's scientific goals, allied with the civic aims of the city's wealthiest citizens, were aided by the close personal and institutional ties he maintained to Columbia University, whose president, Seth Low, was a major figure in the City's political and cultural life. To an extent, Britton and his philanthropic allies shared a vision of the Garden as an instrument of social policy, a view nurtured by the environmentalist beliefs of the nineteenth century that had guided Frederick Law Olmsted in his design

10 Richard Drayton, "Imperial Science and a Scientific Empire: Kew Gardens and the Uses of Nature, 1772–1903" (PhD Dissertation, Yale University, 1993): 16–18.

11 Ibid., 18–20.

FIGURE 2
Nathaniel Lord Britton, 1904.

for Central Park. The individual philanthropists who incorporated the Garden provided the fledgling botanical institution with legal expertise as well as financial resources and helped create the institutional framework that governed the Garden throughout Britton's term as director. That this framework was challenged by the chancellor of today's New York University, Columbia's main institutional rival, should have been unsurprising. That Britton, Low, and their Garden coincorporators were caught off guard says something about their expectations of success and sense of entitlement to the city's cultural wealth.

Although ultimately they met this challenge to their planned institutional structure, the Garden incorporators nevertheless faced even more daunting tasks: raising the necessary subscription money during one of the nineteenth century's bleakest economic depressions and getting the Garden built according to a specific plan (drawn up by Britton) while faced with strong public opposition by powerful Boston-based members of the landscape architecture profession. A detailed recounting of Britton's original 1896 plan for the construction of the Garden, along with a discussion of the nature of the horticultural rivalry between New York and Boston authorities comprise the remainder of this chapter.

Nathaniel Lord Britton (1859–1934)

Nathaniel Lord Britton (FIGURE 2), the crucial figure in the development of NYBG, was born on Staten Island on January 15, 1859 to Jasper Alexander Hamilton Britton, 52, and his second wife Harriet Lord Turner, aged 30. Related to the Vanderbilt family, the Brittons traced their lineage in the New World to seventeenth-century New York. Nathaniel Britton served as the first director-in-chief of NYBG from 1896 until his retirement in 1929; he died on June 25, 1934 and was buried in Staten Island's

FIGURE 3
John Strong Newberry, 1881.

Moravian Cemetery.[12] During his 75 years, he built a career as one of New York City's prominent men of science and established institutional botany on an Americanized British model in his native city.

Educated primarily in geology (and to a lesser degree in botany) at Columbia's new School of Mines in the 1870s, Britton had been a student and assistant of John Strong Newberry (FIGURE 3), himself a "major influence on the development of New York science."[13] In many respects, Britton's own career trajectory resembled that of his mentor. As boys, both men had been passionate natural historians and had gathered botanical specimens near their homes. Both established themselves in the wider context of American science by helping to build the reputation of New York City institutions: Newberry at Columbia and Britton at NYBG. Newberry, trained in medicine, had spent his early working life on various Western expeditions as a government botanist, determining specimens on US Surveys in California, Colorado, Arizona, Utah, and New Mexico.[14] Similarly, while still a student at Mines, Britton had joined the TBC and had begun building what became, by the end of his life, a lengthy bibliography of botanical publications. Upon graduation from Columbia, he had served as Newberry's assistant in geology and, echoing his mentor's early work, served for two years as the botanist to the New Jersey Geological Survey. His 1881 PhD dissertation at

12 Nathaniel Lord Britton Papers and Family Records, Staten Island Institute of Arts and Sciences, Staten Island, New York.

13 Douglas Sloan, "Science in New York, 1867–1907," Isis 71 (1980): 39.

14 Ibid.

Columbia, "A Preliminary Catalogue of the Flora of New Jersey," grew out of this experience.[15]

Further parallels can be drawn between Newberry and his student. The elder man had been a dedicated institution builder. Before Columbia, Newberry had helped rejuvenate the city's Lyceum of Natural Sciences, leading it as it changed its name to the New York Academy of Sciences, and became a modern organization dedicated to fostering specialization (even though Newberry himself remained a classic natural historian with a breadth of interests and experience). Though sometimes criticized for the very breadth of his knowledge, Newberry remained firm in his belief in science's public role and community involvement within the city and the nation, a commitment absorbed by the young Nathaniel Britton in the 1870s and 1880s.[16]

Henry Gleason, NYBG's Assistant Director and a relatively frequent visitor at the Brittons' home not far from the Garden, recalled that the director-in-chief had "epicurean" tastes in food and employed a good cook (whose husband served as Britton's driver) but was indifferent in his dress and personal appearance. In their later years, the Brittons, Gleason noted, rarely entertained, and Nathaniel had, apart from his taxonomic and administrative duties, "no known form of recreation or amusement." Britton's evenings after dinner were usually spent engaged in contemplation of Garden duties. Gleason recalled that Britton spent much of this time smoking a pipe, rarely speaking, and pacing on a carpet in his study, into which his footsteps had worn a groove.[17]

Of Britton's temperament, one contemporary noted, "the least said"... the better."[18] Diminutive in stature, with a stooped posture and a long beard that appeared to comprise the majority of his weight, Britton nevertheless cut an intimidating figure.

15 Biographical, bibliographic, and genealogical information on Nathaniel Britton can be found in: E. D. Merrill, "Biographical Memoir of Nathaniel Lord Britton, 1859–1934," *National Academy of Sciences Biographical Memoirs* 19 (1938): 147–202; *Dictionary of Scientific Biography* vol. 2 (New York: Scribner's, 1973); *Dictionary of American Biography* XI (New York: Scribner's, 1958): 121–122; M. A. Howe, "Nathaniel Lord Britton" *Journal of the New York Botanical Garden* 35 (1934): 169–180; "Nathaniel Lord Britton," *Proceedings of the Staten Island Institute of Arts and Sciences* 7 (1933–1934): 101–108; M. L. Fernald, "Nathaniel Lord Britton, 1859–1934," Proceedings of the American Academy 70 (1936): 504–505; T. A. Sprague, "Nathaniel Lord Britton," *Kew Bulletin of Miscellaneous Information* (1934): 275–279; H. H. Rusby, "Nathaniel Lord Britton," Science 80 (August 3, 1934): 108–111; and Frans A. Stafleu and Richard S. Cowan, *Taxonomic Literature: A Selective Guide to Botanical Publications and Collections, with Dates, Commentaries, and Types, Volume 1: A–G* (Utrecht: Bohn, Scheltema, Holkema, 1976): 331–348. The archives of the Staten Island Institute of Arts and Sciences contain useful records on the Britton family genealogy. See also the vertical files, "Nathaniel Lord Britton," maintained by the New York Botanical Garden Archives.

16 Sloan, "Science in New York," 42, see note 13.

17 Ibid. and Henry A. Gleason, "Thumbnail Sketches of Botanists," 1961, unpublished manuscript, New York Botanical Garden Archives.

18 Gleason, "Thumbnail Sketches."

A contemporary botanical admirer described him as relatively ill-humored, except when engaged in botanical fieldwork, at which time his normally lugubrious face lit up and he became extremely talkative. Such transformations, apparently, occurred only sporadically. When faced with an interesting plant specimen, Britton was known to concentrate and assume an almost dreamlike expression. He was also possessed of a remarkable memory for Latin binomials, as well as the specific details of many of his previous botanical expeditions.[19]

Britton's legendary willingness to bully his staff into completing tasks was summed up in bitterly humorous verse composed after his death by one of his oldest friends, Arthur Hollick, a botanist he'd grown up with on Staten Island:

Lives of men like him remind us
Murder may not be a crime
And to kick at him departing
would indeed be joy sublime

What to him are life's diversions?
A species new his chiefest goal,
while his stern ideas of duty
Call down curses upon his soul

Let us then be up and doing
Ere we go to meet our fate
Hand it to him good and plenty
DO IT NOW IT'S GETTING LATE![20]

Despite the damning portrait by a purported chum, Britton remained an astute judge of ability, a gift that served him well as Garden administrator. Other contemporaries remembered him in a somewhat different light. George Small, son of J. K. Small, a curator of museum exhibits, recalled that Britton expected his staff to put in "full time" and to be on call daily. Members of the Garden staff took turns taking a full day of Sunday duty, during which they were supposed to make an inspection tour of the entire garden and check in with the guards; "it was quite a chore," Small remembered.[21]

"The Garden was [Britton's] baby," George Small recalled. "He put all of his efforts and I know that he left a good bit of money to the Garden when he died, but he ran the Garden [and] he expected people to follow his orders." Small's boyhood

19 José Álvarez Conde, *Historia de la Botanica en Cuba*. (La Habana: Publicaciones de la Junta Nacional de Arqueologia y Etnologia, 1958): 292.

20 Gleason, "Thumbnail Sketches," see note 17.

21 George Small Oral History, New York Botanical Garden Archives.

FIGURE 4
Elizabeth Gertrude Britton, date unknown.

memories of time spent with the Brittons do not reconcile easily with those of Henry Gleason, who served at one time as Britton's second in command. Whereas Gleason remembered Britton's single-minded concern with Garden details, Small recalled visits to the Britton home with his father that were punctuated by friendliness and laughter. The truest portrait of Britton is likely to lie somewhere in the middle.

Elizabeth Gertrude Knight Britton (1858–1934)

Mrs. Britton's own reputation among Garden employees was similarly mixed, though the daughter of another Garden employee remembered her as a bit "warmer" in personality than her husband.[22] Others who worked closely with Elizabeth, however, recounted her as

> a domineering sort of woman. She came in and once she got to the Garden we all knew she was here. Most of the girls were a little afraid of her. She was kind of loud. And she made her presence known. She came to the Garden about ten [o'clock] in the morning, and if anyone saw her coming, they disappeared quickly.[23]

Born on January 9, 1858 to James and Sophie Anne (Compton) Knight, Elizabeth Gertrude Knight Britton (FIGURE 4) had spent a considerable amount of her girlhood on her grandfather's Cuban sugar plantation, where she acquired the fluent Spanish language skills that served her and her husband well during their joint botanizing

22 Lorna Nash Weinberg Oral History, New York Botanical Garden Archives.

23 Lilian Weber Oral History, New York Botanical Garden Archives.

expeditions to the Caribbean in the early twentieth century. Upon graduation from the Normal College (later Hunter College of the City University of New York) in 1875, she secured a position on its faculty. She later published her first botanical paper (on albinism in plants) in an 1883 issue of the *Bulletin of the Torrey Botanical Club*. She met Nathaniel Britton as a member of the Torrey Club, and they married in 1885, while he served as Newberry's assistant in geology at Columbia College.

Throughout the 1880s, Elizabeth Britton supplemented her duties as a faculty member at the Normal College by working to organize Columbia's moss herbarium and initiated an impressive decades-long series of correspondence with a widespread and cosmopolitan set of bryologists throughout the world. By late 1888, she was prepared to publish the first in a series of papers that would eventually comprise a set entitled *Contributions to American Bryology*. That same year, according to NYBG lore, she had, in the wake of a recent trip to Kew with her husband, suggested the establishment of a botanical garden in New York City at a regular meeting of the Torrey Club.[24] In 1894, she also published a series of well-illustrated articles aimed at a popular audience, so that by the time the Garden was officially established and construction was underway in 1896, she had earned a place at the top of her field and in the public eye.[25] Though lacking any official graduate training in botany Elizabeth Britton's reputation and prowess as a student of mosses was such that she served as a graduate mentor to at least one doctoral candidate at Columbia in the early years of the Garden's history. She was chosen as one of only twenty-five charter members of the Botanical Society of America, an honor not granted to any of her credentialed female contemporaries.[26] As the Garden was organized, she also served as the unpaid "Honorary Curator of Bryophytes," a position from which she conducted significant taxonomic research on mosses.

24 Marshall A. Howe, "Elizabeth Gertrude Britton" *Journal of the New York Botanical Garden* 35 (May 1934): 97–104. The idea of a botanical garden for the city, of course, did not originate with Elizabeth Britton (see Chapter 2), but she was a key figure in the 1889–1890 movement that ultimately resulted in the city's reservation of Bronx Park land for the New York Botanical Garden. Records for a Torrey Botanical Club meeting indicate that Elizabeth Britton discussed her recent work at the Royal Botanic Gardens at Kew, England, when she may indeed have repeated the remark her husband attributed to her: "Why can't we have something like this in New York?" See Minutes of the Torrey Botanical Club, October 24, 1888, New York Botanical Garden Archives. On Elizabeth Britton, see also Lee B. Kass, "Elizabeth Gertrude Knight Britton (1858–1934)" in L. S. Grinstein, C. A. Biermann, and R. K. Rose, eds. *Women in the Biological Sciences: A Bibliographic Sourcebook* (Westport, CT: Greenwood Press, 1997).

25 Howe, "Elizabeth Gertrude Britton," 97–104, see note 24.

26 See Nancy G. Slack, "Botanical and Ecological Couples: A Continuum of Relationships," in Helena M. Pycior, Nancy G. Slack, and Pnina G. Abir-Am, eds., *Creative Couples in the Sciences* (New Brunswick, NJ: Rutgers University Press, 1996): 235–240.

By the 1910s, Elizabeth Britton had begun collaborating on crucial portions of an ongoing effort to catalogue North American flora, had been elected president of an active group of botanists interested in the taxonomy of mosses, and had helped inaugurate the organization to which she would devote increasing energy in the late 1910s and 1920s: the Wildflower Preservation Society of America. Her participation in this latter organization was part of a series of efforts she mounted to deter exploitation of native species of either plants or animals, efforts that were also fostered by her experiences trying to halt vandalism at the Botanical Garden (see Chapter 4).[27] She also regularly wrote letters and published popular articles highlighting the necessity of public playgrounds, arguing that by providing designated spaces for active play, the city would spare its unique parks, Central Park and the Botanical Garden, from such inappropriate uses.[28] Elizabeth Britton maintained her status as a public figure while expanding her bibliography of published botanical papers, a list that eventually exceeded 300. Together, the Brittons helped define an institutional agenda that dominated the life of NYBG from its inception in the 1890s up until their retirement from official duties in 1929.

Choosing a Setting for a New Urban Oasis

The borough of the Bronx, initially known as the city's "north side," comprising New York's twenty-third and twenty-fourth wards, had been, before its incorporation into the city, a series of separate suburban towns on the Harlem Railroad line along the Hudson River and a set of villages east of the Bronx River. The City of New York annexed the portion of land west of the Bronx River in 1874, and the eastern portion 20 years later, joining both to Manhattan politically as part of New York County. (The Bronx became its own county in 1914.)[29] The Bronx River itself has its source in New Castle, New York and flows 30 miles southward to join the East River. Until the mid-nineteenth century, it had been known for its purity, although industrial

27 Howe, "Elizabeth Gertrude Britton," 97–104, see note 24. In a series of lavishly illustrated articles in the *Journal of the New York Botanical Garden*, Elizabeth Britton highlighted a species of plant she designated as in danger of extinction from overzealous flower gatherers or thoughtlessly destructive hikers. Her acerbic writings were directed not simply at invasive species or pickers of native wildflowers: in 1919, she assailed the "noisy, dirty, and selfish" alien starlings that had begun driving native New York birds away and stealing gardeners' fruits and vegetables. Derisively, she dubbed this invasive species the "Bolshevik Birds." See letter from E. Britton to *The New York Evening Post*, February 25, 1919.

28 "More Playgrounds, Mrs. Britton's Plea: Our Parks Are Being Ruined for Lack of Them, Says Botanical Garden Director's Wife" *The New York Times*, August 19, 1922.

29 Evelyn Gonzalez, "From Suburb to City: The Development of the Bronx, 1890–1940," in Bronx Museum of the Arts, *Building a Borough: Architecture and Planning in the Bronx, 1890–1940* (New York: The Museum, 1986): 5–13.

development, including the erection of a snuff mill by the Lorillard family in the 1840s, had diminished this water quality significantly by the turn of the twentieth century.[30]

By the early 1880s, city parks advocates recognized that the future of this rather bucolic, northern, New York City suburb was to be an urban one characterized by dense development and the extension of the Manhattan grid system. The two wards' population began growing rapidly, with apartment houses constructed along new north-south railway routes. Between 1890 and 1900, the Bronx population more than doubled to over 200,000 residents. A decade earlier, *New York Herald* editor John Mulally had begun arguing for the purchase by the city government of an interconnected system of Bronx parks to serve as the "lungs" of the new north side. He marshaled statistics demonstrating greatly increased property values in the vicinity of Central Park, with a concomitant increase in the city's tax revenues. By 1884, Mulally's "New York Park Association" had successfully petitioned the New York State Assembly for approval of its park plan. Bronx, Van Cortlandt, Crotona, and Pelham Bay Parks were the tangible results of Mulally's successful campaign.[31]

Mulally's efforts to secure parkland for New York took place within the context of rapid urban development and increasing immigration and poverty. This was the era of Jacob Riis's famous photographic portrait of New York's slum life, *How the Other Half Lives*, as well as the flourishing of a pulp fiction trade, whose tales of urban squalor fed an incipient anti-urban bias and spurred one of the earliest waves of middle-class suburban flight along street car lines. In 1894, a "committee of seventy" was organized ; its mainly Yankee, Protestant, and Republican membership included J. P. Morgan; William Bayard Cutting; Morris K. Jesup; William E. Dodge, Jr.; and the younger Cornelius Vanderbilt. These men were among the incorporators of NYBG in 1891.[32] Unsurprisingly, such tumultuous economic times underscored the appeal to a social elite of what one commentator described as the "profoundly conservative concept of reform" embodied within the environmentalist designs of Frederick Law Olmsted. Large public parks, beginning with Central Park at midcentury and continuing with the Botanical Garden at century's end, comprised "stable institutional props for a nation confronting fast social change."[33] In the minds of these men, Olmsted's

30 Gary Hermalyn, "A History of the Bronx River" *The Bronx County Historical Society Journal* 19 (1982): 1–22.

31 Gonzalez, "From Suburb to City," 10–13, see note 29; Hermalyn, "History of the Bronx River," 11, see note 30.

32 Paul Boyer, *Urban Masses and Moral Order in America, 1820–1920* (Cambridge, UK: Cambridge University Press, 1978): 121–131; 165.

33 Geoffrey Blodgett, "Frederick Law Olmsted: Landscape Architecture as Conservative Reform," *Journal of American History* 62 (March 1976): 870–872. Blodgett argued that Olmsted, although a committed democrat, envisioned a political and social democracy that "remained responsive to the cues of trained and cultivated leadership," one that could devise aesthetic and structural

utopian environmentalism retained its power to exercise a harmonizing, tranquilizing, and refining effect upon the increasingly diverse populace, especially when coupled with the socializing, educational power of a large museum designed specifically to convey a world of natural order.[34]

Britton himself was not immune to this traditional logic, and into the twentieth century, he took pains to describe the Garden's grounds as a form of landscape park, distinct in character and purpose from the proliferating number of city playgrounds. "Public parks, gardens, and playgrounds are areas vitally necessary for the well-being of a large majority of persons resident in cities," Britton wrote. "They provide attractions and inducements for at least a certain amount of outdoor life amid surroundings more or less approaching country conditions, partially devoid of the distracting environment of the street."[35] When speaking of Central Park, he made sure to declare allegiance to Olmsted's vision, a stance with clear implications for his understanding of the Botanical Garden's grounds:

> The functions of parks and of playgrounds are quite different, and it is not possible to use either for the purposes of the other. A park is a place for rest and recreation. A playground is a place for play and violent exercise, which are not conducive to either rest or recreation. A park calls for trees, shrubs, lawns, flowers, paths, driveways, and artistic development, and requires high maintenance expenditures for care and upkeep and for guarding to discourage vandalism. A playground needs none of these features although a few shade trees are desirable where there is space for them. It is the failure of the public to recognize these essential differences between parks and playgrounds which cause much of the worry and trouble of our park authorities. The present movement for more playgrounds as such, is therefore highly important and worthy of all possible encouragement; portions of undeveloped and unimproved park lands are available for playgrounds, which should, when established, be so designated and plainly distinguished from the areas developed as parks.[36]

And yet, although its founders would not have described it this way, Britton's garden in some ways actually anticipated the later development of these "progressive" parks of the early twentieth century, a tradition that would include the construction of playgrounds and other public spaces dedicated to specific socializing purposes. In the case of a public botanical garden such as NYBG, the public space was devoted not only to "refreshing" landscapes, but at times to picnic areas; ice skating; and, crucially,

---

arrangements to ameliorate the social problems concomitant with urbanization and the rise of the new, "crude and materialistic impulses of popular culture."

34 Boyer, *Urban Masses*, 238–241, see note 32; Steven Conn, *Museums and American Intellectual Life, 1876–1926* (Chicago: University of Chicago Press, 1998): 6.

35 *New York Evening Post*, May 28 1910.

36 N. L. Britton letter to *New York Evening Post*, May 28 1909.

scientific education at both popular and professional levels.[37] In this instance, then, the science and nature education might be considered analogs of the regimented programs of active play that took place in early twentieth-century playgrounds.

Britton's view of an urban-based, nationalistic botany was thus informed by the environmentalist social precedents of the nineteenth century and the contemporary problems faced by New York City as he struggled to establish his American Kew. He and his wealthy contemporaries were happily aware of the potential of their exotic museum and conservatory displays to serve as a "social space of emulation in which civilized forms of behavior might be learned."[38] Garden benefactors in the late 1880s and early 1890s were quick to contend that the Garden would have a refining influence on New York's citizens. The nationalism inherent in Britton's conception of his glasshouse collections of botanical exotica, the way in which he anticipated an ability to instill a sense of wonder in his urban audience and allow them the vicarious experience of plant hunting throughout the Western Hemisphere, entailed a degree of environmentalist belief in the ability of physical surroundings to shape public attitudes and behavior. The Botanical Garden was thus partially the product of the social concerns of the city's plutocrats, for whom it would serve, alongside sister institutions such as the Museum of Natural History, as a "pillar of bourgeois self-definition."[39]

Yet, to overemphasize the traditional historiographic elements of "social control" and social engineering involved in the effort to design a public garden and to characterize NYBG as merely a venue for the socialization of a pluralistic citizenry would be to miss the essential aspect of the impetus behind the Garden-building effort: the Garden was designed, primarily, as an institutional venue for scientific botany.

37 See Robert Weyeneth, "Moral Spaces: Reforming the Landscape of Leisure in Urban America, 1850–1920" (PhD Dissertation, University of California–Berkeley, 1984); Galen Cranz, *The Politics of Park Design: A History of Urban Parks in America* (Cambridge, MA: MIT Press, 1982); and Monique Mosser and Georges Teysot, eds., *The Architecture of Western Gardens: A Design History from the Renaissance to the Present Day* (Cambridge, MA: MIT Press, 1991): 466–467.

38 Tony Bennett, *The Birth of the Museum: History*, Theory, and Politics (London: Routledge, 1995): 24.

39 Sven Beckert, *The Monied Metropolis: New York City and the Consolidation of the American Bourgeoisie, 1850–1896* (Cambridge, UK: Cambridge University Press, 2001): 48. The desire to highlight the degree of social control inherent in the rampant institution-building of the nineteenth century has lead to intriguing interpretations of the history of the American Museum of Natural History. Donna Haraway describes the diorama displays at the AMNH as allowing museumgoers "communion with nature through beasts killed" on faraway hunts, expeditions in which visitors might participate vicariously. Haraway also characterizes the elite enterprise of museum building as part of a concerted plutocratic public health effort aimed at "inoculating" the collective population against the urban threat of decadence and moral decay. Such a strict interpretation would be inappropriately limiting for the New York Botanical Garden. See Haraway's *Primate Visions: Gender, Race, and Nature in the World of Modern Science* (New York: Routledge, 1989): 27, 48.

Establishing an Institutional Framework

Grants of city parkland (and funds to develop them) were not easy to come by, and it was fortunate for the Torrey Club that in addition to an active roster of middle- and upper-class female members, the ranks included two respected New York jurists. Charles P. Daly was chief judge of the New York City courts and a former legal advisor to Abraham Lincoln. Addison Brown, a federal judge for the Southern District of New York, was Nathaniel Britton's coauthor of *An Illustrated Flora of the Northern United States, Canada and the British Possessions*. Brown, with the assistance of Daly (and undoubtedly much input from Britton), drafted and then lobbied for the legislation that officially incorporated NYBG. As an accomplished member of the TBC and a key member of the committee on the botanic garden, the 60-year-old Brown represented an earlier era of natural historical endeavor. His official profession did not preclude him from distinguishing himself as a botanist, and he readily collaborated with the Columbia-educated Britton, a man three decades his junior, who had trained as a professional scientist. In addition, as a man of substantial financial means and social standing, Brown not only bridged the realms of amateur natural history and professional science, but also personally embodied an alliance of philanthropy with institutional botany.

Brown and Daly's official Act of Incorporation of the New York Botanical Garden passed the New York Assembly and was signed into law (appropriately enough, by New York's Governor Flowers) on April 28, 1891. The law explicitly charged NYBG incorporators with the responsibility for the

> collection and culture of plants, flowers, shrubs, and trees, the advancement of botanical science and knowledge, and the prosecution of original researches therein and in kindred subjects, for affording instruction in the same, for the prosecution and exhibition of ornamental and decorative horticulture and gardening, and for the entertainment, recreation, and instruction of the people.[40]

The Act provided for the creation of a Garden that would be run by a group of incorporators who, in turn, would elect from among their ranks a board of managers responsible for conducting all business affairs and for the financial management of the organization. The Board of Managers would be composed of nine members of the NYBG Corporation elected for terms ranging from one to three years; two ex-officio members, the Mayor of the City of New York and the Président of the Board of Parks Commissioners; and six Scientific Directors. The 1891 statute stated that these latter managers, the Scientific Directors, would "have the management and control of the scientific and educational departments of said corporation and the appointment

40 The New York Botanical Garden, Act of Incorporation, Chapter 285 of the Laws [of New York State] of 1891, Amended by Chapter 103 of the Laws of 1894.

of the director-in-chief of said institution, who shall appoint his first assistant and the chief gardener, and be responsible for the general scientific conduct of the institution." Finally, a 250-acre parcel of city parkland north of the Harlem River (in either Bronx, Van Cortlandt, or Pelham Bay Park, depending upon the agreement of the incorporators and the Parks Department) was officially reserved for use by the Garden incorporators provided that by 1898 they were able to raise a minimum of $250,000 through private subscription. Once this sum was obtained, city bonds would be issued to provide another $500,000 in matching public funds for the construction of a fireproof botanical museum, laboratories, and horticultural houses.[41]

The 1891 Act was straightforward in its description of the various Garden governing bodies, assigning control of the most essential operations of the new institution, those tasks directly related to the Garden's scientific and educational mission, to the Board of Scientific Directors. According to Brown's plan, this powerful board would consist of the President of Columbia College; Columbia's professors of botany, geology, and chemistry; the President of the Torrey Club; and the President of the New York City Board of Education. These men, of course, would confer on matters of policy with their choice of the Garden's director-in-chief.

Over two years before Brown, Daly, and Britton's legislative handiwork successfully incorporated the Garden, however, one of their most vocal press allies had publicly warned them of the perils of city politics and of the trouble that awaited them if their new scientific garden should be made subject to the vagaries of New York's contentious municipal climate. "Should an institution of this type be established and made part of politics," *The New York Herald* admonished in late 1888, "it will be a failure [for] no one can go into politics and come out clean. [S]uch beautiful and pure things as flowers should never be contaminated by political influences." Instead, in a curious turn of phrase, the paper counseled the Torrey Club that New York's newest cultural institution should be managed by a "gentleman pope," free from the interferences of government and from the dictates of "governor, mayor, [or] alderman." Only the "love he has for the things he cultivates" in the new Garden, the *Herald* claimed, should guide the actions of the man in charge of its operation.[42] Those interviewed for the *Herald* piece, including a "German expert nurseryman" and the owner of the "oldest florist shop" in the city, uniformly asserted that this loving, gentlemanly guidance would be found in a board of college professors and other "distinguished gentlemen who take an interest in the cultivation of flowers." Even members of the city's political establishment seconded the *Herald's* sentiments. Hampden Robb, then president of the Board of Park Commissioners, had argued for control of the botanic garden by the TBC itself, because "the direct management of

41 Ibid.

42 *The New York Herald*, November 26, 1888.

such a garden is a specialty and should be in competent hands." In return for a parcel of public land, Robb said, the Garden should set aside days of free public access.[43]

The men involved with the establishment of the Garden matched these aspirations: Seth Low, president of Columbia College and scion of a wealthy New York family, sat on the Board of Managers and served as a Scientific Director, whereas Britton, himself the descendant of a venerable New York clan, and the *only* professor of botany at Columbia, was clearly a driving force behind the Torrey committee's efforts. Other "distinguished gentlemen" of fabulous wealth were amply represented in the ranks of Garden managers: Cornelius Vanderbilt, President of the New York Central Railroad, served as the Garden's first president with steel magnate Andrew Carnegie as vice-president, and banker and art connoisseur John Pierpont Morgan acting as Garden treasurer. These were the "wealthy, public spirited citizens . . . who [loved] the beauties of nature as they [did] Art" successfully called by the Torrey committee to endow the Garden with social and financial resources.[44] Addison Brown's contribution of $25,000 toward the establishment of NYBG matched that of these more renowned Gilded Age names: Andrew Carnegie, Morgan, Cornelius Vanderbilt, and John D. Rockefeller each pledged the same $25,000 sum. By making the pledges, these prominent New Yorkers entrusted Britton with the responsibility for establishing a grand, new, landscaped public space with the potential to reclaim a portion of the city's park system for genteel pursuits.

Looking to Central Park

Three decades after the construction of Central Park, but perhaps a dozen years after that park's "democratization" by city officials and various ethnic communities, the patrician New Yorkers responsible for funding the grandest city landmarks and cultural institutions sought to recreate a genteel park north of the East River.[45] In an 1890s memorandum exchanged between Britton and Brown, Chairman of the Torrey Club's Botanical Garden Committee, the men agreed that a crucial first step would be to construct driveways through their portion of Bronx Park in order to attract the "driving class" from whom they expected the majority of their financial support.[46] And yet, though designed to be the reincarnated ideal of Olmsted's genteel public space, the Garden was always publicly identified by Britton and other incorporators as built "not for the few, but for the whole people [for whom] easy access is indispen-

43 Ibid.

44 Torrey Botanical Club, "The Need of a Botanical Garden in New York," see note 3.

45 Roy Rosenzweig and Elizabeth Blackmar, *The Park and the People: A History of Central Park* (Ithaca, NY: Cornell University Press, 1992).

46 "Buildings and Grounds Planning, 1891–1898," Addison Brown Papers, New York Botanical Garden Archives.

sable."[47] As a public space, NYBG would satisfy the civic pride of an extremely wealthy class of philanthropists as well as a more modestly endowed middle class of urban garden enthusiasts and scientists. Botanic gardens had long done so: British and Continental public gardens had been convenient venues through which royal and aristocratic affluence and culture could be displayed to a wide audience. Precisely this imperative had led in the early eighteenth century to a competition between the European Houses of Orange and Bourbon for specimens of exotic global flora.[48]

Similarly, the decision to build Central Park in Manhattan, though couched in democratic rhetoric, had been "fundamentally rooted in the interests of the wealthiest citizens," who sought a grand park to raise the cosmopolitan stature of their city. These men were joined by uptown landowners who sought increased real estate prices, as well as vociferous advocates in the elite press. Yet the park's gentlemen boosters nevertheless were careful to claim to represent the public interest. These groups argued for a large park on three grounds: first an environmentalist belief in the ability of such landscapes to foster the moral uplift of immigrants, thus promoting order among the lower classes; second, the cultivation of a cosmopolitan city; and third the improvement of the city's public health. Furthermore, by the 1830s, New York streets that formerly had served as elite promenades (Broadway, the Bowery) had become publicly contested spaces for working class street culture. Consequently, these wealthy men sought to create Central Park as a new, controlled venue for safe outdoor strolls and carriage rides by their wives and families. By the time Olmsted and Vaux's romanticized "Greensward Plan" was completed in the early 1860s, New York's carriage-bound elite once again enjoyed temporarily a relatively uncontested setting for the display of their wealth. By the early 1870s, however, the victorious elite were again in retreat, as New York's ethnic communities began creating their own, "democratized" park, which subsequently became the setting for more active sorts of recreation. Thus New York's public botanical garden, as it was created in the 1890s, was not only a scientific institution, but was also in part an attempt to recapture a public space by elite New Yorkers in the wake of their retreat from Central Park. That attempt, however, was consistently couched in the rhetoric of public inclusion and access.[49]

47 C. F. Cox to N. L. Britton, August 5, 1895, Addison Brown Papers, New York Botanical Garden Archives.

48 Drayton, "Imperial Science," 33–39. see note 10.

49 Rosenzweig and Blackmar, *The Park and the People*, see note 45. Walker and Simo make explicit the link between Olmsted's naturalistic landscape designs and the early-nineteenth-century European Romanticism of Rousseau, Wordsworth, and British garden designers. Arriving in the US, this Romantic ethos assimilated the work of American Renaissance writers Hawthorne, Melville, and Thoreau; the landscape design of Andrew Jackson Downing; and the Hudson River School of painting typified by Frederic Church and Asher Durand. See Peter Walker and Melanie Simo, *The Search for Modernism in the American Landscape* (Cambridge, MA: MIT Press, 1994): 19.

It seemed, then, as if Britton and his fellow TBC members had heeded the *Herald's* advice and had successfully negotiated the contentious political landscape of America's metropolis. With blue-chip backers, a clearly defined organizational structure, and statutorily guaranteed public funding all but assured, the Garden's success seemed secure. And yet the Garden's newly minted gentlemen managers were due for a series of surprising challenges as they began to build New York's great garden, not directly from the often maligned "politicians" but ironically from other "gentlemen" academics, some guided by their own, separate, institutional imperatives, and others seeking to promote particular established interests in botany and landscape design. Through the 1890s, these genteel challengers in fact made adept use of the political process, exploiting connections in the Assembly and the City's Parks Department to delay the actual construction of the Garden itself. The ends they attempted to advance, however, were entirely their own.

Columbia College: An Institutional Model

In mid-February of 1892, Addison Brown received a note from fellow Garden incorporator Charles Anderson Dana, the controversial and highly outspoken editor of one of the city's most popular dailies, *The Sun*. Dana informed Judge Brown that in Albany the State Senate Committee on Cities was about to hold a public hearing on a proposed amendment to the Botanic Garden's Act of Incorporation. He suggested that "someone ought to be there from our side of the question."[50] Within days, the proposed amendment, which would have altered the composition of the Garden's management hierarchy by expanding it to include representatives from each of the city's colleges, was gaining press attention. On February 16, *The New York Evening Post* reported that in response to this potential change, Columbia College, which had pledged $25,000 toward the $250,000 in private donations necessary for the NYBG Corporation to receive city funds and land, had threatened to withdraw its support from the entire project.[51]

Columbia's assistance, both financial and intellectual, was crucial to the Garden's prospects for success. During the middle decades of the century, the college itself had been relegated to the margins of the city's intellectual life due to its persistent adherence to its traditional classical curriculum. By the 1880s, however, with the assumption of the College presidency by Seth Low, it once again enjoyed a social status, intellectual reputation, and degree of political influence unmatched by any other city school.[52] Columbia's ties to Garden incorporators went beyond its clear

50 Charles Anderson Dana to Addison Brown, February 13, 1892, Addison Brown Papers, New York Botanical Garden Archives.

51 *The New York Evening Post*, February 16, 1892.

52 Thomas Bender, *New York Intellect: A History of Intellectual Life in New York City, From 1750 to the Beginnings of Our Own Time* (Baltimore, MD: Johns Hopkins University Press, 1987).

statutory control of the Board of Scientific Directors: Low himself, of course, was a Garden manager, and Britton a Columbia graduate and professor. In addition, the TBC made Columbia its headquarters, and the College's herbarium (compiled and organized by Britton) and botanical library were to become the nucleus of the NYBG collections. Finally, graduate courses of instruction were to be offered at the Garden, with credit toward advanced Columbia degrees.

In the early 1890s, Britton needed to look no further than his alma mater and employer, Columbia College, to gain a strategic understanding of the ways in which the needs of the city, its wealthy citizens, its general population, and the interests of a specific institution could be reconciled to the overwhelming benefit of that institution itself. Among the city's numerous philanthropic figures, Columbia president Seth Low ranked as one of the most important individuals in the early history of the Botanical Garden. Low represented both the sort of urban northeastern "Progressive" whom Britton needed to target as a means of support, as well as one of the region's most adept practitioners of the very sort of institution-building Britton undertook in the 1890s. Whereas Britton possessed a modest (by Gilded Age standards) personal fortune, Low was the son of an extremely wealthy New York merchant family, had served as mayor of Brooklyn (prior to its incorporation in 1898 into Greater New York), and became mayor of postconsolidation New York after his tenure at Columbia. Both men enjoyed working relationships with many of the city's most gilded names: Andrew Carnegie, J. P. Morgan, and the Vanderbilt family (although given his great wealth and public stature, Low naturally dealt with them on a more equal footing).[53] These men, the "power elite" of late nineteenth century America, were those with "the most direct power to influence decisions and alter events, [and] who controlled property and affected the social relationships that flowed from that control."[54]

Born in 1850, Low graduated from Columbia in 1870 and then entered his father's successful mercantile firm. He later founded A. A. Low and Brothers, which gained a leading position in the tea and silk trade with Japan and China. Low also helped finance the Chesapeake and Ohio Railroad and the first trans-Atlantic cable. Later, as Mayor of Brooklyn, from 1881–1885, he had pursued a reformist agenda that included reducing the city's debt, upgrading public schools, and implementing a merit system in public hiring.[55] At Columbia in the 1890s, Low managed to transform the city's patrician, colonial-era college into a modern research university (in 1896). Britton, a similarly gifted, if less prominent, administrative and scientific entrepreneur, sought comparable success in creating the city's botanical garden.

53 Britton apparently did have family ties to the Vanderbilt clan.

54 Thomas McCormick, *China Market: America's Quest for Informal Empire, 1893–1901* (Chicago: Quadrangle Books, 1967): 10–18.

55 John A. Garraty and Mark C. Carnes, eds. *American National Biography* (New York: Oxford University Press, 1999).

In their attempts to realize their respective institutional visions, Britton and Low each took pains to harness the city's late-nineteenth-century tradition of civic boosterism, contending that New York's well-established status as the center of national commerce required a similar metropolitan preeminence in culture and learning.[56] Although distinctions between the values of commerce and culture, the sense that business pursuits and learning were mutually exclusive, had not been made by early-nineteenth-century New Yorkers, nevertheless the city had endured a reputation for excessive commercialism since the colonial era.[57] Appeals to the intellectual and civic pride of the city elite, then, were designed to underscore wealthy New Yorkers' desires to claim for the city a place of leadership in learning analogous to its commercial status. In doing so, Low cast Columbia as the crucial engine of the city's intellectual growth. As Douglas Sloan noted

> It was Low's genius to recognize that New York's massive urban problems could be made a part of the traditional New York booster theme and thus be turned to the advantage of the University. Low was one of the first to advance the notion that what happened in the rest of the country for good or ill was experienced first and in a heightened way in New York City. . . . Therefore, [Low] argued, "from what quarter are trained intellect and consecrated purpose to come which are to grapple successfully with these problems in the coming time, if not out of New York itself and out of the schools and colleges of New York?"[58]

Low's belief in the power of trained scientific expertise and his strategic articulation of these beliefs to advance the cause of his own institution echoed the successful efforts of Britton's British models to place Kew at the center of a far-flung scientific empire. Low, of course, did more than provide a model of institution-building: by adapting the needs of Britton's proposed garden to those of Columbia, he directly ensured NYBG's success, in the process making it in its early years virtually an adjunct to the university.[59]

In Low, then, Britton perceived a model, a mentor, and an ally. It was part of Low's plan to gain for Columbia the facilities for basic scientific research that the Garden and the American Museum of Natural History provided. By positing a cultural symbiosis between the city and the University, Low managed to tap municipal and philanthropic sources of wealth and support, while maintaining private control over the institution and saving enormous amounts of Columbia's money in the process.[60]

56 Sloan, "Science in New York," 37–38, see note 13.

57 Bender, *New York Intellect*, 60, see note 52. The inaccuracy of this perception was explored in Chapter 2.

58 Sloan, "Science in New York," 54, see note 13.

59 So argues Sloan, persuasively, ibid., 53–54.

60 Ibid.

New York University's chancellor Henry MacCracken's attempt to gain similar access to the resources at Columbia's disposal met with Low's and Britton's vigorous opposition, and this challenge to Low's institutional vision for Columbia (and the NYBG) was surmounted successfully through skillful legal maneuverings and the design of the Botanical Garden's organizational charter. Once established, the Garden's close institutional relationship to Low's University meant that Columbia had, according to Sloan, "acquired a fully outfitted and staffed department of botany, its physical plant supported, and its research expeditions underwritten by the city through public funds and philanthropy."[61]

This crucial link between Columbia and the Garden was not entirely Low's work. Aware of the benefits of such a collaboration, Britton proposed the outline of the arrangement to Low in the spring of 1895. Britton argued that the Columbia herbarium (the nucleus of which had been John Torrey's personal collection, which Britton had organized in taxonomic sequence) was of little use to basic undergraduate education. Rather, he proposed, the collection should be moved to the Bronx, to be housed in the new museum and available for graduate-level research to be conducted under the auspices of the Garden and the University.[62] Ultimately, the arrangement between the two New York institutions proved so beneficial that Nicholas Murray Butler, Low's successor as Columbia president, further formalized their relationship, recalling Britton from his emeritus status to hold a seat on the University's Faculty of Pure Science.[63]

Institutional Rivalries: City Colleges Contend for Access to Garden Science

Fifty-two-year-old Reverend Henry M. MacCracken, the man challenging Columbia's supremacy, was then the new Chancellor of the University of the City of New York, today's New York University (NYU). Born in Ohio in 1840, MacCracken had been educated at Princeton and in German seminaries before becoming a member of the Presbyterian clergy. Since 1881, he had been a professor of philosophy at NYU, and upon ascending to the chancellor's chair in 1891, he began working to establish the school's prominence in the city. In the process, he publicly challenged the traditional prerogatives enjoyed by Columbia College.[64] Chancellor MacCracken, by seeking

61 Ibid., 58.

62 N. L. Britton to Seth Low, April 30, 1895, Columbiana Collection, Low Library, Columbia University Archives.

63 This arrangement was analogous to those between Columbia and the American Museum and Columbia and the Metropolitan Museum of Art. The directors of these institutions each held positions on Columbia's faculty. See N. M. Butler to N. L. Britton, December 7, 1907 and N. L. Britton to N. M. Butler, December 10, 1907, Columbiana Collection, Low Library, Columbia University Archives.

64 For convenience and clarity, I will refer to MacCracken's university by its modern name, New York University. See *Dictionary of American Biography*, "Henry H. MacCracken," note 15.

to secure a role in the management of the city's new botanic garden for his own university, sought to undermine his rival institution's exalted status.

The legislation proposed by State Senator Hagan in early 1892 would have altered the Garden's Board of Scientific Directors to include, in addition to the four representatives from Columbia, two members of the NYU faculty, and one faculty member each from St. John's College (today Fordham University), Manhattan College, Normal College, and Barnard College. In response to MacCracken's tactics, Columbia began circulating new Garden subscription agreements to potential philanthropic supporters. These revised contracts stipulated that if Columbia's sole control of the scientific conduct of NYBG was diluted by statute, all subscriptions would be void.[65] The *Tribune* warned that although the necessary financial support for the new botanic garden had seemed assured, the new bill threatened to "spoil the scheme." Angered by this unexpected legislative maneuvering, Brown went on record with a *Tribune* reporter, openly disparaging the qualifications of other institutions to run a botanic garden and asserting that Columbia's long history of botanical research and its willingness to invest a large amount of its own resources in the new garden entitled it to unquestioned authority in its scientific governance. Henry H. Rusby, a physician, professor of botany at the College of Pharmacy, and member of the original Torrey Club's botanic garden committee of 1889, seconded Brown's assertions. Columbia, Rusby maintained, was the "oldest, most widely known, and most responsible" of the city's institutions of higher learning, and its herbarium and library were of high botanical importance. He noted that if any of New York's educational institutions could make a valid claim for inclusion, it would be a medical or veterinary college. These institutions had not done so because, he asserted, they felt their interests would best be served by Columbia's management. Without Columbia's support, Rusby predicted, the entire enterprise would collapse.[66]

MacCracken countered their arguments by noting that, although his institution and the others named in Hagan's bill had been granting city residents baccalaureate degrees for years, they had never been consulted by NYBG planners at any point in the process of securing public appropriations or devising an organizational structure for the new educational garden. MacCracken argued that the Board of Scientific Directors should be larger and more accountable to the public in the way that the city's museums of natural history and art were. The proposed enlargement of the board, the chancellor contended, would help foster scholarly communication among the city's scientists.[67]

65 *The New York Evening Post*, February 16, 1892. St. John's College was the Jesuit institution, later Fordham University, located across the street from the Garden. The Normal College, Elizabeth Britton's alma mater, would become Hunter College, part of the City University of New York system.

66 *The New York Sun*, February 25, 1892.

67 *New York Daily Tribune*, February 17–18, 1892.

Columbia president Low was partially responsible for the contentious state of affairs in the late winter of 1892. Two months before the matter became fodder for the daily press, MacCracken had in fact contacted Low, privately advising him of his intention to secure via legislative action NYU representation on the Garden's Board. Just before Christmas, 1891, Low had replied, disingenuously, that it was beyond his power to endorse MacCracken's proposed change in plans. He characterized the botanical garden as entirely the project of the Torrey Club and insisted that Columbia was essentially along for the ride; the fact that this was patently false did not elude MacCracken, who continued to press his case. Low took the opportunity to assert to MacCracken that the current structure of the Scientific Directors was best, with "scientific control . . . substantially centered in one strong and responsible institution."[68] At MacCracken's urging, NYU officials also privately contacted other NYBG incorporators. J. J. Stevenson, an NYU administrator, wrote Britton in February and March of 1892, protesting the unfairness of Columbia's favored position and astutely citing the relatively trifling expense to Low's college and the enormous expense to the city of setting up the Garden. Charles Daly counseled Britton with the reassuring assertion that "Stevenson can do us no harm. He does not even understand the charter."[69]

If MacCracken was surprised by the degree of opposition to his efforts in February 1892, he did not say so to the city's reporters. Yet, within a week of Brown's public diatribe against the insurgent colleges, MacCracken decided to postpone seeking passage of the proposed amendment, promising instead to gain admission to the Garden's directorate when the time came for the first city funds to be appropriated. Officials from the other city colleges named in the deferred amendment hailed his actions and approved of his proposed course of action, phrasing their arguments in ways designed to garner maximum public support for their future efforts. By expanding the Board of Scientific Directors, they said, the Garden's government would be more "popular, economical, and effective."[70] MacCracken's public postponement did not signify a private retreat from his goal of raising NYU's cultural and intellectual stature, however. During the spring of the following year, he would manage to avoid press scrutiny while actually securing passage of his desired amendment.

Meanwhile, Britton had remained justifiably wary of the Chancellor's intentions, and in a meeting with MacCracken in late February 1893, his suspicions were confirmed. As Britton later recalled, the Reverend candidly informed him that he was "deeply in earnest on this matter [the expansion of the Garden management], for we cannot

68 Henry MacCracken to Seth Low, December 18, 1891 and S. Low to H. MacCracken, December 21, 1891, Addison Brown Papers, New York Botanical Garden Archives.

69 J. J. Stevenson to N. L. Britton, February 4 and March 2, 1892; Charles P. Daly to N. L. Britton, March 11, 1892, Addison Brown Papers, New York Botanical Garden Archives.

70 *New York Daily Tribune*, February 24, 1892.

afford to let a rival institution obtain the prestige which this movement will give it without making a fight for our own recognition."[71] Convinced of MacCracken's tenacity, he admitted anxiously to Brown a week later that they might have to consider compromising. "Even a single representative could be admitted from NYU," he speculated, "without the consent of the subscribers, including Columbia." Possibly, he conceded, "the Columbia trustees [would] have to be brought over by persuasion."[72] If Britton seemed ready to acknowledge MacCracken's seriousness, his instincts proved correct. On May 5, 1893, in an act whose title made no overt reference to the yet unbuilt NYBG, the Board of Scientific Directors was expanded to include representation from NYU.[73] MacCracken had been able legally to install four representatives of his own university into the Garden management structure. Yet in the intervening seven months, as the severity of the national economic crisis stalled the fundraising process, his success remained unknown to Britton and the other Garden incorporators. Not until January 1894 were the Chancellor's stealthy political machinations revealed.

During the first week of 1894, Garden manager Charles Cox was contacted by NYU's Charles Butler. Referring to the May 5, 1893 act of the state legislature, Butler apparently sought a copy of the minutes from that month's Board of Managers meeting, which had been held January 5 at the offices of the American Geographical Society. MacCracken himself followed up Butler's effort by making the same request in a January 15 letter to Garden president Cornelius Vanderbilt. Having never heard of the legislature's act of the previous May, neither Vanderbilt nor Cox had any idea why the NYU representatives should have any right to read those minutes, and they turned to Addison Brown for advice on how to respond.[74]

Judge Brown was furious. Upon reading the revised Act of Incorporation in the printed compilation of the prior year's statues and learning that MacCracken and three of his faculty had been granted their eagerly sought stature within the NYBG hierarchy, he fumed, "it seems almost humiliating to be obliged to deal with such trickery." Brown angrily recalled that in 1892 the NYU chancellor had openly declared

71 N. L. Britton to Charles F. Cox, February 2, 1894, Addison Brown Papers, New York Botanical Garden Archives. In defense of his actions, MacCracken would later claim that at this meeting on February 25, 1893, he had informed Britton of his revived plans to seek legislation that spring. Britton strongly denied this in his letter to Cox.

72 N. L. Britton to A. Brown, March 4, 1893, Addison Brown Papers, New York Botanical Garden Archives.

73 Significantly, however, MacCracken's clandestine efforts of 1893 benefited NYU only. The colleges that had been named in the original 1892 amendment had been omitted from the final form of the bill passed in May, 1893. Though he had skillfully employed populist rhetoric when arguing for "wider public accountability" for the Garden, MacCracken's primary aim had always been the elevation of his own institution's stature.

74 H. H. MacCracken to Cornelius Vanderbilt, January 15, 1894, Addison Brown Papers, New York Botanical Garden Archives.

his intention to postpone his efforts until the Garden, like the city's other museums, made its first annual approach to the legislature for an appropriation. MacCracken's subsequent maneuvers, according to Brown, amounted to treachery. And yet, the irate jurist-botanist conceded, "nevertheless we *are* obliged to deal with [the situation], or abandon the whole enterprise." Brown and Britton feared that the new amendment would nullify the subscriptions they had received to date, and they were not confident that those pledges would be renewed if they had to begin the process again.[75] Bringing his legal training to bear upon the problem and citing relevant case law, Brown felt sure that MacCracken's actions had been unconstitutional, since the true subject matter of the May 5 bill had been concealed by an innocuous title, and thus had been enacted without the knowledge of all concerned. Nevertheless, fearing the effect negative publicity might have on the generosity of the Garden's benefactors, Brown eschewed the possibility of a lawsuit and he advised both Low and Britton to avoid bringing the situation to the attention of the press.[76]

Instead, the Judge contemplated drafting a bill that would repeal MacCracken's act, and he proceeded to contact several members of the Assembly and Senate in Albany. Privately, Low counseled Brown against such action, saying the Garden should not use similar "false tactics," by seeking secret legislation passed "in alternate years."[77] By the end of February, however, a compromise had been reached. With the acquiescence of both Columbia and NYU, MacCracken's legislation would be repealed, but NYU biology professor, W. Gilman Thompson, would be added to the Board of Scientific Directors. On March 13, 1894, at Low's office on the Columbia campus, Low, Britton, Brown, and James Kemp, the Columbia professor of biology, voted to admit Thompson to their ranks as a Scientific Director. The first major controversy in the Garden's history was over.

MacCracken recognized the extent to which Low's backing of the Garden represented a clever institution-building strategy for Columbia. Low understood that by establishing the Garden on a quasipublic, quasiprivate basis, he was securing for his university access to what would eventually grow to become a first-rate set of natural history collections but without the necessity for maintaining them at Columbia or paying for them entirely out of the University's budget.[78] MacCracken, faced with

75 Britton's fears were not unfounded: Columbia's Seth Low, in a letter to Judge Daly soon after the discovery of MacCracken's success, bluntly asserted his unwillingness to accept divided responsibility for the scientific conduct of the Garden with his rival, and, perhaps as a threatening bluff, "offered" to have Columbia step aside entirely. S. Low to C. P. Daly, January 13, 1894, Addison Brown Papers, New York Botanical Garden Archives.

76 A. Brown to S. Low, January 6, 1894; A. Brown to C. F. Cox, January 10, 1894, Addison Brown Papers, New York Botanical Garden Archives.

77 S. Low to A. Brown, February 21, 1894, Addison Brown Papers, New York Botanical Garden Archives.

78 See Sloan, "Science in New York," note 13.

fewer resources, was determined to secure for NYU as many of the advantages that such a project, funded in part with public money, would afford. In the end, though the Columbia–NYBG relationship remained the strongest, MacCracken at least forced Low and the Garden trustees to acknowledge a responsibility to the city's other institutions of higher learning.

Meeting Economic Challenges in the Mid-1890s

The next impediment to Garden plans was neither personal nor institutional but economic: the Panic of 1893, touched off by the stock market crash that followed the bankruptcies of the National Cordage Company and the Philadelphia and Reading Railroad. Publicly, US President Grover Cleveland, backed by the weighty opinion of J. P. Morgan, blamed the nation's use of silver specie, as mandated by the Sherman Silver Purchase Act of 1890, as the root cause of this economic depression. In 1890, Morgan had viewed the reintroduction of silver-backed dollars as a serious misstep in public monetary policy, arguing that because at that time foreign companies and individuals held more than $3 billion of US securities, these investors would lose significant sums if the US government increased the money supply (thus devaluing the dollar). These investors fled the US securities market, taking their gold from the US Treasury with them, eventually exacerbating the economic slide of the mid-1890s (while helping define the contours of a class struggle over monetary policy and launching the career of prosilver populist William Jennings Bryan). Meanwhile, some 192 railroad lines had failed by 1893, and by 1898, fully one-third of the nation's railroad routes were in receivership. Unemployment skyrocketed to 20 percent.[79]

Despite the highly inauspicious economic climate, however, Nathaniel Britton was able to secure the participation and the crucial financial backing of an impressive roster of New York plutocrats. Cornelius Vanderbilt (1843–1899), grandson of the first Cornelius ("the Commodore") and president of the New York and Harlem Railroad, had agreed to serve as the Botanical Garden's first president. Andrew Carnegie, noted philanthropist and author of the widely read essay "Wealth," in which he had declared that "the man who dies . . . rich dies disgraced," had agreed to become the Garden's first Vice-President. Carnegie's cooperation, though crucial, was not necessarily a surprise, given his belief in the power of parks such as the Botanical Garden to serve as "ladders upon which the aspiring can rise [and] means of recreation by which men are helped in body and mind. . . ."[80] The son and grandson of Scottish social activists and reformers, Carnegie had been a perfect candidate for Britton's approach. After increasing US steel production beyond that of Britain by the late nineteenth century,

79 Jean Strouse, *Morgan: American Financier* (New York: Harper Collins, 1999): 305, 319-320; and Boyer, *Urban Masses*, 125–127.

80 Andrew Carnegie, "Wealth," *North American Review* 148 (June 1889): 653–664.

Above
FIGURE 5
The Bronx River gorge in winter, date unknown.

Left
FIGURE 6
Glaciated landscape in the hemlock forest, ca. 1913.

Carnegie had prepared himself to begin giving his vast fortune away. Finally, J. P. Morgan, banker, art connoisseur, presidential advisor, and guarantor of American capitalism, served, appropriately, as treasurer. Other philanthropists populating the new institution's Board of Managers included William E. Dodge, Dwight O. Mills, William Schermerhorn, Charles F. Cox, Charles P. Daly, and Samuel Sloan.

On March 17, 1895, New York's *Daily Tribune* happily announced that despite the recent economic hardships, the planners of NYBG were continuing their efforts to bring world-class botany to the city. "The admirable project is prospering," which to many citizens, the *Tribune* conceded, "was probably an agreeable surprise." Although "the long financial depression has disappointed expectations for raising funds," nevertheless, the paper predicted, once the garden began to "demonstrate its practical utility," it would "surely succeed." Despite the heavy financial burdens then faced by the city, the Garden, the *Tribune* contended, had "a strong claim not only to private support but [also] to public endowment."[81]

That spring, as efforts to secure the required $250,000 in private donations continued, Low and Britton prepared to confer with the City Parks Commissioners regarding their choice of Bronx Park as the site of the soon-to-be-built Garden. The Commissioners assigned Samuel Parsons, the Park Superintendent, and Calvert Vaux, then the City's landscape engineer, to visit the newest city parks north of the Harlem River with Low and Britton and to report their recommendations.[82] On May 18, Britton and Parsons traveled to the Bronx River, and examined the soil and "vigor of the vegetation" of the surrounding areas. Hiking through the approximately 650 acres of glaciated landscape that comprised the entirety of Bronx Park, Britton informed Parsons that the southern portion, relatively inaccessible and containing little forest growth and much marshy land, would ill suit the needs of a public garden.[83] In the

81 *New York Daily Tribune*, March 17, 1895.

82 Samuel B. Parsons, Jr. (1844–1923) figured prominently in American horticultural affairs and New York landscape design during the nineteenth century, serving first as Secretary of the Horticultural Department at the Philadelphia Centennial Exposition and then as New York City's Landscape Architect at the time the NYBG was founded. A graduate of Yale's scientific school, he had been interested in the development of model farms; had entered his father's (S. B. Parsons, Sr., 1819–1907) famous Flushing, New York, business as a nurseryman; and had become a partner in Calvert Vaux's design firm. His career in landscape architecture included design commissions for the cities of Birmingham, Alabama and San Diego, California as well as plans for parks and cemeteries throughout the US. He had also been active in the attempt to revive the New-York Horticultural Society in the 1880s and an active horticultural writer, contributing seven books to the expanding canon of nineteenth-century American horticultural literature. See Charles Birnbaum, *Pioneers of American Landscape Design* (Washington, DC: National Park Service, 1993): 93–95 and "A Visit to Parsons' Nurseries" *The Horticulturist* 30 (September 1875): 17.

83 This portion would later be granted to the New York Zoological Society for the creation of the Bronx Zoo. Britton and Parsons would also visit Van Cortlandt Park at the end of May, 1895, but

Northern reaches of the park, however, the soil, "deep and loamy," was more to Britton's liking. There, over centuries, the modest Bronx River had cut a picturesque gorge (FIGURE 5) through which a steady stream of fresh water tumbled before finding its ultimate outlet into the East River.

Britton was also impressed by the hemlock forest (FIGURE 6), a grove of old-growth trees that predated European settlement of the region.[84] The future director-in-chief's preference for this northernmost portion of the Park was reinforced by the lack of private real estate development on its northernmost boundary. It was his hope that protected from such "border nuisances" as the construction of taverns and other unsavory establishments, the Garden's beautiful natural scenery would be able to fulfill its mission as a "refining" influence.[85]

Upon returning from his visit with Parsons to the Bronx River gorge, Britton recommended to the Scientific Directors the choice of the northern parcel of Bronx Park land, asserting that it "combines all the desirable elements of ready accessibility, rich and varied soil, grand natural scenery, dense natural forest and freedom from obnoxious surroundings." Yet, he added that he did not feel he and his committee members (W.G. Thompson and J. F. Kemp) could arrive at a satisfactory plan for the location of the garden's buildings "without advice from competent landscape engineers." He had been unable to discuss the matter with Calvert Vaux at that point and conceded that, although he had a general set of principles by which he hoped to see the Garden organized and the Museum constructed, the details of all such efforts would "require prolonged study."

Before the particulars of the Garden's layout were considered, however, the finance committee was able to report to the Managers that the long sought-after figure of $250,000 had finally been reached by mid-June, 1895. The June 19 edition of the *Tribune*, long a garden booster, cheerfully informed readers that the Garden was now "assured." Judge Charles Daly, undoubtedly as relieved as the rest of the Garden incorporators that the Torrey Club's seven-year-old idea seemed to be nearing fruition, assured *Tribune* readers that though the first object of the new institution would be scientific research, its second priority would be to serve as a venue for the "recreation of those who live by their labor." For Daly and his cohort of philanthropists and botanists, the now seemingly certain creation of a grand civic institution at a time of widespread financial uncertainty represented the achievement of a laudable

---

Britton remained convinced of the superiority of the northern end of Bronx Park.

84 The NYBG tract included several easily discernible characteristics of a glaciated landscape, including polished and striated bedrock outcrops, potholes in rock formed by eddying glacial meltwater, and large erratic boulders deposited by the retreating glacier. See Gary Hermalyn, "A History of the Bronx River" *Bronx County Historical Society Journal* XIX (1982): 1–22.

85 Minutes of the Scientific Directors, May 24, 1895, New York Botanical Garden Archives.

goal. Yet, they acknowledged that by reaching this stage, they also courted the risk of public censure. Prudence required that the Garden be justified in more universal terms than the mere claims of pure scientific research. By emphasizing the degree to which the proposed new public space would serve the needs of the largest proportion of the city's population, Daly sought widespread acceptance at a crucial moment in the Garden's history. The "contentment [of those laborers] with their lot is of the highest importance to the maintenance of public order," New York's esteemed jurist asserted. "It is especially so at this period when we hear so much about the conflict between capital and labor, of the hostility to the building up of colossal fortunes, [and] the loud complaint of the anarchist of the inequality between the rich and poor." Daly's comments reflected the concerns of his class of New Yorkers, those with the financial means (despite the depression) to help endow an institution whose future nevertheless was contingent upon public financial and, as Daly presumed correctly, *political* support.[86]

Designing the Garden

With private and municipal funds secured, the NYBG incorporators began building a botanical garden. On June 17, 1896, the Garden Managers delegated the particulars of site design to a commission consisting of director-in-chief Britton; R. W. Gibson, the Museum architect; John Brinley, a civil and landscape engineer; Scientific Director Lucien Underwood; and Samuel Henshaw, head landscape gardener. Lincoln Pierson, of the greenhouse design firm of Lord and Burnham, creators of a conservatory for railroad financier Jay Gould at his Hudson River Valley estate in Tarrytown, New York, added his expertise to this group. This group was to work in conjunction with the previously assembled Committee on Plans. Both bodies conferred on Oct. 9, 1896 and composed a general outline for the Garden's physical development. By late autumn, they were prepared to present their ideas to the assembled Managers.[87]

Britton's November 30, 1896 report was extensive, and presented his vision of the general layout for the Garden (FIGURE 7). He sought to reconcile this conception of a botanic garden with the public imperatives of park design that had governed landscape architecture since Olmsted and Vaux had reified Andrew Jackson Downing's adaptation of English naturalistic landscape ideas in the creation of Central Park 35 years earlier. The report began with a defense of the selective removal of trees for

86 Minutes of the Board of Managers, June 18, 1895, New York Botanical Garden Archives; *New York Tribune*, June 19, 1895.

87 "Report of the Plans Commission, November 30, 1896" Bulletin of the New York Botanical Garden 1 (January 1897): 23–24. Britton was indubitably the guiding hand behind the extensive document and was the lead signatory as Director-in-Chief, a position he had held since his official appointment on July 1, 1896. Subsequently cited as "Britton Plan."

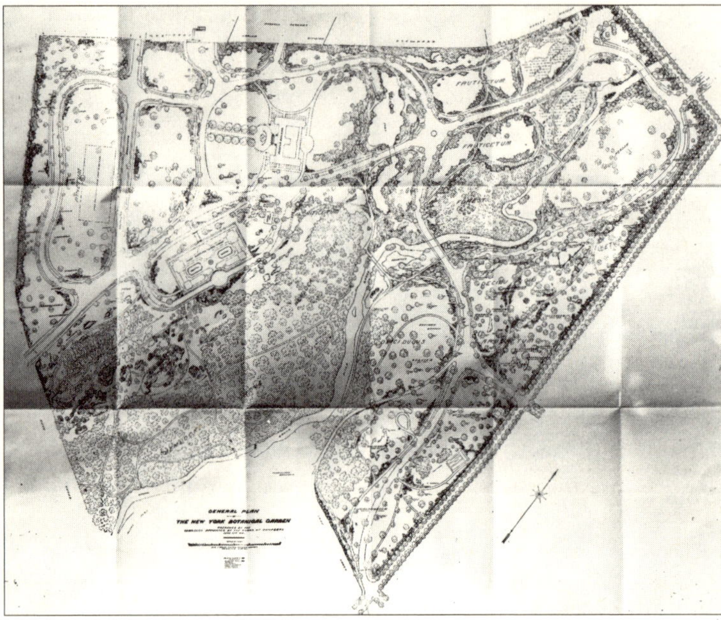

FIGURE 7
Britton's original general plan for NYBG, 1896. Note that the Conservatory was not constructed at the site depicted here, but at the location reserved for "a second horticultural house" (see FIGURE 8).

the provision of gently curving carriageways. Asserting that "it is a matter of congratulation," Britton maintained that his plans commission "endeavored to preserve the natural features of the tract so far as possible without prejudicing the location of driveways, buildings and the scheme of planting." They succeeded in largely retaining the general natural beauty of the park, he reported, since "the natural features of the area . . . lend themselves to artificial improvements without any loss of beauty or effect." Although some healthy trees were removed where road grading and "the rounded character of the curves which are imperative in a public park" made such actions necessary, "the number so selected," he wrote, "is so insignificant . . . that we feel that they need not be seriously considered, especially as all of them are abundantly represented at other points in the Garden."

Future Garden visitors seeking to use these graceful drives could arrive by means of the Harlem line of the New York Central Railroad, Britton noted. In addition, those arriving by Webster Avenue trolley would have only a one block walk eastward to the Garden entrances, which were located opposite the Bedford Park Station and, for carriage drivers, at Mosholu Parkway (which had been designed to link the entire system of the Bronx parks, from Pelham Bay to Van Cortlandt). Finally, Southern Boulevard, "a fine driving road," provided direct access to Bronx Park from the south.

Increased access from the north and east was anticipated with the opening of other roads in coming years. In November 1896, Britton had also planned to construct a series of pedestrian and carriage entrances around the northern and eastern borders of the Garden, with kiosks at each of these points designed to shelter gate attendants.

Britton and Brown concurred that a crucial first step would be to construct driveways through their portion of Bronx Park in order to attract the "driving class" from whom, despite Judge Daly's strategic egalitarianism, they expected the majority of their financial support.[88] These carriageways had been laid out by Vaux himself before his death by drowning in November of 1895, and Britton's planning commission adopted the famous landscape designer's plan for the portion of the Garden west of the Bronx River with only one minor revision.[89] The commission added to Vaux's plan a set of internal roadways on the eastern side of the river, creating a unified system designed to incorporate the Southern Boulevard, a broad pedestrian plaza at the Bedford Park railroad station, and an elevated extension of the Mosholu Parkway that would carry carriages on a bridge spanning the New York Central train tracks.[90] This latter bridge, Britton anticipated, would comprise "the most notable driving entrance to the Park for many years."[91] Within the Garden itself, a road from 35 to 40 feet in width was planned to extend from the Bedford Park station to the northern end of the park, following the natural contours of the landscape, except where it would cross a proposed stone arch erected over a man-made lake. This lake, in turn, would replace a bog that occupied a depression between the elevated site of the Museum and the area designated for the planting of a fruticetum.[92] Britton's committee also designed the layout of footpaths as well as roadways. His system of 10- to 25-foot-wide pathways, he wrote, was designed to permit "the ready distribution of crowds of people to the several portions of the Garden." In this way, he hoped, pedestrians would be kept as "much as possible" to the paths within the decorative grounds, although such spatial restrictions, he added, need not be as rigidly enforced within the arboretum, pinetum, or other forested areas.

The Museum building, as designed by commission member Gibson, was to be slightly over 300 feet across, with two symmetrical wings each comprising a third of that width. At its widest, the Museum would be 50 feet deep, with a "central feature

88 "Buildings and Grounds Planning, 1891–1898," Addison Brown Papers, New York Botanical Garden Archives.

89 Although rumors of Vaux's despondency led to speculation that the death was a suicide, Vaux's most recent biographer lends credence to the efforts of his son to dispel such speculation. See Francis R. Kowsky, *Country, Park, and City: The Architecture and Life of Calvert Vaux* (New York: Oxford University Press, 1998): 320.

90 Constructed in 1868, Southern Boulevard at that time ran within the Garden grounds, north of the current site of the Haupt Conservatory.

91 Britton Plan, 28, see note 87.

92 Ibid., 39.

projecting beyond the main structure, both to the front and rear." It was described by Britton as an "Italian Renaissance" edifice, with symmetrical arrangements of columns, entablature, and pilasters on the first two floors, and a third floor distinguished by dormered windows with steep mansard-style roofs. The central feature of the building, Britton promised, would be its domed reading room. The commanding height of the imposing building, he asserted, would nevertheless be mitigated in visual effect by the harmony of its lines with the woodland surroundings of the garden.[93]

Within this large structure, the Scientific Directors proposed a three-part plan of organization. The first floor would be devoted entirely to exhibits describing the economic usefulness of plants used for foodstuffs; drugs; textile manufacture; timber; and the production of oils, resins, gums, waxes, and sugars. Each display would be illustrated with plant specimens, photographs, models, and drawings, all arranged to illustrate the commercial products of these plants and the processes of manufacture associated with each. The second floor was reserved for the maintenance of a general botanical museum, in which plant specimens were to be displayed systematically from "the most lowly organized to the most complex plants;" a physiological museum, in which a collection would be designed to convey the "phenomena and processes of plant life;" a paleobotanical collection, in which a selected set of fossil plant specimens would trace "the sequence of the appearance of the various groups of plants upon the earth and their distribution in time;" and finally a display intended to convey the diversity of plant life in the New York City region. This latter museum area was to be modeled on a similar collection of local flora maintained at the British Museum. Each "natural family" type in the general museum was to be illustrated by actual plant specimens as well as the inclusion in each display of fruits, seeds, roots, leaves, bark, along with photographs and drawings.[94] Finally, an 800-seat lecture hall was placed on the western end of the building, nearest the railroad line.[95]

In his report, Britton took pains to justify the choice of the Museum site in terms of topography and ease of access. The NYBG planning commission, Britton wrote, had repeatedly examined the entire 250-acre garden parcel and had even entertained several suggestions that had been "informally received," but had unanimously agreed that the hill approximately 1000 feet east of the Bedford Park Station was "incomparably better than any other within the tract," because it was near the rail line and the Parkway and provided solid ground for constructing the building's foundation. In addition, Britton noted, this location would allow for considerable expansion of the Museum building in the future. The commanding hilltop position

93 "Report of the Secretary and Director-in-Chief for 1897" *Bulletin of the New York Botanical Garden* 1 (1897): 167. Subsequently cited as "Report of the Secretary and Director-in-Chief for 1897."

94 Britton Plan, 51–52, see note 87.

95 Gibson's Museum building and its constituent exhibits are discussed in detail in Chapter 4.

was also chosen for its proximity to a plateau approximately 900 feet to the southeast upon which he wished to erect the first horticultural greenhouse. This arrangement would allow the two buildings to share a powerhouse for heat.

Britton's initial choice of location for the first Lord and Burnham greenhouse thus was driven by both pragmatic and aesthetic consideration: one powerhouse for two buildings, as well as "abundant tree growth" that "permit[ted] a very beautiful treatment," with "enough large and perfect trees to partially mask the building and protect it from winds." Britton predicted that when completed, the Lord and Burnham edifice would be "one of the most beautifully located glass houses in the world." At that time, he planned for the first horticultural complex of 13 houses to cover an area of at least 45,000 square feet, with two east-west running wings extending from the central dome of the 90-foot-high palm house and then turning southward to form an enclosed court. In this protected space between glasshouse wings, Britton planned to install aquatic plantings and "other ornamental features." The Conservatory framework itself would be essentially fireproof, as it was to be constructed mainly of an iron superstructure, with glass panels and a bluestone facade for the foundation. The expansion-permitting casings that held the panes of glass would be the only major wooden component of the greenhouse. The entire structure was to be painted white. Overall, Britton described the style as "modernized Italian renaissance."[96] The heating system was designed to permit a minimum temperature of 70 degrees Fahrenheit even in winter, and heating mains were to be submerged beneath the Conservatory in trenches. Radiators would be screened from view by plantings. The powerhouse was still to be situated at a distance from the Conservatory, so as to avoid "the nuisance and unsightliness always incidental where the boiler power is placed in proximity to the houses."

Walks through the greenhouse displays were designed to be "commodious," with twisting paths through the larger houses and straight walkways in the smaller ones. The decision to construct a large number of differentiated horticultural houses, Britton noted, meant that a highly diverse set of plant life could be displayed. He planned to collect and maintain for display within these varied settings a cactus garden, a fern collection, an economic collection of tropical fruits, and an aquatic exhibition.[97] In addition to the Museum and the great glasshouse, several smaller structures were to be dispersed throughout the remainder of the 250-acre site, including a residence for the Brittons east of the river and cottages for the first and second gardeners. Finally, the planning commission's report called for the erection of propagating

96 Britton Plan, 165, see note 87.

97 "Report of the Secretary and Director-in-Chief for 1897." Though identified in this report as the most desirable site for Lord and Burnham's planned work, the great glasshouse ultimately would be constructed at a different location, along the Southern Boulevard. This change will be discussed below.

houses; a stable; and a series of three small, "rustic," rain shelters. Around each of these buildings, on an estimated 25 acres in all, Britton proposed plantings of a "purely decorative" nature, with no attempt at arrangement by scientific classification."[98]

If Britton and his plans commission had devoted a great deal of effort to planning the Garden's structures, they devoted an equal degree of consideration to the layout of the plantings. First, Britton endeavored to preserve as many healthy specimens of trees as road building would permit, even in areas not designated as forest or arboretum. Within the plots reserved for systematic planting of conifers, deciduous trees, and decorative plants, Britton's commission spared hale trees from the saw, even if they did not fit into the scheme he had devised for those areas. In addition, approximately 30 acres of the southwestern portion of the garden were to be given over to a pinetum, or conifer plantation, upon which Britton planned to grow between 90 and 100 species of pine in a "natural sequence, allowing for from two to five permanent individuals, the larger ones to stand not less than 100 feet apart." A tract east of the river, "much diversified in character [and] affording a great variety of soil and exposure," would be reserved for the Garden's collection of hardwood trees, 275 species of which were to be arrayed, alongside the extant specimen trees, in a "nearly natural sequence."[99] As in the case of the pinetum, Britton planned to plant in excess of the number of individuals desired, because not all were expected to grow to maturity. Paths through this 70-acre deciduous arboretum were to be laid in such a way as to "bring individuals of every natural family into view." As a complement to the arboretum, Britton called for a nearby 15-acre tract to be cultivated as a fruticetum containing approximately 850 shrubs. Within the fruticetum, the commission advocated a "very elastic arrangement" of space for a "nearly natural" sequential arrangement of families, which would trace a "curving line around and through the middle of the tract," allowing "considerable areas of greensward between each of these families." In this way, Britton hoped to allow the shrub collections eventually to double, without a disruption in the ordered arrangement.[100]

For the systematic collection of herbaceous plants, Britton's plans commission reserved an eight-acre parcel of land in the Garden's southern end: an unforested sloping meadow bordered by marsh land and a portion of forest, which would provide ample space for the cultivation of plant species suitable for aquatic, meadow, hillside, rock, and shade habitats. He proposed to arrange in "natural sequence," "every species which will probably exist in the open in our latitude." (In addition to the rock-loving specimens located in the herbaceous plantation, Britton decided to include a rockery "mainly for decorative effect.")

98 Britton Plan, 34, see note 87.

99 Ibid., 34–35.

100 Ibid., 36. They also planned for a systematic display of woody vines in the viticetum, while reserving the ability to deploy vines for merely decorative effect elsewhere in the Garden.

The 1896 report also outlined the creation of several wetland areas, which were designed to utilize existing water sources for the cultivation of moisture-loving plants. A five-acre bog garden, in which wetland trees, shrubs, and herbaceous plants could be grown, was designed to occupy the northwestern segment of the park. Here Britton did not plan a strictly systematic arrangement, opting instead merely to plant species of the same genus together. In addition, by strategically utilizing two small local streams that originated outside the Garden, the planners hoped to transform a natural bog lying north of the elevated site reserved for the Museum into two adjacent lakes, which ultimately were to be separated by a north-south driveway. Overall, Britton proposed a total water area for NYBG of approximately six acres, excluding the Bronx River. As part of these wetland plans, the river itself would be lowered by two feet by removing a portion of the old Lorillard mill dam. This alteration would facilitate drainage while minimizing the presence of the stagnant pools in which disease-carrying insects might breed. This restoration of the river level nearer to its original depth would also, the commission reported, enhance the aesthetic character of the landscape. Britton even planned to add to the front of the newly lowered dam an irregular pile of boulders, "which would break up the even character of the . . . fall into a foaming cascade, which would be in harmony with the forest and much more beautiful than the present unnatural mill dam."[101]

Finally, in addition to including provisions for tapping the city's water supply via the construction of large mains on the Garden property and providing specific ideas regarding drainage, sewage disposal, lighting, and grading, the plan designated an acre of Garden land for the cultivation and display of plants selected for their economic value. These products, Britton wrote, had "usually found [a] place in botanical gardens, and we regard the feature as a desirable one." This collection would be arranged by product rather than primarily by botanical classification. In concluding the report that accompanied the 1896 plan, Britton noted that his outline of prescriptives for remaking the Garden landscape had "purposely been prepared in a very elastic way," so that later modifications and unforeseen contingencies might be met as construction proceeded. Such contingencies, in the form of political and scientific controversies, soon reappeared and, as in the case of the earlier controversies over the composition of the Board of Managers, were eagerly chronicled by the New York press.

The Garden Design Controversy

On December 12, 1896, the editors of *The New York Times* questioned the validity of the Britton landscaping plan. The *Times* belittled the members of the Garden's plans commission, asserting that they lacked the standing of Frederick Law Olmsted.[102]

101 Ibid., 46.

102 *The New York Times*, December 12, 1896. The *Times* failed to recognize Calvert Vaux's role in designing the Garden's roadways.

That afternoon, *The New York Evening Post* echoed the critical refrain, urging the Parks Department to postpone approval of Britton's plans until more "expert" opinions might be heard. The *Post* particularly bemoaned the fact that the NYBG plans had been "kept secret" from public scrutiny and had been sent to the city Building Department *before* being presented to the Commissioners of the Parks Department. The paper argued that Boston's Arnold Arboretum represented the proper reconciliation of scientific purpose and landscape design and, thus, served admirably in its role as "public pleasure ground" as well as research institution.[103] The creators of New York's new public garden, they implied, would do well to follow such a distinguished precedent. Within days, *The New York World* joined the *Times* and the *Post* in urging a delay so that Britton's original plan might be evaluated by "the best experts in landscape gardening and botany."[104] In the midst of this public controversy, Judge Brown, as chairman of the Garden's Committee on Plans, presented his own official report to the entire Board of Managers. He was careful to defend Britton's work.

In endorsing the Britton plan, Brown stressed the uniqueness of every botanical garden and asserted that no particular botanical garden could "be taken as a simple model for another." Each garden, he argued, is "a special study, having reference to its site and topography," and in the case of Britton's report of November 30, Brown said, such careful study had been obtained "on the spot" from "competent experts in the various departments [of the new Garden]." Each of these expert staff members, Brown maintained, "were able to give, in the fullest degree, constant personal attention and repeated instruction on every point." These commission members, he contended, had in fact made repeated visits since 1891 to the "principal gardens of Europe," in an effort to examine the designs of each. In addition, he continued, those involved with the preparation of the planning commission's report had in fact "sought the advice of the persons in this country believed to have had the most experience in such enterprises, so far as they were accessible to the Committee." Brown thus publicly defended Britton's plan and his competence in landscape design, while at the same time asserting that outside expert advice had already been secured. The "valuable aid," of the nation's premier landscape architect, Frederick Law Olmsted, however, was unavailable due to poor health. In the late Calvert Vaux, designer of the Garden's road system, the commission had had the benefit of "the highest expert skill," and his last work (i.e., on the NYBG driveways) "largely controlled the plan of all the other landscape designs," Brown reported.[105]

103 *The New York Evening Post*, December 12, 1896.

104 *The New York World*, December 14, 1896.

105 "Memorandum to Board of Managers from Standing Committee on Plans," December 14, 1896, Addison Brown Papers, New York Botanical Garden Archives.

The decision *not* to cluster all of the principal buildings near the southern end of the Garden adjacent to the grounds of St. John's College (today Fordham University), Brown noted, was based upon the advice of Daniel Morris, assistant director of the Royal Botanic Gardens at Kew, and formerly the director of the imperial Public Gardens in Jamaica. At Kew, on a single day during the summer of 1895, Morris had reported, approximately 400,000 people had entered the garden. Surely, Brown insisted, New York's fine garden, in the opinion of its strongest boosters soon to rival even Kew in prestige and popularity, must make allowances for similar numbers of visitors. "To make possible the proper management and proper supervision of such masses," he wrote, "and to prevent them from becoming dangerous to each other and destructive to the Garden, an easy and natural division of any such crowds, by a separation of the buildings of most interest, is a prime necessity."[106]

Brown recommended approval of Britton's plan by the entire Board of Managers, concluding that it was "in all material respects as good as can be framed." Further, Brown insisted, he was most gratified that Britton's commission had managed to preserve all the "unique natural charms" of the Garden tract, and congratulated the Garden's designers for appropriating "almost every part of the diversified grounds to the most apt and beautiful uses." Yet Brown, aware of past public controversy and attuned to the likelihood of ongoing press scrutiny, closed his report on a conciliatory note, asserting that the plan easily admitted "a large elasticity in treatment, and abundant room for modifications in detail as may be hereafter desired." Suggestions for modifications were readily forthcoming.

Charles Sprague Sargent and a New York–Boston Rivalry

During the 1890s, one of the more influential of the numerous American periodicals devoted to gardening and landscape design was New York's *Garden and Forest*, edited by a New York City Park Commissioner, William A. Stiles, and published under the direction of Charles Sprague Sargent, director of Harvard University's Arnold Arboretum in Boston. Stiles and Sargent had followed closely the efforts of Britton and the Torrey Club to establish a large public garden. In the early part of the decade, they aided Britton by chronicling his success in garnering funds and offering advice to the Garden incorporators. In 1891, Sargent asserted to readers interested in the actions of New York's garden builders that the ultimate value of such an endeavor would depend upon "the capacity of the man who controls it."[107] In the spring of 1894, as fundraising efforts continued despite the negative effects of the economic depression,

106 See Minutes of the New York Botanical Garden Board of Managers, December 14, 1896, New York Botanical Garden Archives.

107 *Garden and Forest*, April 8, 1891.

Sargent further counseled that the task of choosing a director for the new Garden would be one of the most important decisions the NYBG incorporators would face. According to Sargent, the NYBG's Scientific Directors needed to appoint a chief with the integrity of the Hookers at Kew—a man who could face varied challenges sure to confront the new institution, including public pressure to construct merely a "pretty flower garden," as well as the power and influence of the city government.[108]

Decades later, the NYBG's Henry Gleason suggested that Sargent, by the 1890s the successful head of the Arnold Arboretum, was lobbying for the job himself. In light of Sargent's position at Harvard, his family ties to elite Bostonians (many of whom made major bequests to the Arboretum during Sargent's tenure), and his large suburban estate just outside of Boston, this appears unlikely. Sargent biographer S. B. Sutton notes that in 1879 Sargent already considered the Arboretum to be his life's work, upon which he would ultimately expend approximately $225,000 of his own fortune.[109] Rather, Sargent's animosity to the ultimate form NYBG took was due in large part to Britton's personal unwillingness to consult the Olmsted Brothers' firm, a snub that represented an essential challenge by New Yorkers to Boston's purported cultural authority in matters of both landscape art and botanical science. Sargent and Britton's prickly personal relationship had also been exacerbated by Britton's leading role in efforts to reform the International Code of Botanical Nomenclature.[110]

Charles Sprague Sargent, of eminent Boston parentage (he was a cousin of painter John Singer Sargent) and influential authority on matters botanical and architectural, was no stranger to the Brittons, who had dined along with historian Francis Parkman at the Sargent estate, "Holm Lea," in suburban Brookline, Massachusetts in the spring of 1889. In fact, despite the later decades of rancor, it appears that Sargent and other Boston botanists had had a close corresponding relationship with their New York brethren in the late 1880s and early 1890s; their

108 *Garden and Forest*, May 9, 1894.

109 S. B. Sutton, *Charles Sprague Sargent and the Arnold Arboretum* (Cambridge, MA: Harvard University Press, 1970): 57. To suggest that this was unlikely, however, is not to assert that it was impossible. It should be noted that Sargent did have ties to New York State, including familiarity with the landscaped Hudson Valley estates of mentors H. H. Hunnewell and H. W. Sargent and his service as chairman of the first Adirondack Park Commission. Griswold and Weller noted 18 separate major bequests from personal friends of Sargent to the Arnold Arboretum. See Mac K. Griswold and Eleanor Weller, eds. *The Golden Age of American Gardens: Proud Owners, Private Estates, 1890–1940* (New York: H. N. Abrams, 1991): 23.

110 See Sutton, *Charles Sprague Sargent and the Arnold Arboretum*, note 109 and W. Trelease, "Charles Sprague Sargent," *National Academy of Sciences Biographical Memoirs* 12 (1929): 247–258. For a convincing portrait of the social background of elite Bostonians such as the Sargents, see Tamara Plakins Thornton, *Cultivating Gentlemen: The Meaning of Country Life among the Boston Elite, 1785–1890* (New Haven, CT: Yale University Press, 1989).

discussions of nomenclatural questions had been, up until that time, rather cordial. In 1887, for instance, Sargent had written Britton regularly as he worked upon the genus *Crataegus* and had exchanged herbarium specimens with the New Yorker. These friendly exchanges of specimens continued into the following decade, as Sargent established the nation's leading horticultural periodical, *Garden and Forest*, to which Britton contributed reviews. When Britton began his career-long quest to reform nomenclature, Sargent had admonished him gently that strict adherence to some of Britton's proposed rules, would lead to work taking "a lifetime."[111]

Yet, by the mid-1890s, the design of the new public garden in New York, coupled with Britton's strenuous commitment to nomenclature reform, had led to an irreconcilable division between New York and Boston botanists, with Britton and Sargent helping to lead their respective camps. Sargent had begun voicing his objections to Britton's plans for Bronx Park in 1896, when they had been unveiled. Sargent had in fact personally examined the Bronx Park site for the Garden, walking through the parcel accompanied by Britton and Garden and Forest editor (and New York Parks commissioner) Stiles in the autumn of 1896. Sargent repeated this tour of the proposed NYBG site in the summer of 1897 as part of an "expert commission" appointed at Stiles's insistence by the Department of Public Parks to evaluate Britton's proposed design.

"The proposed site," Sargent maintained after visiting the park, "does not seem to be an ideal one . . . either in situation or topography. The surface of the ground is very much broken by ledges, so that the different parts of the Garden must be necessarily considerably removed from each other," he contended. Sargent also took issue with the overall size of the parcel of Bronx Park land set aside for the Garden, maintaining that it would outgrow the space allotted it. In addition, he argued that it would be comparatively inaccessible to residents of the southern, western, and eastern portions of the greater metropolitan area. Finally, he contended that only by grouping the Garden's buildings in one end of the parcel would an aesthetically pleasing result be reached, and land would be left over for the planting of trees.[112] (Sargent's antipathy to the site chosen in Bronx Park may have been increased by the fact that during his walk with Britton, he had severely injured an ankle while jumping over a fence.)

Britton countered Sargent's criticisms by noting that placing the new Garden at the geographic center of Greater New York was impossible, and thus that *any* location was bound to be relatively inaccessible to some regional residents. He also contended that the land set aside by the city grant was ample for both garden and park purposes and that any more might be difficult to administer. Finally, Britton disparaged Sargent's

111 C. S. Sargent to N. L. Britton, March 21, 1889; December 9, 1889; and January 10, 1891, Arnold Arboretum Archives, Harvard University.

112 C. S. Sargent to A. Brown, November 27, 1896, Arnold Arboretum Archives, Harvard University.

own institution, claiming that Sargent's assertions regarding tree planting were less than accurate since "many of the trees [in the Arnold Arboretum] have done poorly," and that the site of the Arnold's administration building was worthy of "serious" criticism.[113] Britton, wary of Sargent's prestige and eager to avoid yet another controversy involving the nascent public garden, privately characterized the Bostonian's objections to the proposed Garden layout as "wild" and inconsistent. Although, as expected, Sargent had been critical of the proposed layout of the Garden's buildings, he had acknowledged, Britton said, the high quality of the Bronx Park soil and admitted that there was more space in the parcel of parkland than he had previously supposed. According to Britton, Sargent's main criticism seemed to be that, "nobody who knew anything about things in general had been consulted." When Sargent disapproved of the planned width of the NYBG's pathways, however, Britton retorted that Calvert Vaux himself had been responsible for setting the proposed dimensions. Britton angrily characterized Sargent as "grumbling," "howling," "offish to the whole scheme," and willing to "practically object to everything." Privately, Britton confided to Judge Brown that Sargent's attitude toward their efforts stemmed from Sargent's desire to "give the younger Olmsteds a job," as well as the "old notion that nobody outside of Boston knows anything."[114]

Sargent, unwilling to shy away from this debate, had utilized some of the most powerful tools at his disposal in the attempt to influence the design of New York's public garden. He had published a thinly veiled attack on the NYBG planners' vision for Bronx Park in *Garden and Forest*, where he had editorialized that "we have said that planning and planting by novices are, as a rule, worse than unsatisfactory, for if there is any art which needs original aptitude, special training and long experience for the highest success, it is the art of planting public and private grounds." Nevertheless, he contended, most people disregarded this advice and considered that any "botanist is considered qualified to plan a public garden. Places of remarkable natural beauty, when once they are acquired by a city, are often put under the control of commissioners," who might be gifted with business acumen but utterly lacking, Sargent said, any facility in landscape art. He concluded that "for the treatment of any piece of ground, public or private, the counsel of a landscape gardener of recognized standing should

113 N. L. Britton to A. Brown, December 5, 1896, Arnold Arboretum Archives, Harvard University. Britton's comments about the ability of the NYBG to administer the its expansive landscape proved prescient: given the budget, Garden administrators had a difficult time discouraging vandalism in the various plantations throughout the 1910s (see Chapter 4).

114 N. L. Britton to A. Brown, November 18, 1896, New York Botanical Garden Archives. For Brown's account of the extended controversy involving the Boston authorities' objections to the site of the Garden, see Addison Brown, *Judge Addison Brown: Autobiographical Notes for His Children* (Boyce, VA: Carr Publishing Company, 1972).

at once be taken." Only such "true artists," Sargent maintained, would be able to work in harmony with Nature by practicing a "broad and catholic art which alone is useful in the preservation of the Yosemite Valley or the scenery of Niagara as it is in planning a pastoral park."[115] Only professional landscape designers, Sargent maintained, possessed the requisite training to build a public garden that would incorporate the natural features of the surrounding landscape. Botanists such as Britton, Sargent worried, were concerned only with the practical institutional needs of a scientific botanical institution, to the exclusion of an informed aesthetic sensibility.

Stiles, as treasurer of the Commission that ran the City's Department of Public Parks and editor of *Garden and Forest*, remained a staunch Sargent ally. He held that Britton's plan threatened to interfere with "the natural scenery of the park." In a clear echo of the animating spirit of Olmsted's Central Park 40 years earlier, the Parks commissioners, in the wake of Sargent's assessment, asserted that

> the primary purpose of a rural park within the reach of a great city is to maintain that rest and refreshment of mind and body which come from the tranquilizing influence of natural scenery. All other additions to the attractions of the people's place should be subordinate to the controlling purpose in a design and maintenance of such pleasure grounds. Anything which interferes with the restful quality of the scenery in so far destroys the highest value of the park.[116]

Sargent, again using the pulpit of the *Garden and Forest* editorial page, similarly invoked a Romantic park ideal to undermine Britton's vision for the Botanical Garden. "Whenever . . . a city possesses park lands with exceptionally beautiful features, their preservation . . . means the saving of capital; it is the husbanding of a source of health which is the source of all wage-earning power. . . ." He continued

> When we consider the almost universal admiration and even affection among civilized men and women for broad natural landscapes . . . and consider its special restoring effect upon those who suffer from the nervous strain of city life, we have the one justifying reason for large urban parks. Small parks . . . are valuable for other purposes, but for the highest rest and refreshment nothing will fill the place of stretches of beautiful natural scenery. Artificiality, the needless intrusion of buildings, anything which interferes with seclusion and the actual contact and communion with pure nature, defeats to some extent the highest purpose of such parks.

He went on to identify Britton's plans for building in Bronx Park as an explicit example of such folly: "the City spent money like water to create some charming natural

115 "Art and Nature in Landscape Gardening," *Garden and Forest* May 19, 1897.

116 Report of the Commissioners of the Department of Public Parks to Mayor William L. Strong, July 31, 1897, Mayoral Records, Municipal Archives of the City of New York.

FIGURE 8 Aerial view of the Museum building and the Conservatory, 1921.

scenery in . . . Central Park," Sargent wrote. "Now that it has come into possession of scenery far more beautiful and ready-made, it would be worse than folly to destroy it."[117]

The Political Context of the Design Controversy

The opposition of New York City's Parks Department, in the form of Stiles's alliance with Sargent, was a formidable obstacle for Britton and the Torrey Club to overcome. Since the annexation of Bronx land and the designation of the parks north of the Harlem River, Stiles's department had been an extremely powerful influence in the development of the new city wards. As late as 1890, the Parks Department controlled the construction of bridges, streets, sewers, and other improvements throughout the Bronx. Only when such improvements were slow in coming had "north side" residents secured state legislation in the early 1890s that created a separate department to address Bronx development concerns.[118] Given the Parks Department's power and the prestige of the Sargent and Olmsted names, it is unsurprising that by June of 1897 several of Sargent's "expert" committee's criticisms had been addressed, and that the Botanical Garden as laid out by Britton's November 1896 plan had been modified.[119]

117 "Natural Beauty in Urban Parks," *Garden and Forest*, June 30, 1897.

118 Gonzalez, "From Suburb to City," 10, see note 29. Gonzalez notes that original development plans for the Bronx had been drawn up by Olmsted, Sr., to include a series of "rural" suburbs characterized by winding streets. The decision to extend Manhattan's grid pattern northward across the Harlem River was made to accommodate the desire for greater density of development.

119 Minutes of the Committee on Plans, May 28, 1897, New York Botanical Garden Archives. The "expert committee" called for by Charles Dana's *New York Sun* and convened by the Parks

Banished from the Garden grounds were the planned homes of the director and head gardeners. Most significant, however, was the relocation of the planned site of the great Conservatory, which Britton had hoped to erect on a commanding rise adjacent to the Museum building, to the site it occupies today along Southern Boulevard (FIGURE 8). The Boulevard itself would also be rerouted to its present location between the campus of Fordham University and the western boundary of the Garden.[120] Nevertheless, the commissioners did not give in to Sargent's suggestion or the pressure of *The New York Sun* editor Charles Dana to group all of the Garden's buildings into one corner of the park, leaving the rest in what Sargent considered a "natural" state. (Ironically, this Sargent ally was the same Dana, who, in 1892, had helped Britton fend off the challenge by NYU's Henry MacCracken.) By late July 1897, the City's parks commissioners, with the exception of Stiles, had approved Britton's revised plans for the Garden and had requested the City's Board of Estimate to issue bonds to finance its construction.[121]

Despite this victory, Britton's plans for the Garden were not yet assured. The final private subscriptions to the Garden's inaugural fund had been received in the mid-1890s, at a time of both economic and political upheaval in the city's history. Consolidation of New York and Brooklyn into Greater New York was slated for January 1898; in the meantime, reformers had alternated with Tammany figures in the mayor's office during the 1890s. Lawyer and real estate developer Hugh Grant had become the city's first Irish-American mayor in 1888. He had been in office as the 1891 Act of Incorporation passed the New York State Legislature, kicking off the Garden's official subscription drive. Grant had been ousted in 1894 by William L. Strong, a reform candidate who became a crucial supporter of Judge Brown and Nathaniel Britton's efforts to secure the city appropriation in 1897. Strong, in turn, was followed by Robert Van Wyck, another Tammany figure, in 1897.[122]

Department had included, besides Sargent, Thomas Hastings, Charles Eliot, John C. Olmsted, and Samuel Parsons. Even with the agreed upon changes, Stiles refused to vote in favor of Britton's plan, casting the only dissenting vote against it among the Parks Board commissioners. It is important to note that Dana's animosity to Britton's plans stemmed from his close personal ties to Sargent. In the 1880s, when Sargent and Stiles, at the request of northwestern railroad interests, had journeyed west to conduct a survey of western forests, they had been accompanied by Dana's son, Paul. See Sutton, *Charles Sprague Sargent*, 93–94, note 109.

120 Britton's original choice for the site of the conservatory corresponds roughly to that occupied in 2006 by the native plant garden.

121 Minutes, New York City Department of Public Parks, July 19, 1897, Mayoral Records, New York City Municipal Archives.

122 David Hammack, *Power and Society: Greater New York at the Turn of the Century* (New York: Russell Sage Foundation, 1982): 116–117 and Beckert, *The Monied Metropolis*, 320, see note 39.

Mayor Strong, though faced with the vociferous opposition of *The Sun*, found support in the Bronx's own newspapers, whose editors were eager for the sort of "uptown" development that the Garden and its neighbor the Bronx Zoo (erected in the southern portion of Bronx Park that Britton had rejected as unsuitable for the Garden) represented. *Dickson's Up-Town Weekly* fretted that the "wealthy, worthy gentlemen" who had been "public-spirited" enough to pledge funds for the Garden's construction might be driven away if the controversy continued. In response, *Sun* publisher Charles Dana, on his deathbed, countered with an editorial decrying "the businessman in politics," accusing the Strong administration of seeking to run the municipality as a business and discarding the advice of experts, presumably the sort represented by Stiles, the Olmsted firm, and Sargent. These "trained officials," in *The Sun's* estimation, were the necessary public check on a "rampant business spirit," and ensured the "reasonable application of skill in the expenditure of public money."[123]

Despite the criticism of the *Sun*, city officials appropriated municipal funds for construction by October 1897. Notwithstanding the substantive changes to the original 1896 design, Britton and his allies could consider the resulting layout of the Garden a victory for their vision of the new institution.[124] The crux of that vision was a practical one, not necessarily at odds with the earlier, romantic vision of Olmsted's parks or the reformist impulse behind it, but in keeping with a desire to create a scientific botanical institution in New York to rival Old World institutions. The Garden, as finally organized and built, represented the culmination of a plan begun in the deliberations of the Torrey Club in 1888 and in conversations held decades earlier by botanists John Torrey and Asa Gray; horticulturists Thomas Hogg and George Nash; and of course, the civic-minded plutocrats who financed its creation. The controlling vision of the Garden's mission remained firmly in the hands of men like Britton and

123 Such a stance was not as ironic as it might appear. Although it would seem that the reformist coalition that had elected Strong to office over the Tammany machine candidate represented a social stratum in New York directly analogous to the elite Boston world of the Sargent family, such an interpretation, David Hammack has shown, suffers from a lack of nuance. As Hammack noted, the various electoral coalitions that formed around New York City municipal elections in the 1880s and 1890s were, contrary to traditional interpretations, surprisingly diverse. To win the mayoralty, candidates could muster the simultaneous support of Tammany and non-Tammany Democrats, the New York County Republican organization, and/or various citizen movements such as the Citizens' Union. Each successful coalition, then, was composed of a varied set of social actors, including wealthy merchants, bankers, professionals, union officials, lawyers, and professional politicians. See Hammack, *Power and Society*, 118–129, note 122.

124 In a fittingly bitter aftermath to the site and design controversy, the members of Sargent's "expert" committee submitted individual bills for their services to the City's Parks Department. Sargent billed $100 and Olmsted over $200, all to be charged to the fund for the erection of buildings in the Botanical Garden. See Minutes of the New York City Department of Public Parks, October 11, 1897, Mayoral Records, New York City Municipal Archives.

Henry Rusby, who were committed to international standards of scientific botany. The New Yorkers' iconoclastic stance in the nomenclature controversy only emphasizes this point: their commitment to professional, and preferably "logical" standards of taxonomic practice outstripped even the most established of European university botanists.

Thus, while never neglecting aesthetic considerations, Britton's paramount concern in laying out the Botanical Garden was practicality: the necessary growing conditions for each family of plants and the need for a comprehensible layout to convey elementary principles of taxonomic botany to New Yorkers. Rusby, in a letter to the city administration during the controversy over the design of the Garden, criticized the Bostonian opponents as "landscape people and editors," prone to offering "all sorts of crazy theoretical ideas." In a somewhat ironic twist, in light of the nomenclature controversy, Rusby and Britton made repeated and opportunistic recourse to the opinions of the administrators of established European gardens when arguing for their own garden plan.[125] "Landscape features [had] been carefully studied," Rusby assured the city's leading politicians, "but had not been allowed, in trivial details, to override experience in working affairs." He criticized the Sargent partisans as carpetbaggers, "alien to the city and the state," and lacking the local authority to determine the city's course of action in its own parks system. It is interesting to note, in light of the difficulties between Sargent and Britton, however, that the Arnold Arboretum shipped to New York, at the height of the controversy over Britton's design, 300 specimens of trees and shrubs, a largesse that might be explained as an effort by Sargent to influence the Garden's design or simply the willingness of even "difficult" members of the botanical community to recognize the importance of the New Yorkers' efforts.[126]

The Power of Harvard Botany

In March of 1897, as Britton and Sargent contested the design of the botanical garden, NYBG incorporator C. F. Cox confided to Addison Brown that he found it "humiliating to have Boston vested with a veto power on a purely New York enterprise." Cox feared that the Garden administration was on its way to becoming mere "servants of the Park Board," operating at the behest of Boston landscape architects.[127] Rusby, another incorporator and curator of the Garden's economic botany collections, complained

125 Britton's personal conception of the purposes of a public botanical garden will be discussed in further detail in Chapter 4. See H. H. Rusby to Francis M. Scott, Corporation Counsel to the City of New York, September 17, 1897, Mayoral Records, New York City Municipal Archives.

126 Minutes of the Board of Managers of the New York Botanical Garden, November 12, 1896, New York Botanical Garden Archives.

127 C. F. Cox to A. Brown, March 5, 1897, Addison Brown Papers, New York Botanical Garden Archives.

to City officials that Sargent and members of the Olmsted Brothers firm were trying to usurp control over the Garden's development.[128] By 1898, Charles Cox and Rusby need not have worried about continued interference from Boston's cultural authorities, given the change in the board's composition that came with consolidation. The creation of Greater New York on January 1 of that year entailed a change in the bureaucratic contours of the city's park administration. With consolidation, the mayor began appointing a Park Board of three people, one for the Bronx, one for Brooklyn and Queens, and one for Manhattan and Staten Island. As had the preconsolidation Board that included William Stiles, this new Board could award contracts and set basic rules, but it would not retain the authority to set day-to-day policy. Though each of these post-1898 commissioners retained strong control over parks within their boroughs, each would be entirely dependent upon the goodwill of the mayor, who had the power to remove them from office. Subsequent to consolidation, the occupants of these posts tended to be politically active (often Democratic) partisans, unlike the genteel individuals (including the *Garden and Forest* editor) who had held the posts previously. The latter had been staunch allies of Charles Sargent, the Olmsted firm, and the city's landscape architect Samuel Parsons, Jr. However, Parsons himself did not last long in the postconsolidation era; he was removed from office early in the mayoralty of Strong's successor Robert Van Wyck.[129] Rather, given the postconsolidation alignment of power over parks, it became clear that to navigate the intricacies of New York politics, Britton needed to remain adept in advancing the Garden cause before all manner of city administrations, either reform or machine-controlled.

The objections of Cox, Rusby, and Britton to Sargent's involvement in the Garden's formation, however, cannot be construed as either an outright break from the genteel ideals of park building or a repudiation of the bourgeois norms that provided the context of Gilded Age institution-building. Nathaniel Britton's views on landscape design and architecture, though comparatively untutored, remained firmly in the camp of Olmstedian/Victorian, middle-class, aesthetic ideals, beliefs that were tied to environmentalist beliefs in the power of "refreshing" physical surroundings to mold public behavior. In one sense, the controversy over the Garden's design in the late 1890s was simply the result of a personal rivalry between members of opposing camps in a contemporary contest over scientific standards for naming plants. The Britton-Sargent clash was a personal one, exacerbated by two famously prickly personalities. Britton's imperious nature was legendary; his obstinate, single-minded performances in the decades-long nomenclature wars underscore his stubbornness.

128 H. H. Rusby to F. M. Scott, Counsel to the Corporation of the City of New York, September 17, 1897, Mayoral Records, New York City Municipal Archives.

129 Frans Stafleu, unpublished history of the New York Botanical Garden, 21 (typescript in New York Botanical Garden Archives).

In Charles Sprague Sargent, he found a formidable opponent. Intimate of Harvard presidents, relative of wealthy merchants and cultural icons, and scion of a powerful Boston family, Sargent was accustomed to being asked advice. Prone to arrogance, Sargent was known to "utter opinions as though they were facts and facts as though they were axioms." He would, his biographer S. B. Sutton writes, "give assistance generously unless he detected hostility . . . in those circumstances he shifted abruptly to the role of critic, and the sting of his criticism knew no moderation."[130] As an established expert with service on the Adirondack Park commission and US forestry panels, and the publisher of an influential horticultural journal, Sargent surely *expected* to be consulted at length regarding the construction and organization of New York's new botanical garden. One can only imagine his pique at not being granted a greater degree of influence over its design.

The critical factor in these struggles, of course, was the residual power and prestige of Harvard's botanists, who considered themselves to be heirs to Asa Gray.[131] After ending his apprenticeship under John Torrey in New York, Gray had spent his career in Cambridge, Massachusetts making Harvard a North American adjunct to the Royal Botanic Gardens at Kew. This was a savvy career move for himself and his institution, because, as contemporary American botanists knew, Kew's power at that time was unrivaled. By following Kew standards in nomenclature and classification and incorporating them into his authoritative *Manual of Botany*, Gray harnessed that institution's formidable prestige for Harvard. In the 1870s, Gray offered the only graduate-level training available in botany in the United States, and maintained the only American laboratory capable of hosting research comparable to that conducted at Kew and on the Continent.[132] In his efforts to build a national network of botanists (answering to his authority) and to seize for Americans the power to write their own natural history, Gray was Britton's most important predecessor. With analogous support from local Boston plutocrats and similar connections to Washington (Gray allied himself with Joseph Henry, Secretary of the Smithsonian), Gray's success at consolidating his control over American botany was a model Britton sought to emulate,

130 Sutton, *Charles Sprague Sargent*, 293, 312–313, see note 109.

131 Betty Smocovitis noted that reliable indictors of Harvard's botanical power were the number of degrees in the discipline that it granted and the prestige associated with the degree holders. Years after Gray's death, in the first decades of the twentieth century, Harvard remained the single largest producer of botanists in the U S. Nevertheless, the institution was not without its problems, plagued as it was during the late-nineteenth and early-twentieth centuries by internal competition among fragmented botanical departments. See Vassiliki Betty Smocovitis, "Botany and the Evolutionary Synthesis: The Life and Work of G.Ledyard Stebbins, Jr." (PhD Dissertation, Cornell University, 1988): 36–37, 42–43, 324–326.

132 Richard Overfield, *Science with Practice: Charles E. Bessey and the Maturing of American Botany* (Davenport: Iowa State University Press, 1993): 15.

even as that level of control grew more difficult as more US institutions contended for resources and prestige. Ultimately, by coordinating various nineteenth-century surveys of North American flora and demanding specimens be sent to his office at Harvard for naming and classification, Gray in a very real sense managed American botany. His death in 1888 coincided with Britton's own rise in the botanical circles and, perhaps serendipitously, with Elizabeth Britton's call for an American Kew to be built in New York.[133]

To emulate Gray's success, Britton openly challenged his Harvard successors. He was not the first to do so: in the 1870s and 1880s, federal botanists had begun, in historian Philip Pauly's words, appropriating Gray's project "to identify North American organisms and . . . assist in their exploitation." It was at this time that the US federal scientific bureaucracy, specifically the US Department of Agriculture's (USDA) various divisions, took the lead in research into horticultural and agricultural problems, including plant culture, pest control, pathology, crop breeding, and plant exploration and introduction. Federal plant scientists had begun, Pauly wrote, to "see in botany the potential key to the reorganization of the nation's food production system." Such federal projects involved larger staffs and better resources than those available even at Harvard. Nevertheless, in the late 1880s and into the 1890s, Harvard, with its German-educated scientists and close ties to Kew, remained, especially in the opinions of its own staff, at the center of American botany.[134]

Organized within the decade after Gray's death, Britton's NYBG thus presented yet another direct challenge to Harvard botany. By exploiting a growing network of correspondents in Washington, both Brittons developed close working relationships with botanists in the USDA and the Smithsonian. Working alliances with men like Joseph N. Rose and Frederick Coville enabled NYBG to challenge the Arnold Arboretum and Harvard's Botanical Garden, institutions harboring their own imperial and national ambitions in forestry, horticulture, and systematics. Britton, an outsider in training and geographic origin and the author of a new book on northeastern American plants organized with a novel plant-naming system, had begun undermining Gray's legacy. Given this context, Sargent's objections to the New Yorkers' efforts were unsurprising.[135]

Britton's challenge to Boston's cultural authority also generated controversy

133 A. Hunter Dupree, *Asa Gray: 1810–1888* (Cambridge, MA: Harvard University Press, 1959) and Philip Pauly, *Biologists and the Promise of American Life: From Meriwether Lewis to Alfred Kinsey* (Princeton: Princeton University Press, 2000): 17–33.

134 Pauly, *Biologists*, 33, 44, 81–84, see note 133.

135 On Sargent's reverence for Asa Gray and Harvard botanists' general role in cultivating Gray's reputation into the twentieth century, see Sharon Kingsland, *The Evolution of American Ecology, 1890–2000* (Baltimore, MD: Johns Hopkins University Press, 2005): 48–57.

because by the late nineteenth century the landscape architecture profession had become so established that to ignore its reputation for expertise courted wide public censure. Bostonians such as Sargent and the partners of Olmsted's firm had become nationally recognized authorities on aesthetic landscape development, whose opinions were sought from California to New Jersey. Furthermore, the Bronx setting chosen for the Garden was particularly contested because uptown districts of the northward-growing city were seen by contemporary urban planners as the most likely scene of a planned utopian landscape. By the 1890s, of course, it was generally recognized that the colonial-era landscape of lower Manhattan was beyond the power of central planners to alter. But in the northern districts, upon whose landscape parks, streets, and other components of an urban infrastructure were just beginning to be installed, a course of planned, rational development might be closely followed. The development of one such prime cultural component such as the Botanical Garden, situated as it was within the confines of the most central of Bronx parks, was bound to be contested by various actors seeking to influence the design of a modern New York City.[136]

Navigating City Politics

Finally, it should be recalled that the contest between Britton and Sargent occurred within the context of 1890s New York, a city whose mayoral administrations alternated during the decade between self-conscious "bourgeois" reformers and loyal political operatives of the Tammany machine. Britton's ties to the city's philanthropic elite made him a natural ally of men like William Strong, but the exigencies of the municipal appropriations process and the powerful tradition of New York park politics had made him perhaps equally adept at negotiating the demands of machine politics and securing the financial support of the city's economic elite. Sargent and Britton's respective attempts to control the construction of the Garden took place within a political context that had allowed for the growth of a counterelite: a Tammany machine, allied to real estate speculators for whom the development of any uptown landscape/utopia free of an urban grid (as Olmsted had originally planned for the Bronx) was anathema. For this latter group, rational planning of the city's northern development offered the opportunity to rationalize real estate profits as well as the distribution of political patronage.[137] This was the political context of consolidation-era New York that Britton and his Torrey Club allies had to negotiate in order to secure municipal financial support, even as they fended off the challenge from self-appointed Boston aesthetic authorities.

136 David Scobey, "Empire City: Politics, Culture, and Urbanism in Gilded Age New York" (PhD Dissertation, Yale University, 1989): 352.

137 Ibid., 423 and Hammack, *Power and Society*, 117, see note 122.

Historian David Scobey portrayed 1890s New York City "bourgeois urbanism," a late nineteenth-century version of the 1850s Olmsted environmentalist ideal, as a failure in its struggle to counter this Tammany-led rationalized ideal of real estate development. The efforts of Brittonian allies in the city's philanthropic elite, Scobey asserts, to gain a measure of control over "uncivil" working class street culture were unsuccessful, and a mid-nineteenth-century middle- and upper-class model of civic and cultural stewardship was defeated. Rather, by the 1890s, Scobey contends, the ideals of this bourgeois urbanism had long been lost, and defeat conceded to the powers of the real estate market. Instead of uplift, Scobey writes, the aim of 1890s elite reform had already become the Progressive crusade to "expose, regulate, and ameliorate." But it was not possible to eliminate, in Jacob Riis's words, "how the other half lived."[138] Now, instead of bringing the working class out of tenement squalor to a suburbanized [or] passive rural arcadia, this last generation of nineteenth-century reformers attempted to carry cleanliness and order *into* urban slums, in the briefcases of housing inspectors, social workers, and playground attendants.[139]

As noted above, Britton's vision of the Garden did share in part with Progressive parks a notion of public spaces as *active* molders of public behavior; the Garden's public education programs and, after WWI, its short-lived soldiers' horticultural training program, for example, may be considered analogs of the regimented programs of play and instruction that took place in urban playgrounds. Yet Scobey places the demise of "bourgeois urbanism" far too early in the 1890s. Britton and his philanthropic cohort understood the Garden's public space in terms that directly recalled the mid-nineteenth-century vision of uplifting public landscapes in the tradition of Central Park. (Britton's vision of NYBG's public purposes, of course, was even further nuanced by his conception of the Garden as primarily a scientific, institutional setting for botanical research.) Contrary to Scobey's portrait of the demise of a bourgeois urbanism, the Garden's creation from 1895 through 1900 underscores the *persistence* of a still-potent environmentalist understanding of public space, one directly descended from Olmsted and Vaux's mid-century vision and that grew out of the desires of the city's philanthropic elite to promote New York City's status as the cultural and intellectual capital of North America. Their success is chronicled in the following chapter.

138 Ibid., 463.

139 Ibid.

Chapter 4
An Educational Garden in a Public Space[1]

By 1900, Britton's zealous stewardship had brought a basic infrastructure for the new botanic garden nearly to completion. Begun in 1898, the Museum building was occupied by the Garden's scientific staff and ready for the installation of planned exhibits. The Conservatory ranges, to comprise some 15 connected glasshouses when finished, were also well along with several houses, including the central dome, already in place. Newly acquired tropical and tender plants had been moved from temporary storage in borrowed greenhouses at Columbia, to new semipermanent displays.[2] Grading of garden pathways and roads according to the approved 1890s plans continued with paving materials quarried from a site not far from the Museum itself.

As the physical garden took shape, Britton began implementing a plan for the New York Botanical Garden (NYBG) that included a dual approach to education as well as bold ideas for the botanical reconnaissance of the United States' expanding domain of political influence. His vision of the Garden as an educational venue should be considered a response to Victorian assumptions regarding the proper characteristics of a traditional natural history museum. With its plutocratic New York benefactors, grand civic architecture, and municipal endowment of land and operating funds, NYBG resembled its local institutional elders, the American Museum of Natural History and the Metropolitan Museum of Art. Also, as the Western Hemisphere's aspiring counterpart to the great botanical research institutions of European empires, NYBG could claim the pedigree of natural history's grand tradition of scientific inquiry. Indeed, the Arnold Arboretum's Charles Sprague Sargent, Britton's sometime political and scientific nemesis, had explicitly asserted in the pages of the influential horticultural weekly, *Garden and Forest*, that he considered botanical gardens to *be* museums. The museum collections of the Garden were granted prominence within

1 A version of this chapter was originally published as "A Scientific Garden in a Public Space: The New York Botanical Garden, 1900–1910," *Studies in the History of Gardens and Designed Landscapes* 22 (2002): 193–213.

2 N. L. Britton, "Report of the Secretary and Director-in-Chief for 1897," *Bulletin of the New York Botanical Garden* 1 (1898): 164–165. Cited subsequently as "Report of the Secretary and Director-in-Chief for 1897."

Britton's vision of the educational mission of the new institution. Yet, ever the eager institutional entrepreneur, the director-in-chief considered Sargent's assessment of the role of the garden as a mere museum of static displays to be an unnecessarily limiting one. Rather, Britton planned for NYBG to serve not only as a venue for the educational display of living and preserved plants and plant products, but also as a cutting-edge research institution, one with a strong link to the city's most prestigious university and its scientific departments.[3]

In devising an ambitious strategy of institutional development, Britton concentrated his considerable energies in several directions. Beginning in 1900, he sought to complete the physical garden landscape, including indoor and outdoor plant collections designed to educate and decorate. Concurrently, he oversaw the completion of the Museum and the installation of its constituent collections, while creating laboratories, a library, and a herbarium of dried plant specimens for advanced botanical training and research. Britton also devised programs of public education for both adults and schoolchildren, using the Museum and the various plantations to host lectures and tours. Finally, as part of these educational efforts, he established a set of publications, edited by NYBG staff and affiliated scientific professionals, to disseminate in both popular and scholarly formats the research conducted under NYBG auspices.[4]

Creating the Garden Landscape: Setting Priorities

As he prepared his overall plan for the Garden in 1896, two years before construction began in earnest, Nathaniel Lord Britton had had cause to consider just what type of public institution he hoped to create and what priorities he wished to set. In August 1896, as Vice-President of the American Association for the Advancement of Science (AAAS) and Chairman of that organization's botanical section, Britton had announced his views on botanical gardens in an address to members gathered in Buffalo, New York. He recalled the botanic garden's Western origins in the medieval apothecary gardens of European monasteries and noted their development alongside universities at Padua, Leiden, Bologna, Uppsala, and Paris. He cited the gardens' growth during a period of European exploration and noted a concomitant increase in the "ornamental and decorative taste for planting." Gradually, he noted with approval, these gardens tended to become public institutions, open to interested visitors. The modern botanical garden, he said, consisted of four main elements: "the economic, the aesthetic, the scientific, and the philanthropic." Each characteristic, Britton recognized, "had been given different degrees of prominence, depending upon local conditions."[5]

3 Charles Sprague Sargent, *Garden and Forest*, April 8 1891.

4 Britton's expansive taxonomic research agenda, which would be designed to make the Garden into *the* US center for Western Hemisphere botany, will be discussed in Chapters 5 and 6.

5 N. L. Britton, "Botanical Gardens: Origins and Development," Address to American Association for

The AAAS address gave Britton a public opportunity to articulate his priorities as the newly appointed director-in-chief of New York City's new botanic garden. The order of the list he enumerated to his listeners, among whom sat allies as well as potential detractors in his battle over botanical nomenclature and the controversy over the layout of Bronx Park, was determined by his interpretation of the historical development of the botanical gardens of Europe. And yet his enumeration of "economic" and "scientific" components was given priority in the body of his speech. Britton thought it proper to include within the scope of economic responsibilities of a botanic garden "facilities for the display and investigation of all plants directly or indirectly useful to man and their products." He included forestry, pharmacognosy, agriculture, pomology, pathology, bacteriology, and organic chemistry as useful economic duties of a modern botanic garden. And he advocated the installation of museum displays illustrating each subject (in which products would be grouped by industry not botanical affiliation), as well as indoor and outdoor economic gardens, in which useful plants were displayed. In all of these investigations, the work of the scientific staff would be essential, Britton insisted, due to the "necessity of knowing just what species is under investigation, its structure, distribution and literature, and in order that the most approved and exact methods may be used in the research."[6]

Britton acknowledged, however, that not all of these areas of economic botany should be carried out by a botanical garden. Research in agriculture, pomology, and plant pathology, he conceded, were "so well organized in America, through our National Department of Agriculture and our numerous agricultural colleges and schools," that there was no need to duplicate such research to a great degree by garden staff. (He did assert, however, that a well-endowed garden able to prosecute additional work in those fields would undoubtedly unearth important knowledge. Such work did, by the 1920s, become an important part of the Garden's research.) That left forestry (Sargent's domain) and pharmacognosy as necessary branches of economic research for botanic gardens. "No such liberal allowances have been made with us" for these investigations, Britton stated, and such work "must prove of the greatest benefit to the country."[7]

To undertake such valuable research, Britton noted, the scientific facilities of a modern botanical garden needed to be "thoroughly equipped." A well-stocked library, properly equipped laboratories, and a comprehensive herbarium, he maintained, would be the "sources whence exact information regarding the name, structure, habits, life

the Advancement of Science, 1896, copy in New York Botanical Garden Archives.

6 Ibid.

7 Ibid. The forestry work Britton eventually helped foster would in large part take place in Puerto Rico, an insular possession of the US whose high population density had led to a pronounced degree of deforestation. See Chapter 6.

FIGURE 9 Museum building near completion, autumn 1901. The black bunting hanging from the windows memorialized US President William McKinley, who was assassinated in September 1901.

processes and products of plants are derived." He conceded that it was "practically impossible for any one library to have all the literature of botany and related sciences, any one herbarium to possess an authentic and complete representation of all species of plants, or any one museum to be thoroughly illustrative." Yet he asserted "the more closely it is approximated the better the results." With this expansive conception of the possibilities for future botanical gardens, Britton foreshadowed the ambitious research agenda he would pursue at his own institution. A properly conducted program of botanical inquiry, he asserted, would include research in taxonomy, morphology, anatomy, physiology, and paleontology, with proper laboratory facilities "for their successful prosecution." Included within this broad set of scientific duties, garden researchers would devote part of their energies to the arrangement systematically of museum displays and plantings in order to illustrate "natural relations of [botanical] families as understood at the time." These displays, Britton noted, should remain flexible enough "to admit of subsequent modification as more exact information relative to those relationships is obtained." Finally, he concluded, the caliber of research would be aided by the maintenance of a close relationship between modern botanical gardens and universities.[8]

8 Ibid.

With his agenda clearly in mind, Britton turned his attention to the actual construction of the New York garden. Work on the Museum building began first, because it would house the Garden's administrative offices, herbarium, library, and laboratories in addition to educational exhibits. Britton approved architect Robert W. Gibson's design of the Museum as an "Italian Renaissance" edifice (see Chapter 3) (FIGURE 9). The choice of Italianate architecture was unsurprising. The style, with its conscious evocation of European nobility, had been in use since the mid-nineteenth century by builders eager to appropriate the cultural authority of Renaissance Italy for New York's public spaces. Ever since A. T. Stewart's enormous, marble-clad "cathedral" of commerce, one of the city's first department stores, was erected on Broadway at Chambers Street in 1846, the region's architects had been eager to eschew their prior republican reliance upon austere Greek Revival architecture for forms that bestowed a grander cultural legitimacy upon new public works.[9] Thus, the central feature of the Museum building was a dome, which spanned the Library's reading room, admitting light to the desks and catalogs below and adding a sense of grandeur to the goings-on within. The flooring immediately beneath the dome consisted of thick glass tiles, designed to allow some of the dome's light to filter into the rotunda of the second floor below. Elsewhere on the uppermost floor, skylights were installed, providing additional illumination to the library stacks, the herbarium, laboratories, and staff offices. The overall structure was, like the Conservatory ranges, designed to be fireproof: wooden finishings were "banished," and furniture would be as nearly fireproof as possible.

Initially, Gibson's plans for the Museum edifice had been even grander. In May of 1896, he had proposed to Britton the construction of a glazed gallery in the rear of the building that would complete the circuit of exhibition halls, form an enclosed courtyard, and provide space for the display of living plants within the Museum itself. This glass-roofed gallery was to have been constructed of glass substantial enough, Gibson said, for a man to "walk upon."[10] Though financial constraints likely prevented him from executing this more audacious plan, Gibson remained ambitious in his design, conferring with the designers and caretaking staff of the city's other major natural history institution, the American Museum.[11]

Actual construction of the Museum building began on January 6, 1898, as the John H. Parker Co., the winning contractor, began marking the corners of the

9 See Mona Domosh, *Invented Cities: The Creation of Landscape in Nineteenth-Century New York and Boston* (New Haven, CT: Yale University Press, 1996): 55.

10 R. W. Gibson to N. L. Britton, May 26, 1896, N. L. Britton Administrative Correspondence, New York Botanical Garden Archives.

11 R. W. Gibson to N. L. Britton, Oct 26, 1896; Britton to Gibson, December 19, 1896; N. L. Britton Administrative Correspondence, New York Botanical Garden Archives.

building's foundation. The collections housed within this building were the crucial components of Britton's plan to create within the Garden an institution of public education. In his 1896 AAAS Vice-Presidential address, the director-in-chief had characterized botanical gardens as "philanthropic agencies" that, through their public educational facilities offered an "orderly arranged institution for the instruction, information and recreation of the people" that was "more efficient for these purposes than a park" alone would.[12] Emphasis on the educational benefits to accrue to the public, of course, had been behind Britton's efforts to secure the grant of public parkland to begin with; he adhered to the plan in his efforts to create and publicize the educational displays in the new Museum.

Though Britton was influenced in his conception of the Garden by several working trips to the Royal Botanic Gardens at Kew, England, these elements of Garden design were also indebted to decades of nineteenth-century American horticultural practice. American estate owners had refined a local tradition of "Italian garden" design, reconciling in the process formal elements such as terraces with naturalistic features derived from the cultivated English landscape. Overall, the decision to fashion these principal NYBG settings after an Americanized Italian ideal was intended to evoke an association with "history, culture, and achievement," a suitable set of ideals for an ambitious scientific institution. Britton also chose Italianate design for the formal grounds immediately surrounding both the Museum and the Conservatory. A fountain was installed in front of the Museum's main entrance, and a set of walls, upon whose columns were placed urns serving as planters, was erected around a broad lawn leading up to the building. These walls enclosed yet another fountain.[13]

Britton, however, met with opposition to his efforts to erect a neoclassical public structure within a large park that many considered 'picturesque'. In response to the criticism of horticultural writers, he asserted that the terra cotta and white brick façade of the Museum would weather to a light gray "most harmonious with the foliage surrounding."[14] In this way, Britton and his architect attempted, of political necessity, to reconcile the demands for civic pride with the still powerful, Olmstedian, design ideology of naturalism. Terra cotta, a material that in the wake of 1893's Columbian Exposition was shunned by most self-respecting architects in favor of marble, does get dirty, but would not enable the building to blend into a picturesque background as Britton predicted. As this bit of public relations spin illustrates, the

12 N. L. Britton, "Botanical Gardens: Origins and Development," see note 5.

13 Rebecca Warren Davidson, "Images and Ideas of the Italian Garden in American Landscape Architecture," (PhD Dissertation, Cornell University, 1994): 16, 39, 40–41. These latter walls and the second fountain were dismantled after Britton's tenure as Director ended.

14 N. L. Britton, "Description of the Museum," *Bulletin of the New York Botanical Garden* 1 (February 15, 1898): 167.

FIGURE 10 Cutting hay on the Garden's grounds, August 1917.

efforts to build a public garden containing a museum and a scientific research center took place within a late-nineteenth-century context of continuing cultural tension over the uses of public space. Olmsted's venerable ideal regarding the uplifting effects upon citizens of artfully "natural," Romantic landscapes still made sense to a significant portion of the city's elite.[15] Yet the ornamental imperatives of a setting organized around formal architectural elements along with the requirements of scientific education challenged the NYBG's creators in a way unique to builders of large botanical gardens. The chosen setting in Bronx Park added particular difficulty as Britton attempted to reconcile each of these requirements in a coherent manner. In the eastern section of the Park, the Bronx River had cut a scenic gorge and tumbled over falls at the site of a defunct early-nineteenth-century mill. Surrounding the river was a 50-acre tract of ancient hemlock forest, a notable botanical anomaly at New York's latitude and yet another scenic resource. The entire Garden parcel was composed of former farmland, working hay meadows (FIGURE 10), a previously harnessed water source, and both old- and second-growth forest. To the extent that any of these features, especially the "natural" ones, were to be maintained, they would become as much parts of the artfully constructed Garden as the formal parterres installed around the Conservatory's entrance.[16]

15 Exceptions to this general tendency could be found, for instance, in the desire of some elites for a carriage speedway in Central Park.

16 In fact, the old mill's dam was modified for scenic and practical purposes, and the Bronx River used to irrigate certain aquatic plants near the riverbed.

Ironically, given the degree of public criticism leveled at Britton's planned layout for the Garden as a whole, little note seems to have been made of the actual aesthetic and practical failure of the Museum building's final design. As erected by architect Gibson, the building was essentially a poorly executed and unharmonious pastiche of exterior design elements. It is unclear exactly *why* Gibson won this commission, though it is likely that he was known to the Garden's founders as the designer of several churches and banks throughout the Hudson Valley and New York City.[17] Gibson was certainly no architectural newcomer. Trained in his native Britain in the 1870s, he had designed numerous public and private buildings, receiving the commission in 1905, for example, to build a private townhouse that stands today as the Cartier Building on Fifth Avenue in Manhattan. In Albany, New York, he had designed the Cathedral of All Saints Episcopal Church in 1884; the lodge and chapel for the Albany Rural Cemetery, 1882-84; and the Romanesque-style National Commercial Bank in 1887.

Despite this resume, Gibson apparently lacked a mature understanding of neoclassical design, the architectural style that, in the years following the triumph of Olmsted's "White City" at the 1893 Columbian exposition, Americans considered the proper aesthetic model for public institutions.[18] The Garden's exhibition spaces were relegated to the lower floors of the building and illuminated by gas lamps and afternoon light from the building's front, south-facing windows, rather than installed on the top floor under the optimal natural light of the Museum's dome and skylights. There were also no balconies designed to allow maximum exposure of the exhibits to overhead natural light. Thus, it appears that the displays may not have been adequately illuminated, the southern exposure and tall windows notwithstanding. Britton justified these choices by noting that visitors arriving through the building's main entrance would find themselves at all times only one flight of stairs either up to the

17 Besides Gibson's entry in the design competition for the Museum, Britton had received eight others. All NYBG Managers and Scientific Directors still residing in the city that summer had been given an opportunity to examine the nine sets of submitted plans for the Museum. Gibson's work had been judged "the best adapted to the work of the Garden," and these were subsequently transmitted to the Parks Department for approval. In addition, a total of three firms had competed for the commission to design the Conservatory. See Minutes of the Committee on Plans, June 17, 1896, New York Botanical Garden Archives.

18 In addition, his final product lacked the didactic features incorporated into the very structure of English predecessors such as the Oxford Museum, where local stone was used in construction of walls and pillars in an ingenious and nationalistic display of native geology. See Carla Yanni, *Nature's Museum: Victorian Science and the Architecture of Display* (Baltimore, MD: Johns Hopkins University Press, 1999): 82–83; http://www.hvnet.com/TOUR/al/hist/natregis.htm [29 March 2001] and Cornelia Brooke Gilder, "Robert W. Gibson, Master of Many Styles," Preservation League of New York State Newsletter (May–June 1984): 4–5. I am indebted to Carla Yanni for her insights on architect Gibson's failure to achieve his neoclassical design goals.

second story or down to the basement from all exhibitions.[19] Finally, terra cotta (mentioned above) represented a curious choice of building material, though this was most likely a function of Britton's cost-cutting priorities. It must be noted, however, that these overall design flaws were not necessarily specific to New York. Contemporary, American, natural history museums were commonly constructed in a similarly grand style and incorporated elements that often compromised the needs of public display.[20]

Britton's Contemporary Precedents: Natural History and a "New" Botany

In evaluating Britton's choices for the construction of the Museum and the development of the Garden's landscape, it is important to recall that Britton served as a bridge between the older forms of nineteenth century natural history and the newer mode of scientific inquiry gaining prominence in the early decades of the twentieth century. Completing his graduate education at Columbia in the early 1880s and trained by John Strong Newberry in geology (and to much lesser extent in botany), Britton had come of age in disciplines that readily lent themselves to public articulation in time-honored, decades-old formats. Such methods included the display of accumulated specimens in labeled lighted cases arranged in rows along the hallways of grand civic buildings. If natural history was focused on objects that were readily presented for such display, and the collection of such objects in a systematic way was what had come to differentiate the practice of science from mere collections of "curiosities," then Britton's priorities and his decisions regarding the public presentation of NYBG science are readily understandable.[21]

Nineteenth-century natural history museums had assumed leadership in the production of knowledge about the natural world. Whereas the colleges of the mid-nineteenth century were mired in a traditional curriculum of theology and classical languages, museum curators had "assumed intellectual leadership because

19 Britton, "Description of the Museum," see note 14.

20 Sally G. Kohlstedt, "International Exchange and National Style: A View of Natural History Museums in the United States, 1850–1900," in Nathan Reingold and Marc Rothenberg, eds. *Scientific Colonialism: A Cross-Cultural Comparison* (Washington, DC: Smithsonian Press, 1987): 173-74. Kohlstedt cites windows placed for architectural design rather than optimal lighting as an example. In fact, Gibson himself may have enjoyed a reputation for submitting cost-conscious designs that would have appealed to the cautious financial instincts of Britton. Gilder, "Robert W. Gibson," 4, see note 18. Over two decades later, when the Garden sought advice on renovations from the Olmsted Brothers firm, Frederick Law Olmsted, Jr. openly disparaged the structure. See Chapter 7.

21 Yanni, *Nature's Museums*, 3, see note 18. For further discussions on the relation of natural history objects and their display in museums, see Steven Conn, *Museums and American Intellectual Life: 1876–1926* (Chicago: University of Chicago Press, 1998) and Kim Kleinman, "The Museum in the Garden: Research, Display, and Education at the Missouri Botanical Garden Since 1859" (PhD Dissertation, Union Institute, 1997).

they fostered original research through the careful and systematic way they dealt with objects."[22] For an example of these principles in action, Britton had only to look southward, to Manhattan, where his contemporary Henry Fairfield Osborn (1857–1935) would within a decade take charge of the American Museum of Natural History.[23] Before Osborn's tenure as president of the American Museum began in 1908, however, the institution was run by Morris K. Jesup, a New York philanthropist and a firm believer in the educational value of "nature cabinet" exhibits of stuffed birds, reptiles, and mammals for a city filled with schoolchildren who had never ventured beyond the metropolitan limits into the countryside.[24]

Jesup's collections had been built up over the course of a decades-long nineteenth-century fascination with fossils. This enthusiasm for collecting material evidence for British geologist Charles Lyell's "uniformitarian" theory was both popular and scientific. On a popular level, finding fossilized remains of prehistoric creatures was, among an educated section of the American middle class, a passion analogous to contemporary "botanizing" enthusiasms. At the level of museum-sponsored science, fossil hunting became, from the 1860s to the 1890s, a serious, competitive business that fostered legendary academic rivalries (most famously between E. D. Cope and O. C. Marsh) and provided Jesup's American Museum with the raw material for spectacular exhibits of reconstructed giant skeletons.[25]

Though Jesup's stewardship of the museum from 1881 to 1908 had placed it on firm financial footing, Osborn considered Jesup's emphasis upon education through the use of "natural history cabinets" outmoded. Reflecting the viewpoints expected of one of his social station, Osborn sought instead to use the museum Jesup left behind as a "'positive engine' for the 'propagation of socially desirable views.'"[26] Of patrician American birth (and the nephew of J. P. Morgan), Osborn had nurtured his vocation in vertebrate paleontology while accompanying scientific expeditions to the American West during undergraduate summers at Princeton. He later continued studies in embryology at Cambridge and comparative anatomy in London under T. H. Huxley,

22 Conn, *Museums and American Intellectual Life*, 16, see note 21.

23 On Osborn and the American Museum, see William Gregory, "Biographical Memoir of Henry Fairfield Osborn, 1857–1935," *National Academy of Sciences Biographical Memoirs* XIX (1937): 53–88; John Kennedy, "Philanthropy and Science in New York City: The American Museum of Natural History, 1868–1968," PhD Dissertation, Yale University, 1968; R. Rainger, *An Agenda for Antiquity: Henry Fairfield Osborn and Vertebrate Paleontology at the American Museum of Natural History, 1890–1935* (Tuscaloosa: University of Alabama Press, 1991); and Donna Haraway, *Primate Visions: Gender, Race, and Nature in the World of Modern Science* (New York: Routledge, 1989).

24 Kennedy, "Philanthropy and Science in New York City," 153–154, see note 23.

25 Rainger, *An Agenda for Antiquity*, 9–13, see note 23. In the wake of Cuvier's insistence that the fossil record proved the existence of prior extinct species, Lyell had demonstrated that present day forces continued to cause changes that characterized natural history. Ibid.

26 Kennedy, "Philanthropy and Science in New York City," 153–154, see note 23.

who introduced the young American to Darwin. Ultimately convinced after much study, however, that his talents were not those of the field naturalist, Osborn had, after the late 1880s, devoted himself to the study of paleontology, setting his scientific goal as "the discovery of new principles from generalization" and shunning "the mere assemblage of facts," a process he characterized as "an extremely painful and self-denying" though "fascinating" pursuit.[27]

Britton's and Osborn's projects coincided in an important way: i.e., they both safeguarded an "exceptional" place for their respective institutions well into the twentieth century, past the point at which many museums had ceded priority as knowledge producers to university departments, the US Department of Agriculture, pharmaceutical companies, and independent research institutes. Osborn, who allowed Britton to list his name as a member of the New York Botanical Garden Corporation, had used his social prestige to secure continued philanthropic backing for his own paleontological efforts. He succeeded in his goal of maintaining relevance for a quintessentially "nineteenth century" natural history museum because of his wealthy connections and his ability to articulate the social concerns of his peers.[28] Britton, though not nearly as socially well-connected or as wealthy, nevertheless managed to do likewise: his 33-year tenure as director-in-chief was characterized by a series of often successful personal appeals to the city's wealthy to underwrite the Garden's growth and the continued relevance of its particular brand of botany up until the 1920s. Britton's efforts to position the Garden at the forefront of US botanical research were also aided by the fact that botany itself was undergoing a reinvigoration as a biological discipline, one that was sparked in 1900 by the rediscovery of Mendelian genetics and the subsequent infusion of government dollars into botanical endeavors. The "exceptional" status of botany, i.e., that it could continue to be practiced professionally *beyond* the walls of a university (though undeniably in conjunction with university departments), was underscored by the explicit reference to botanical gardens in an analogy used by Cope to illustrate his ideal of a museum-university partnership. He admonished his colleagues at Philadelphia's Academy of Natural Sciences to safeguard its institutional relevance by forging a relationship with the University of Pennsylvania: "the Academy . . . should bear the relation to the University," he said, that "the Jardin des Plantes does to the medical and other schools of Paris."[29]

Britton was also well aware of contemporary developments in botanical studies beyond New York and was in correspondence with J. M. Coulter at the University of Chicago; Charles Bessey of the University of Nebraska; and, of course, C. S. Sargent and his Harvard colleagues at the Arnold Arboretum and the Gray Herbarium. During

27 Gregory, "Henry Fairfield Osborn," 59–62, see note 23.

28 Rainger, *An Agenda for Antiquity*, 22–23, see note 23.

29 Cited in Conn, *Museums and American Intellectual Life*, 61, see note 21.

the late 1890s, Coulter was engaged in an ambitious taxonomic study of the plant family Umbelliferae while organizing the University of Chicago's department of botany. As the new department head, Coulter oversaw the construction of the Hull Botanical Laboratory, while encouraging his faculty to continue a series of morphological and ecological studies. By 1898, Coulter presided over a Chicago botanical department equipped with a 3,000-volume library, greenhouses, a completed Hull Lab, and outdoor gardens in two city parks.[30] At the Arnold Arboretum, Sargent continued work on his *Silva of North America* while administering that institution's 6,000-volume library and its herbarium. Under Sargent's direction, the Arboretum at the turn of the century began sponsoring numerous expeditions by one of the nation's more famous plant hunters, E. H. "Chinese" Wilson, to Asia in order to secure numerous species for introduction into North America.[31] The work of these two institutions was thus traditional, in that it fit into the paradigm of taxonomic field botany established by Hooker *per et fils* at Kew, but also incorporated elements of what had come to be known as the "new botany."

First in Nebraska and later in Iowa, Charles Bessey had continued writing textbooks and promulgating the tenets of this "new botany," which entailed a shift in focus from (merely) plant collection and description to evolutionary classification, structural physiology, pathology, and ecology. Bessey's efforts at adding these new dimensions to American botany had been stimulated by the development of evolutionary theory and a subsequent reevaluation of the various natural sciences. Reformers like Bessey questioned not only what to study, but how to study it, and had begun arguing for a larger role for laboratory study of physiology and pathology vis-à-vis fieldwork in botanical science. Bessey, Britton's ally in the nomenclature debates (see Chapter 5) had in fact become a leader in this movement. From his Midwestern posts at state land grant colleges, he helped foster a new relationship between university studies and national agricultural practice, and had done much to focus the US Department of Agriculture's attention on the problems identified by this new scientific methodology. "As the study of living organisms," Bessey's botany, then, would "differ little from zoology in what it studied: physiology, the life processes involved in obtaining food, assimilation, growth, and reproduction; morphology, the structures (leaf arrangement, vascular systems, roots, reproductive organs) that allowed the organism to carry out its life processes; and geography, the environment in which the organism operated." Bessey never underestimated the

30 A. D. Rodgers, *John Merle Coulter: Missionary in Science* (Princeton, NJ: Princeton University Press, 1944): 157–159.

31 Ida Hay, *Science in the Pleasure Ground: A History of the Arnold Arboretum* (Boston: Northeastern University Press, 1995): 115–140. On Wilson's life as an itinerant plant collector, see Roy W. Briggs, *"Chinese" Wilson: A Life of Ernest H. Wilson, 1876–1930* (London: HMSO, 1993).

importance of continued taxonomic study; he argued only for a different emphasis and different uses for this work, i.e., the uncovering of phylogenetic (evolutionary) relationships among plants.[32]

Given his familiarity with the work of his peers, Britton was able to draw upon a clear set of institutional precedents in organizing the Garden. He also sought to establish the Garden's relevance within a context of changing botany and the emergence of alternative venues in the US for botanical research. These natural history museum models were found on both sides of the Atlantic, yet from Kew to the Upper West Side of Manhattan (site of the American Museum), the arrangement of these collections had been informed by Francis Bacon's injunction to Western scientists to collect, describe, and classify objects in a meaningful, narrative manner.[33] This mandate to instruct through the regimented display of artifacts was, of course, what characterized the traditional natural history museum throughout the nineteenth century.[34] Such a system of display has been dubbed an "object-based epistemology," which Steven Conn defines as an "intellectual architecture . . . predicated upon the assumption that objects could tell stories to the untrained observer." Designers of such exhibits, Conn writes, believed that all museum artifacts on display should speak for themselves on sight, and that the meanings contained in these displays would be apparent to anyone studying them closely. Victorian administrators believed that these objects, if displayed effectively, would function as representations of the entire body of knowledge about the natural world. Furthermore, the knowledge presented in such "visual sentences," Conn reports, generally represented the "metanarrative of evolutionary progress":

> A trip through the galleries followed a trajectory from simple to complex, from savage to civilized, from ancient to modern. The form that museums developed in the last half of the nineteenth century made this lesson inescapable to anyone who strolled their galleries. Museums functioned as the most widely accessible public fora to underscore a positivist, progressive and hierarchical view of the world, and they gave that view material form and scientific legitimacy.[35]

Such a portrait of natural history museums is informed by an interesting consideration of the polyglot urban settings in which they were established: contemporary museum builders, faced with the "daunting task of educating and/or controlling" a swelling population of immigrants, justified their belief in an object-based epistemology with

32 Richard A. Overfield, *Science with Practice: Charles E. Bessey and the Maturing of American Botany* (Ames: Iowa State University Press, 1993): 74.

33 Conn, *Museums and American Intellectual Life*, 22–25, see note 21.

34 Ibid.

35 Ibid., 45. On object-based epistemology, see also Soraya de Chadarevian and Nick Hopwood, eds., *Models: The Third Dimension of Science* (Stanford, CA: Stanford University Press, 2004).

a firm belief that "knowledge conveyed visually was not culturally or linguistically bound."[36] This confidence in the efficacy of display led to the employment of the characteristic feature of nineteenth-century natural history museums: row upon row of glass specimen cases, in which were meant to be exhibited representations of the sum of human knowledge. In this way, the physical format of these natural history museums was related intimately to the content of the information they sought to convey: "the form that these museums took," Conn asserts, "made it possible for the visiting public to understand the meaning of museum displays."[37]

It was thus to the paradigm of the museum case that Britton and his curatorial team turned as they planned the series of exhibits to be housed in three floors of Gibson's neoclassical Museum building. What would enable Britton to depart from the constraints that the natural history museum ideal imposed, however, were the exceptional characteristics of the botanical garden itself. Though Britton eagerly harnessed the pedagogical power of the museum case, he managed to retain institutional relevance not by simply updating or rotating those displays. Indeed, the *other* functions intrinsic to the mission of the botanical garden (research of taxonomic and economic import; exploration; publication; and the provision of an open, landscaped, public space—a *garden*—within an urban center) were all responsible for the survival of NYBG into the twenty-first century.

With faith in the new Garden's importance for science, for the City of New York, and even for the nation at large, Britton and his staff designed a set of museum collections that conformed to the nineteenth century tradition of "object-based epistemology." To carry out the Garden's public educational mission "most efficiently," Britton noted, the various museum displays and indoor and outdoor exhibits had to be "arranged to present the principle facts in such a manner as to be capable of comprehension by persons not acquainted with the scientific aspects of the subject."[38] Reflecting the research priorities of the director-in-chief, these new museum collections

36 Conn, *Museums and American Intellectual Life*, 6, see note 21. Conn's analytical framework begs modification because, although careful to disavow an overreliance upon the familiar story of museum decline (and university ascendance), he nevertheless places the locus of the natural history tradition squarely in the realm of the museum, which, he asserts, consequently was relegated to the status of intellectual backwater by the early twentieth century. Yet Conn's account omits any consideration of botanical gardens. If his inquiry were extended to include these exceptional incarnations of the natural history museum, his conclusions about that decline would require reevaluation.

37 Ibid., 8. Such a method for displaying "ordinary" but significant objects also served, Conn notes, to distinguish the great institutions of the late nineteenth century from their "Barnumesque" antebellum predecessors. These natural history museums, with their emphasis upon taxonomic order, replaced the "freakish" with the "representative and ordinary," underscoring in the process a prevalent natural theology.

38 "The Laboratories," *Journal of the New York Botanical Garden* 1 (1900).

FIGURE 11 Systematic botany exhibits in the Museum building, 1905.

were divided into three main divisions: a taxonomic collection, an economic collection (including plants crucial for the production of drugs), and a paleobotanical collection.

The Museum Collections

The Systematic Collection. The systematic collection (FIGURE 11), designed to illustrate the principles of taxonomy to a visiting lay audience, was housed in over 120 wood and glass cases that took up the entire second floor (approximately 28,000 square feet) of the Museum building.[39] Each of these cases contained a "concise definition of the group of plants there represented" centered above the doors.[40] This overall collection comprised several subsets of exhibits as well, each designed to convey standard botanical principles to visitors as they marched from case to case. The primary category of exhibits on the second floor consisted of a general collection of dried herbarium specimens mounted on sheets of paper with accompanying labels giving detailed information regarding the name, provenance, and taxonomic affiliations of each sample. The names given on the labels, in keeping with Britton's stated

39 See *Bulletin of the New York Botanical Garden* 5 (1907) and *Journal of the New York Botanical Garden* 9 (1908): 172–79.

40 J .K. Small, "Report of the Curator for 1904," *Bulletin of the New York Botanical Garden* 4 (1904): 38.

position, conformed to the American Code of nomenclature as it had been first promulgated at Rochester in 1892.[41] These mounted examples were set upon glass shelves within the display cases and were frequently surrounded by drawings (sometimes beautifully rendered watercolor portraits illustrating plant structures) and photographs taken of the live plant in the field by the Garden's expeditionary parties.[42] In addition, specimens were sometimes further explained by plant models; by fruit samples; and, when possible, by the display of fossil plants closely related to the representative specimen in order to convey a sense of plant evolution.[43] In this way, the "synoptic" collection of exhibits, as its name suggests, was meant to represent the larger realm of the world's flora.[44] Certain plants were chosen to stand in as representative members of the more comprehensive taxa to which they belonged: a single specimen indicated characteristics typical not only of that plant's species, but of its genus and family as well. The flat thick sheets of cardboard on which the herbarium specimens were mounted were chosen for their resistance to fading; other samples were arranged in specimen jars. Half of 1900 had been spent by workers of the Garden's own printing press preparing labels for these exhibits.[45]

In conjunction with the general synoptic collection, Britton and John Kunkel Small, the chief curator of both the Museum collections and the Herbarium, established a microscopic exhibit and a display of local New York flora. A short, heavy man with a gruff demeanor, Small had been one of the Garden's earliest employees, joining Britton's staff in 1898, three years after taking his doctorate in botany as Britton's student at Columbia.[46] While still a graduate student, he had organized the University's herbarium, the core of which had been John Torrey's personal collection and which would also form the nucleus of the Garden's Herbarium, into the sequence called for by European botanists Adolph Engler and Karl Prantl. This was the first American herbarium to be so arranged. The Engler and Prantl system, published in Germany between 1887 and 1915, was the first widely used system to be based upon contemporary conceptions of evolutionary progression.[47] This taxonomic

41 See Chapter 5 for a discussion of Britton's role in the nomenclatural debates that characterized botany at the turn of the twentieth century.

42 New York Botanical Garden, *Descriptive Guide to the Grounds, Buildings, and Collections*, 1907, 1909, 1916, 1920, 1934.

43 *Journal of the New York Botanical Garden* 1 (January 1900).

44 See Conn's discussion of this function of museum objects, *Museums and American Intellectual Life*, 4, note 21.

45 J .K. Small, "Report of the Curator for the Year Ending 1900," *Bulletin of the New York Botanical Garden* 1 (March 1900): 316.

46 Henry Allen Gleason, "Thumbnail Sketches of Botanists," 1961: 42–43, unpublished manuscript New York Botanical Garden Archives.

47 Scot Medbury, "Taxonomy and Arboretum Design" Arnoldia 53 (1993): 14–15.

organization was extended beyond the Herbarium to the Museum displays and to accompanying exhibits of live plants in the economic garden and systematic plantations themselves. Small's devotion to the systematic museum collections and the Herbarium was reputedly such that he often spent his own funds to sponsor collecting expeditions or acquire new specimens.

His commitment to taxonomy was equally absolute according to contemporaries. "If he ever learned anything about the content, or even the existence of any kind of botany except taxonomy," Henry Gleason recalled, "he never betrayed the fact in his speech or his writings." Small, along with his mentor Britton, tended to be a taxonomic "splitter," quick to designate plants that deviated even a small degree from type specimens as new species.[48] According to Gleason, a Garden student who later served as Britton's assistant director, there was "little indication that [Small] ever thought of such taxonomic matters as hybridization or mutation, or that he ever considered that some of his species were merely ecological variants. If a plant was just a little different from a described species, it was to Small a new species and was promptly described as such."[49]

Small's microscopic display consisted of a series of 24 stations at which specimens were mounted under microscopes in order that museumgoers could examine plant structures at the cellular level. The microscopes, donated by NYBG incorporator William E. Dodge, were housed in specially constructed plate glass cases, which covered the entire mechanism save the upper portion of the draw tube, which protruded from the glass box and allowed the visitor to peer below at plant structures magnified approximately 70 times their actual size. To keep these displays in focus, Small and his assistants affixed metal clamp rings to the upper portion of the tube in order to hold it firmly in place. Small typically selected for display in the microscopy exhibit such representative specimens as slime molds, freshwater diatoms (algae), seaweeds, and the spores of various, fungal, plant parasites.[50]

48 Interestingly, there is evidence to suggest that Elizabeth Britton, unlike her husband, exhibited the tendencies of both a taxonomic "lumper" (someone inclined to recognize fewer species within families of plants) and a "splitter." In 1912, as she prepared a portion of the bryological volume of the Garden's massive publishing project, the North American Flora, she reported that she and her co-authors had managed to reduce substantially the number of moss species within various families. In three of 13 cases, however, families of mosses were enlarged by the splitting of species into multiples. Elizabeth Britton was the author of all three families. See *Journal of the New York Botanical Garden* 13 (February 1912).

49 John Kunkel Small records, internal New York Botanical Garden employee file, New York Botanical Garden Archives; Gleason, "Thumbnail Sketches," 42–43, see note 46.

50 Marshall A. Howe, *The Microscopical Exhibit at the New York Botanical Garden* (Rochester, NY: Bausch & Lomb Optical Co., 1902).

The local flora exhibit of herbarium specimens completed the second floor displays. Designed as an inventory of plants growing within a 100-mile radius of New York City, this popular attraction was the result of a collaborative effort with the Torrey Botanical Club (TBC), whose weekly collecting excursions to locales within that same radius had begun in the nineteenth century and continued into the twentieth. Garden staff assistants also supplemented Small's local flora exhibit with the results of a year spent collecting in local woods and fields in 1900.[51] Once these exhibits were installed, they were available to visitors eight hours a day, seven days per week.[52]

The Economic Collection. The second major component of the Botanical Garden's Museum, the economic collection, was housed on the main floor and was placed under the curatorial direction of Henry Hurd Rusby (1855–1940), a physician, member (and later chair) of the Garden's Board of Scientific Directors, President of the TBC, and subsequently Dean of the Columbia University College of Pharmacy. Rusby had since his youth been an active member of the city's botanical scene, garnering honors for his personal herbarium at the Philadelphia Centennial Exposition and joining the Torrey Club in 1879. As a medical student in the 1880s, he had collected plants in the southwestern US for both the Smithsonian Institution and the pharmaceutical manufacturer Parke, Davis, and Company. Upon graduation, he had explored remote regions of Bolivia, Columbia, and other South American locales, initiating a lifelong series of collecting expeditions in search of economically and medicinally important plants, including coca and cinchona, the plant from which quinine, a key treatment for malaria, is derived.[53]

Rusby's zeal for his curatorial responsibilities matched that of Small, and he did not allow a host of other professional obligations to deter him from building up a considerable collection of products with which he planned to demonstrate a multitude of economic uses for plants. Early in 1898, as the Museum was still being constructed, he conferred with Britton on possible methods for securing the necessary material to flesh out his vision of an economic collection. While conferring on these plans, Britton and Rusby decided to establish a collection that would rest "upon a more distinctly and systematically educational basis than is customary in institutions of this kind." The director-in-chief hoped to avoid the typically monotonous arrays of economic products found in most botany museums and instead attempted to fashion displays

51 Small, "Report of the Curator," see note 45.

52 *Journal of the New York Botanical Garden* 1 (1900).

53 David Williams and Susan Fraser, "H. H. Rusby: The Father of Economic Botany" *Brittonia* 44 (1992): 273–279; C. W. Ballard, "Obituary: Henry Hurd Rusby" Science 93 (1941): 53–54; H. H. Rusby, "Guide to the Economic Museum of the New York Botanical Garden," *Bulletin of the New York Botanical Garden* 41 (1921): 3–4. Specimens from this trip were those subsequently taken for identification at Kew's herbarium by Nathaniel and Elizabeth Britton in 1888.

that would clearly connect the featured plants and products with the industrial processes that were responsible for their manufacture.[54] In doing so, he and his curatorial staff would lend legitimacy to a view of nature as an often exotic set of exploitable resources.[55]

Because Garden funds for acquisitions were scarce, Rusby decided that he would create exhibits using plant-derived products donated by various manufacturers, as well as through exchange of specimens with similar institutions. In order to secure material suitable to entice other museums to send along medicinal and economic plants and herbarium specimens, Rusby urged Britton to send collectors out into the Garden's vicinity to gather useful bark and roots.[56] He also realized that the Garden could secure material by agreeing to pay the expenses incurred by any field collector willing to venture to remote regions, provided that that collector's salary was paid by another institution. In practical terms, this meant that NYBG would frequently agree to terms with explorers whose overall expeditions were conducted under the auspices of other museums, organizations, or government agencies. Rusby took pains to cultivate relationships with other botanical and scientific institutions, secure in the knowledge that other gardens and museums would be "eager to secure material of absolutely authenticated origin [since] the greater portion of [specimens] ordinarily coming to hand lacked this [authenticity] in some way." In order to guarantee a supply of botanical specimens for exchange that had such an authenticated provenance, Rusby and Britton decreed that all specimens destined for exchange must be gathered by trained and reliable botanical collectors who would properly record the origin of each specimen.[57] "The more valuable portions of the economic collections" were assembled this way, Rusby asserted, with their provenance and botanical identities thus securely established. This was, he boasted, what made the NYBG economic collection "unique."[58] As a result of this strategy, Rusby's botanical agents negotiated agreements with institutions as diverse and as distant as the Mexican National Institute of Medicine, the Technological Museum of Sydney, Australia, and the British colonial government of India.[59]

Rusby did not ignore the possibility of building his economic collection with actual donations and set out systematically to improve the Garden's chances for

54 H. H. Rusby, "Report of the Honorary Curator of the Economic Collections," *Bulletin of the New York Botanical Garden* 1 (April 13, 1899): 253.

55 See Carla Yanni, *Nature's Museum: Victorian Science and the Architecture of Display* (Baltimore, MD. Johns Hopkins University Press, 1999).

56 See H. H. Rusby to N .L. Britton, March 23, 1898, cited in Williams and Fraser, "H. H. Rusby," 276, note 53.

57 Rusby, "Report of the Honorary Curator," see note 54.

58 Williams and Fraser, "H. H. Rusby," 277, see note 53.

59 Rusby, "Report of the Honorary Curator," see note 54.

securing such industrial largesse. He prepared a list of 100 of the "leading vegetable product industries" and compiled an accompanying list of articles he needed in order to properly illustrate the manufacture of each. He then sent copies of these lists to representatives of the industries he had identified, including his old employers Parke, Davis as well as the tobacco industry and International Paper. With this written request, he included a printed circular in which he justified the Garden's request for assistance. To stoke the interest of parsimonious executives, he delineated the proposed organizing principles behind each economic exhibit and invited each industry to contribute photographs, models, and other materials. Although the response was not overwhelming, the materials received in the late 1890s, including a display illustrating the manufacture of spool cotton thread, and the promise of a model paper mill, he stated, had fully repaid "all the expenses and trouble involved in the attempt."[60]

Though the overall program devised by Rusby and Britton proved to be a labor-intensive process (as any program of specimen exchange and donation-seeking would), Rusby managed to fill 210 display cases with nearly 10,000 objects that he arranged according to their economic value and uses.[61] Within each of these economic categories, he organized plants in taxonomic sequence, from lower, less complex organisms to higher plants. Samples of foods (along with drug plants, the largest category represented) and fibers used in the manufacture of clothing occupied the western hallway of the main floor of the Museum. Included here were such fibrous plants as cotton as well as fruits and cereal grains.

Beyond these cases, in the westernmost wing, Rusby arranged displays illustrating tobacco products, spices, beverages, oils, sugar, starches, wood, paper, and cork. In the eastern hallway he placed a series of cases containing plants from which various drugs were derived. Finally, the eastern wing of the main floor was given over to a display devoted to North American woody plants, from which, the accompanying labels explained, were derived valuable resins, tars, pulp, and dyes.[62] These exhibits, like their counterparts in the second floor's systematic collection, made frequent use of photographs, drawings, and fruit samples. In addition, highly descriptive labels accompanied each specimen, bearing the common and scientific names of the plants as well as information regarding their geographic origins, cultural requirements, and industrial transformation into salable products. Rusby took special satisfaction in the successful exhibition of the economic collections. He asserted in 1921 that though "it had been a most laborious and expensive operation" to send NYBG explorers into remote regions "often destitute of roads" with necessary preservative equipment (including glass jars and formaldehyde solution), nevertheless "the result had more

60 Ibid.

61 Williams and Fraser, "H. H. Rusby," 277, see note 53.

62 *Bulletin of the New York Botanical Garden* 2: 27–28.

than repaid the sacrifice." The economic exhibits, like the various systematic displays, Rusby proclaimed, had "brought within the reach of the people of this metropolis the study of many things not otherwise to be seen." He further stated that although the primary purpose of the economic exhibits was to illustrate practical values of plants, they had nonetheless added to scientific knowledge of morphology, taxonomy, and physiology.[63]

Rusby's success in creating an intelligible economic collection was not complete, however, until he could be reasonably certain that its pedagogical objectives were being met. Within five years of its opening, he worried that his displays, although serviceable to a trained botanical observer, would not enlighten those lacking previous knowledge of the subjects on display. "The casual visitor," he warned, might find the exhibits "unintelligible," and he advocated a more expansive set of explanatory labels as well as the distribution of guiding pamphlets to museumgoers. "Probably a small charge for such literature," he believed, "would add to its appreciation by its recipients."[64]

The Paleobotanical Collection. The third and final component of Britton's Museum was the collection of fossil plants that had been gradually accumulated by his lifelong friend and fellow "botanical geologist," Arthur Hollick. Born and raised on Staten Island near the Britton family, Hollick had joined the Torrey Club along with his friend Nat while both were still students in the mid-1870s. After an undergraduate education at Columbia College, he earned a PhD in botany at George Washington University.[65] Hollick's scientific interests eventually brought his work far beyond the borders of the Garden: he spent a good portion of his active career on official leave from NYBG in order to work with the US Geological Survey in Alaska. In granting these official absences, Britton made sure to insist that his old friend collect fossilized and other plant specimens for the Garden. Housed in the basement of the Museum, the paleobotanical collections were arranged in order to describe a narrative of floral evolutionary "progress" that was comprehensible to a lay audience. The cases of fossil plants were organized according to the sequence of geological eras in which they appeared.[66] The "main object" of this collection, Hollick noted, was to enable "the casual visitor to obtain a general idea of the significance of the [displays]."[67]

63 Rusby, "Guide to the Economic Museum," see note 53.

64 H. H. Rusby, "Report of the Honorary Curator of the Economic Collections," *Bulletin of the New York Botanical Garden* 4 (1904): 49.

65 Charles Arthur Hollick, "Torrey Botanical Club Reminiscences," *Memoirs of the Torrey Botanical Club* 17 (1918).

66 New York Botanical Garden. *Descriptive Guide to the Grounds*, 1904, 1920.

67 *Journal of the New York Botanical Garden* 9 (1908): 214.

Public Education

Nature Study for Schoolchildren. Though the garden plantations and each of the three Museum collections were designed to be intelligible to an unaccompanied lay visitor, Britton decided within several years of the Garden's opening to provide docents free of charge. They also compiled an official NYBG guidebook available for purchase by visitors. Initially, not many of the daily visitors took advantage of the free tours of the grounds and Museum, and Britton determined to advertise the service using prominent signs.[68] Among the several exhibits, the most consistently popular were the microscopic displays and the local flora displays. Annually, thousands of the Bronx's public schoolchildren, escorted by their teachers, walked through the Museum's spacious and austere second floor, examining the labels and illustrations surrounding samples of local plants and trees. Many of these displays were housed using an arrangement that Britton had admired at the Metropolitan Museum of Art. These consisted of a series of swinging frames, enclosed in removable glass plates, each attached to a central upright pole and into which were inserted flat objects for the visitors' examination.[69]

In 1905, Britton and curator Marshall A. Howe decided to provide a "nature study" education for as many of the borough's grade-school children as possible. On April 14, 1905, Howe inaugurated a series of regular lectures and guided tours to fourth and fifth grade classes of public school students. That day, 750 children, representing 14 Bronx schools, were seated in the basement auditorium of the Museum building to hear Howe discuss "Flowerless Plants." His lecture was illustrated with colored lantern slides. After the talk, the children were ushered in groups of 50 onto the second floor where they were escorted through the systematic exhibits. Howe explained what seaweeds were and then passed each group on to William A. Murrill, the NYBG mycologist, who gave a tour of the fungi collection.[70]

Next, the students were met outdoors by other Garden staffers, who conducted them on an annotated walk through the NYBG forest, stopping to reinforce a point that Howe or Murrill had made regarding the various plants on display. Each group of children completed their field trip to the Garden with a walk through the public conservatories, again accompanied by a staff botanist or horticulturist. The subjects chosen had been designed to correspond with the students' botanical curriculum.[71]

68 Minutes of the Scientific Directors of the New York Botanical Garden, December 8, 1906, New York Botanical Garden Archives.

69 N. L. Britton to R. W. Gibson, December 19, 1896, N. L. Britton Administrative Correspondence, New York Botanical Garden Archives.

70 "Nature Studies in Park: How City Gives School Children Lessons in Trees and Flowers," *New-York Daily Tribune*, August 20, 1905.

71 Ibid. and "Practical Nature Study," *New York Evening Post*, April 15, 1905.

Ever cognizant of ways to garner positive publicity for the new public Garden, Britton invited a reporter from the *New York Evening Post* to witness the inaugural set of nature study lectures. Favorably impressed, the newsman proclaimed to the city the following day that "the successful manner in which the programme [sic] was carried out was highly gratifying to all concerned. Much interest was manifested by the children in all they saw and heard, although the average age of the boys and girls was only about eleven years." Britton asserted to his press audience, a regular presence at the Garden during the publicity conscious Britton years, that the series of lectures, though attempted in an "unsystematic, desultory way" in certain European nations, was actually an innovative format being tried by his staff. Each class would return at different days for the remainder of the spring for follow-up lectures by Britton on plant taxonomy and by head gardener George Nash on horticulture. Britton planned to extend the lecture series to a total of approximately 2,500 students that year.[72] The nature study classes proved extremely successful and, within a few years' time, began attracting classes from public schools in Manhattan. In succeeding years, Britton expanded the lectures' scope to encompass a wider variety of topics, enlisting Rusby, for example, to give a talk on "Industries Depending on Forests," which provided an introduction to the economic collections.[73] Other members of the staff followed suit, instructing schoolchildren on their areas of specialty. By 1906, 10,000 Bronx and Manhattan schoolchildren were being escorted through the Garden grounds for nature study lectures; in 1909, the figure had increased to 17,000.[74]

At the turn of the twentieth century, the nature study movement aimed to instill in children an inquisitive and loving attitude toward the natural world. Britton's plan for the Garden's nature study program was aimed at urban children in local public schools. Others in this national movement, including popular science writer Anna Comstock and horticulture professor Liberty Hyde Bailey, both of Cornell University, targeted rural children, who, they feared, would forsake life in the country for overcrowded cities.[75] Nature study was taught primarily by female teachers, who sentimentalized plants and animals in an effort to foster children's appreciation of (rather than control over) nature. Control remained the prerogative of male professionals, who used the information they gained from experiment and observation to synthesize new knowledge. By the early twentieth century, though popular as a form

72 *New York Evening Post*, April 15, 1905.

73 New York Botanical Garden Circular, "Nature Study of the 5B Grade," April 16, 1907, New York Botanical Garden Archives.

74 N. L. Britton, "Report of the Director-in-Chief," Bulletin of the New York Botanical Garden 6 (1906): 195–211; Britton, "Report," Bulletin of the New York Botanical Garden 9 (1909): 153–163.

75 Pamela Henson, "'Through Books to Nature: Anna Botsford Comstock and the Nature Study Movement," in Barbara T. Gates and Ann B. Shteir, eds., *Natural Eloquence: Women Reinscribe Science* (Madison: University of Wisconsin Press, 1997): 126, 137.

of elementary pedagogy that combined book learning and direct "field" experience, nature study was coming under attack as unscientific from this same male-dominated profession.[76]

At a regular meeting of Garden staff and students in late 1908, Mary Anderson, then studying ferns on scholarship at NYBG, defended the continued necessity of nature study.[77] Though conceding the existence of conflict between "scientists" and "naturalists" over the proper grade school curriculum, she noted that a widely agreed upon course of study was still in the early stages of development. The key tenet of nature study, she maintained, would be an applied focus in which, for example, beetles would be examined not merely as insects but as potential destroyers of crops whose development should be monitored and chronicled. Similarly, the parts of a flowering plant were of interest to children engaged in proper nature study to the extent that the functions of each plant organ in reproduction were discovered and conveyed. The guiding question in nature study education, Anderson asserted, must be "how does this affect the lives of the children?" To the extent that their interest was aroused, she held, "assimilation would be rapid." To this end, the grade school curriculum should not emphasize classification, but instead concentrate the students' attention on the "immediate environment of the normal child": e.g., pets, house plants, local birds, etc. Gradually, Anderson told her botanical colleagues, through the knowledge of basic facts of the immediate domestic surroundings, the children's "circle of knowledge" would widen to encompass an understanding of the larger processes of life.[78]

Despite the potential for controversy between partisans of this brand of "nature study" and those of a more "masculine" science, Nathaniel Britton remained committed to the public educational role of the Garden. Speaking once again before a convention of the American Association for the Advancement of Science in 1909, Britton characterized his botanical garden as an important adjunct of public education and a "highly specialized park in which the plantations [had been] formed and arranged primarily with regard to botanical facts and theories."[79] His belief in the educational mission led him to continue regular lectures to local grade-schoolers and to devote NYBG staff (male and female) and resources to the task. Of course, those Garden lectures in fact *did* aim to convey a rudimentary understanding of botanical classification (a window on the world of a professional, masculinized science), and

76 Ibid.

77 Mary Perle Anderson had been an investigator of the geographic distribution of Japanese ferns at the NYBG since 1906. Educated at Mount Holyoke and MIT, she had conducted other research at Woods Hole, Chicago, Columbia, Kew, and the Jardin des Plantes. She had also served as a high school instructor of biology in New England and a college instructor at Mt. Holyoke. She had won the first prize in the Garden's initial Wildflower Preservation essay contest in 1904. She also taught nature study at the Horace Mann School. See *Journal of the New York Botanical Garden* 10 (1909): 56.

78 Ibid., 56–60.

79 Journal of the New York Botanical Garden 11 (1910): 25–30.

they were aided in this attempt by the very design of the various collections of living and preserved plant specimens.

Lecture Series for Adults. Ever the institutional entrepreneur, Britton's vision of public botanical education was not exhausted by children's nature study. As part of his plan to raise the profile of the Garden and make it as integral a part of the public educational landscape as the city's other publicly funded museums and scientific institutions, Britton instituted a regular series of Saturday afternoon lectures, to be delivered to an adult audience in the Museum building's auditorium. These lectures in fact antedated the completion of the Museum; the earliest NYBG-sponsored talks open to the general public had been held in meeting rooms at the American Museum of Natural History (AMNH) beginning in December of 1895. Daniel Morris, Assistant Director of the Royal Botanic Gardens at Kew, England, had delivered the inaugural lecture there, giving the audience of over 500 an account of the history of Kew and its development as an integral part of the British Empire. Cornell's horticulturist Liberty Hyde Bailey followed Morris in the spring of 1896 with a discussion of "Natural Scenery and Landscape Gardening."[80]

Once the Garden's own building had been completed, Britton was quick to print and circulate flyers announcing sets of six lectures in the spring and six more in the fall of each year. Included on each lecture announcement was a note from Britton giving the most convenient Harlem Railroad train scheduled to arrive in time for the discussion. These talks, aimed at a general audience, were given by Garden staff members and by professional botanists. In 1905, the same year that NYBG began its nature study lectures, for instance, US Department of Agriculture botanist Frederick V. Coville discussed "The Indian and His Uses for Plants," and Francis Lloyd, a regular NYBG explorer, spoke on the "Pines and their Life History." Meanwhile, Garden employees D. T. MacDougal, Howe, Murrill, and Rusby each presented reports on portions of their research to an audience of New Yorkers interested in botany and horticulture. The lecture series was successful almost as soon as the Museum's new auditorium was opened. Held from April through June and again from October through November, 17 such lectures had been offered during 1900 and each attracted audiences of 75 to 500.[81]

Professional Education

Within several years of the opening of the Museum building to a curious public, the various collections began growing at a rapid rate, taxing the ability of Small and the other curators to find adequate space, not only for the exhibited specimens, but for

80 *Bulletin of the New York Botanical Garden* 1 (1) (April 15, 1896): 17.

81 N. L. Britton, "Report of the Director-in-Chief to the Board of Managers," *Bulletin of the New York Botanical Garden* 2 (6) (1901): 19.

the ever-expanding Herbarium. Begun with the nucleus of the old Columbia College herbarium, the NYBG collection of specimens (many of them types) increased by an average of several thousand annually. The spoils of Garden-sponsored expeditions, shipped on steamers from various Western Hemisphere ports, continued to arrive at New York piers by the thousands.

Proud of his new institution's ability to create such an indispensable botanical tool and eager to demonstrate a certain level of success to his expanding list of philanthropic backers, Britton ensured that each copy of the monthly Garden Journal had a comprehensive list of Museum, Herbarium, and Library accessions. Britton was determined that the set of botanical resources he had coaxed into existence would be utilized in the most efficient manner possible. In 1902, with a Museum, Herbarium, Library, and botany laboratory established and basic equipment secured, he persuaded his Scientific Directors to appropriate funds to establish a series of small scholarships (of approximately $50 per month) to attract researchers from around the nation to NYBG. The "greater majority of working botanists on whom the general progress of botanical science rests in America," Britton told the Board, "are instructors and professors in colleges, curators of museums, and members of the staffs of experiment stations, in which the facilities and time for research are extremely limited." Granting these NYBG fellowships, Britton argued, would greatly aid the production of necessary scientific knowledge by enabling competent investigators to take paid leaves of absence from their official posts and pursue research using the NYBG's new facilities. "The general obligation which rests upon an institution of this character," Britton told the Scientific Directors in 1902, "to widen the limits of botanical knowledge in all directions is emphasized by the constant appeals made to us to undertake the solution of problems embracing questions of the highest scientific importance . . . which can only be dealt with by the aid of trained investigators." University fellowships were generally bestowed upon an "immature generation of workers," Britton noted: young men and women seeking professional status. By contrast, the Garden's scholarships, if granted to researchers already credentialed, would directly "aid the development of the science of botany."[82]

Such an infusion of scholars into the Garden community, Britton realized, would help the Columbia graduate students in residence and would provide a steady supply of publishable research for the Garden *Memoirs*. The Garden would be assuming its proper status as a sponsor of cutting-edge botany, a role, Britton insisted, that was generally beyond the purview of even Federal agencies devoted to agricultural and scientific topics.[83]

82 Minutes of the Scientific Directors of the New York Botanical Garden, May 6, 1902, New York Botanical Garden Archives.

83 N. L. Britton to Board of Scientific Directors of the New York Botanical Garden, November 18, 1902, N. L. Britton Administrative Correspondence New York Botanical Garden Archives.

Britton's scholarship program provides an insight into the level of professionalization in American botany at the turn of the century. Though he himself had been educated primarily in geology, not botany, and had collaborated rather famously and profitably with Judge Addison Brown on a classic work of nineteenth century systematic botany, by the early 1900s, he generally came to consider only botanists working within those organizational structures he named in his petition to the Scientific Directors to be true scientific botanists.[84] Of necessity, he and his staff continued to correspond with a host of amateur "botanizers," accepting and identifying specimens sent by mail and offering encouragement to schoolteachers and other avocational botanists around the country. The Garden also continued to employ reliable field collectors who may not have had any formal scientific training. And yet in devising the Garden scholarship program in 1902, Britton in a very real sense acknowledged the importance of university training and professional status within the botanical ranks. NYBG became an institutional promoter of the scientific professionalizing process.

The Garden's official relationship with Columbia University signified the extent to which Britton considered his institution a training ground for these scientific professionals. A firm believer in the necessity of close ties to academic researchers, Britton in 1900 secured an arrangement with Columbia's Department of Botany (of which, several years earlier, he had been sole faculty member) through which advanced courses of botanical research by Columbia students would be conducted at the Garden's facilities. Work completed under the direction of Britton's senior NYBG staff would be accepted toward completion of graduate degrees in botany awarded at the University. By late 1900, 20 students, including some from outside Columbia, had enrolled in graduate courses at the Garden.[85] Such close ties, of course, had been the intention of Seth Low at the Garden's inception in the 1890s, when the Columbia president had recognized in the Garden an opportunity to secure indirect public support for a very expensive adjunct to the University's scientific program (see Chapter 3). During this first decade of the Garden's history, Low and Britton (and Low's successor, Nicholas Butler Murray) each understood the degree to which collaboration would benefit their respective institutions and acted accordingly.

84 N. L. Britton and A. Brown, *An Illustrated Flora of the Northern United States and Canada, from Newfoundland to the Parallel of the Southern Boundary of Virginia, and from the Atlantic Ocean westward to the 102d Meridian*, 2nd edition (New York: Dover Publications, 1970).

85 Britton had intended to include within these arrangements a special course in scientific forestry and had conferred in 1895 and 1896 with Gifford Pinchot and the faculty of Columbia's School of Mines to do so. This part of the plan, it appears, did not come to fruition. See N. L. Britton, Administrative Correspondence with Columbia University Department of Botany, 1892–1896, New York Botanical Garden Archives.

The Garden faculty available to mentor matriculated students included both Britton and his wife, Elizabeth; Lucien Underwood; Marshall Howe; P. Rydberg; G. V. Nash; J. K. Small; Francis Lloyd; E. S. Burgess; and, until he left for the Carnegie Institution in 1903, Daniel MacDougal.[86] The roster of NYBG-Columbia alumni grew to include many who would make significant contributions to taxonomic botany in the first half of the twentieth century, and several of these new PhDs would go on to staff Britton's garden long after he retired. Most notably, Henry Gleason completed his dissertation at the Garden and would later serve as acting director.[87]

Also included among the many students in residence at the Garden laboratories was a sizable contingent of women. During the first two decades of the Garden's graduate program, at least 35 of the 85 matriculated students were female. Though many left after taking MAs in order to pursue posts in secondary-school education, two women, Gertrude Burlingham and Winfred Josephine Robinson, received doctorates in 1908 after submitting dissertations on the taxonomy of the Agaricaceae (mushrooms) and pteridophytes (ferns), respectively.[88] By January of 1908, a total of 24 men and women, including matriculated graduate students pursuing Columbia degrees and senior botanists receiving Garden fellowships, were in residence at NYBG. Among this group were J. C. Arthur, a professor of physiology at Purdue University and a recipient of NYBG funds for three consecutive years; Mary P. Anderson, the Columbia student and nature study teacher at the Horace Mann School; and Homer D. House, expert on the genus Ipomoea (morning glory) and later an administrator of George W. Vanderbilt's "Biltmore" estate in North Carolina.[89] Overall, facilities for research and NYBG faculty expertise enabled Britton to offer a set of subjects for investigation at the labs that, while heavily weighted in favor of taxonomy, nevertheless included an impressive number of subjects: the taxonomy of algae, fungi, bryophytes, pteridophytes, and spermatophytes; morphology and experimental morphology; "developmental taxonomy;" paleobotany; plant geography; physiology of cells; chemical physiology; and physiological anatomy.[90] Between the inception of the Garden's research program in 1897 and 1908, totals of 71 men and 63 women registered to use the Garden's research facilities, and the subjects of their investigations ranged from mycology to zoology.[91]

86 David L. Lentz and Marlene Bellengi, "A Brief History of the Graduate Studies Program at the New York Botanical Garden," *Brittonia* 48 (1996): 404–412.

87 Ibid.

88 Ibid.

89 *Bulletin of the New York Botanical Garden* 5 (1908): 160–161.

90 *Journal of the New York Botanical Garden* 10 (1909): 130–131.

91 The "complete" listing of investigations conducted at the NYBG during this period includes: taxonomy, mycology, paleobotany, embryology, morphology, cytology, zoology, experimental morphology, physiology, desert botany, biology, nature study, phytoecology, lichenology, sociology,

The Laboratories. Crucial to the success of this program of professional education were the series of Garden laboratory facilities established when the Museum building opened and run under the direction of Daniel MacDougal.[92] An accomplished botanist and field worker, MacDougal had been Professor of Botany at the University of Minnesota until Britton lured him away in late 1898 with the promise of first-rate research equipment. Britton and MacDougal had known one another through contacts in the American botanical community for some time. In mid-1897, MacDougal, perhaps with an eye toward career advancement, had written Britton in the Bronx, congratulating him on the consummation of his plans for the new botanic garden. He followed up his note the following year with a humorous but friendly jibe at Britton's early plans for an extended program of Garden-sponsored botanical exploration. "Congratulations on your annexation of Porto Rico," MacDougal wrote, "I hope you will not devastate the native flora to such an extent as to render the island unfit for a tropical lab."[93] Within weeks, Britton had offered MacDougal the job as head of the laboratories, at a salary of approximately $2,100 annually (slightly less than half of Britton's) and responded in kind to MacDougal's joke with his own warning that the philanthropic responsibilities of Garden administrators included "crawling through sewers" and "visiting millionaires as we all do to pass the time."[94]

Within a month, MacDougal's appointment had been approved by the Board of Garden Managers. He was anxious to begin work and suggested to Britton that in equipping the new research facilities they look for a model to the laboratories at Berlin's botanic garden. MacDougal also noted with approval Britton's plans to expand the Library, predicting that a solid collection would strengthen the Garden's research program even though, in a veiled reference to Harvard's facilities, he conceded that NYBG would lack "glass flowers."[95] On his way east from Minnesota in 1899,

---

bryology, soil bacteriology, sanitary biology, agriculture, nutrition, tropisms, pedagogy of science, phytogeography, anatomy, artificial parthenogenesis, economic botany, ecology, forestry, local floras, museum administration, experimental evolution, and pathology. See *Journal of the New York Botanical Garden* 9 (April 1908).

92 On MacDougal, see Sharon Kingsland, "The Battling Botanist: Daniel Trembly MacDougal, Mutation Theory, and the Rise of Experimental Evolutionary Biology in America, 1900–1912," *Isis* 82 (1991): 479–509, and Sharon Kingsland *The Evolution of American Ecology, 1890–1900* (Baltimore, MD: Johns Hopkins University Press, 2005).

93 D. T. MacDougal to N. L. Britton, August 19, 1897 and October 7, 1898, N. L. Britton Administrative Correspondence, New York Botanical Garden Archives.

94 N. L. Britton to D. T. MacDougal, December 5, 1898, N. L. Britton Administrative Correspondence, New York Botanical Garden Archives.

95 D. T. MacDougal to N. L. Britton, January 20, 1899, N. L. Britton Administrative Correspondence, New York Botanical Garden Archives. MacDougal was referring to the Harvard botany department's

MacDougal stopped at St. Louis to tour the facilities at the Missouri Botanical Garden. There, he met with the director William Trelease, who, he informed Britton, might pose something of a competitive threat in the rush for botanical resources. Trelease was determined, according to MacDougal, to keep his own library at the "front" rank and was attempting to establish a similar laboratory. [96]

By August of 1899, MacDougal was ready to estimate the cost of furnishing the laboratories with necessary equipment: $4,465.50; a month later he hastily appended an additional request for unforeseen costs of $800.[97] Included in these figures were over $1,900 for optical equipment, including several Leitz microscopes, lenses, and photographic apparatus. Furniture, MacDougal noted, would cost over $200 more, and the morphological laboratory alone would require almost $700 to outfit. The largest single outlay of almost $1,100 would have to be for physiological equipment, which comprised a "Roentgen ray apparatus," a heliostat, pumps, batteries and gas fixtures.[98] Thirteen rooms within the Museum were set aside for MacDougal's purposes, including a constant temperature chamber; embryological, physiological, and morphological laboratories; a classroom; a photographic darkroom; and a chemistry laboratory. [99]

Despite the costs involved, MacDougal's ambitions for the NYBG laboratories extended beyond the Bronx. His allusion to a "tropical lab" in his jocular correspondence with Britton in 1898 gives a clue to his priorities. A laboratory set up within the Western Hemisphere tropics was one of the great desiderata of American botany around the turn of the twentieth century. Though in 1898, MacDougal thought that the newly acquired American territory of Puerto Rico was the most logical location, this laboratory would eventually be established in 1903, under NYBG auspices, at Cinchona on the island of Jamaica. Situated at an elevation of 5,000 feet in the Blue Mountains outside of Kingston, the laboratory and its 1,500 acres had been established

---

collection of hand-blown, life-size, and anatomically accurate, glass flower and plant models. This "Ware Collection of Blaschka Glass Models of Plants," a group of 847 individual glass sculptures, was the work of a German father-son team of glass blowers, Leopold and Rudolph Blaschka. The exhibit, commissioned by Boston philanthropists Elizabeth C. Ware and Mary Lee Ware to serve as scientifically accurate instructional models, still draws fascinated visitors to Cambridge. See Lorraine Daston, "The Glass Flowers" in L. Daston, ed.*Things that Talk: Object Lessons from Art and Science* (New York: Zone Books, 2004): 223–254 and Richard Evans Schultes and William A. Davis, "Glass Flowers Bloom Forever in Unique Harvard Collection," *Smithsonian* 13 (1982): 100–107.

96 D. T. MacDougal to N. L. Britton, March 6, 1899, N. L. Britton Administrative Correspondence, New York Botanical Garden Archives.

97 D. T. MacDougal to N. L. Britton, August 22, 1899, N. L. Britton Administrative Correspondence, New York Botanical Garden Archives.

98 D. T. MacDougal memorandum, Sept. 26, 1899, New York Botanical Garden Archives.

99 *Bulletin of the New York Botanical Garden* 1 (March 30, 1900).

by the British colonial government in the nineteenth century to cultivate cinchona trees, the bark of which supplied quinine to combat malaria. The colonial Jamaican government leased the land and its buildings to the Garden for a period of 10 years beginning in 1903. Britton planned to make Cinchona's facilities (a dormitory, three offices, two greenhouses, and two laboratories with bench space for 12 researchers) available to students and senior scientists interested in tropical botany.[100] It would also serve as a base of operations during the expeditions necessary to compile the Garden's *Flora of Jamaica*. Winifred Robinson, one of the three women to receive a doctorate while studying at the Garden during Britton's tenure, recalled her 1904 trip to use the Cinchona facilities as a "delightful part" of her botanical education, although she noted that although glassware and reagents were provided by NYBG, visiting students needed to bring their own microscopes (a commonplace at the time).[101] While it was in operation, the Cinchona laboratory proved to be an extremely popular venue for American research, attracting rising scientific talent such as Forrest Shreve, who conducted an investigation into "physiological plant geography" by measuring rates of transpiration above 4,500 feet.[102]

The Cinchona tropical station, Britton and MacDougal agreed, fulfilled a "great need" for a laboratory in the tropics of the Western Hemisphere in which plants could be propagated in an outdoor setting before being transported to North American botanic garden conservatories. Thus, Cinchona would serve in a capacity similar to that of the network of tropical botanic gardens of the British empire: it would act as a way station for the NYBG tropical collections. "Larger plants collected in the tropical forests," Britton noted, could be transported to temperate latitudes "only with difficulty and considerable loss unless they have been again rooted in the tropic and sent north in pots or tubs."[103]

Britton's initiative was roundly hailed throughout the American botanical community; botanists and horticulturists throughout North America enthusiastically endorsed his Jamaican plans. Investigators from numerous American institutions availed themselves of the facilities Britton and MacDougal had built in both the Bronx and at Cinchona. In 1902, A. P. Anderson of the Columbia University botany department accomplished pioneering work on edible starches, which, MacDougal proudly reported to the Garden Managers, yielded "new products of great economic value as foods."[104] In total, by 1904, 45 researchers (including 25 women) were registered workers at the

100 *Journal of the New York Botanical Garden* 7 (November 1906).

101 W. J. Robinson, "A Visit to the Botanical Laboratory at Cinchona, Jamaica," *Journal of the New York Botanical Garden* 5 (October 1904): 187–194.

102 *Journal of the New York Botanical Garden* 7 (August 1906).

103 *Journal of the New York Botanical Garden* 5 (1904): 1–2.

104 D. T. MacDougal, "Report on the Laboratories," *Bulletin of the New York Botanical Garden* 2 (1902): 452.

NYBG laboratories, and eight more had conducted investigations at Cinchona.[105] MacDougal himself had used the laboratories to begin the collaboration on the mutation theory with Dutch scientist Hugo deVries that would bring him his greatest fame. (Though this work was ultimately discredited, MacDougal confidently reported in 1903 that he had succeeded in forming new species in "the only known instances where new forms of [that rank] have been observed to originate under circumstances properly guarded against error.")[106]

Very few American scientific institutions were willing to play a cooperative role in building the tropic laboratory by pledging financial support, despite Britton's concerted attempt to form a cross-institutional network upon which the responsibility for funding it would rest. In 1906, as Cinchona hosted a record number of researchers, Britton circulated a letter to 45 universities, gardens, and museums in which he requested that each contribute funds to support the appointment of a permanent resident investigator at the tropical laboratory.[107] Only five of the letter's recipients responded with a pledge of money, however, and Britton had to abandon the plan.[108]

Despite this setback, even the British scientific establishment, from whose colonial apparatus Britton and MacDougal had secured the Cinchona facilities, took positive notice. In the spring of 1904, the editors of *Nature* lauded the enterprise of their American "cousins," who, though possessed of "numerous experimental stations" refused to be satisfied. "Vast as is their territory," the editors wrote, "they have invaded British territory, in a most genial way it is true, but still they have annexed, with our consent, a portion of the island of Jamaica." Lamenting Britain's own relative lack of progress with regard to the "practical" applications of botany and agriculture, the *Nature* editorialists concluded that while they did not "question the utility of ironclads and cruisers as protectors of commerce," it was becoming "obvious to those who [were] watching the proceedings of [their] neighbors and rivals that if [Great Britain] did not largely extend scientific training and induce [its] wealthy citizens to follow the example of their American brethren in endowing science, the necessity for protection will vanish, and that not slowly." Within several years of founding NYBG, then, Nathaniel Britton, his staff, and New York's philanthropic elite had accomplished a remarkable feat: they had managed, if in a small way, to reverse the centuries-old direction of trans-Atlantic Anglo-American scientific influence, becoming in the

105 D. T. MacDougal, "Report on the Laboratories," *Bulletin of the New York Botanical Garden* 4 (1904).

106 D. T. MacDougal, "Report on the Laboratories," *Bulletin of the New York Botanical Garden* 3 (1903): 222–223.

107 Britton, "Report of the Director-in-Chief," *Bulletin of the New York Botanical Garden* 6 (1906): 195–211.

108 N. L. Britton, "Report of the Director-in-Chief," Bulletin of the New York Botanical Garden 7 (1907): 213.

process a target of British imperial envy![109] NYBG would help give American botany a further imperial cast in the coming decades.

The Library and Herbarium. The Library and Herbarium collections, as crucial components of the Garden's research-conducting abilities, received as much careful attention as the Museum exhibits and the laboratories. The Library began filling with valuable accessions, eventually straining its storage capacity beyond levels anticipated by its architect. The Garden's book collections began with an 1896 agreement Britton forged between the Garden and Columbia College to deposit the College's 5,000-volume library in the new Museum building. Just as the College herbarium, containing personal collections of John Torrey, became part of the Garden's valuable reserve of scientific information, so also did Columbia's botany library. Other book collections deposited with the new NYBG Library shared the venerable pedigree of the Herbarium: included in an 1899 accession from the New York Academy of Medicine were volumes that had been part of David Hosack's personal library.[110]

Botanist Anna Murray Vail served as the Garden's first librarian. At the end of 1903, she reported to Britton that over 1,600 volumes had been added to the collection that year, bringing the Library's total holdings to 14,591.[111] The book collections increased at such a steady rate in part due to Britton's direct stewardship. When authorizing staff members' trips to scientific capitals in Europe and elsewhere, Britton suggested they scour local bookshops on behalf of the Garden. Vail and others undertook such book-hunting missions. In addition, worried that the supply of complete sets of crucial botany texts was rapidly diminishing due to the growth of public library collections in the English-speaking world (a trend made possible to a large degree by the Garden's own Vice-President, Andrew Carnegie), Britton contracted with a Geneva book dealer to acquire all botanical titles published between 1700 and 1870.[112] With this aggressive approach to collection building, Britton compiled as comprehensive a collection of the literature of botany as his budget allowed. In

109 "Laboratories for Botanical Research," Nature 69 (April 7, 1904): 538–539. Lacking financial support from other institutions, Britton was forced to let the lease for the Cinchona laboratories lapse in 1913. By 1914, they were being used by the British Association for the Advancement of Science, which allowed American students continued access to the facilities. British imperial envy, it appears, was short-lived. See "Report of the Secretary and Director-in-Chief for 1914," *Bulletin of the New York Botanical Garden* 9 (1915): 11.

110 Bernadette Callery, "Collecting Collections: Building the Library of the New York Botanical Garden," Brittonia 47 (1995): 46 and Minutes of the Scientific Directors of the New York Botanical Garden, January 9, 1899, New York Botanical Garden Archives.

111 Anna Murray Vail, "Report on the Library," *Bulletin of the New York Botanical Garden* 3 (1903): 224.

112 N. L. Britton to Board of Managers, December 20, 1898, N. L. Britton Administrative Correspondence, New York Botanical Garden Archives.

1902, for example, among many Library accessions, the Garden acquired a 1450 edition of a rare herbal entitled *De Virtutibus Herbarium*; a 1581 reproduction of the same volume; a "profusely illustrated" 1614 *Hortus Floridus*; and a six-edition set of Blackwell's *Herbarium Selectum* that had been printed and bound with richly colored plates of botanical illustrations in 1715.[113]

To pay for this dedicated program of library building, Britton inaugurated, with the sanction of the Garden Managers, yet another special fund: he sought an appropriation of $4,000 for book purchase in 1899 and at least $1,000 annually thereafter. Within several years of the Library's opening, he had secured regular subscriptions from Garden officers and benefactors to continue the acquisition program.[114] Vail noted that of over 1,800 volumes added to the Library in 1905, 860 had been purchased using "Special Book Fund" dollars.[115] Britton continued overseeing the Library's expansion throughout his career, taking advantage of various opportunities to acquire sizable European collections whenever they became available. In 1923, for instance, he raised $14,000 to purchase from the Geneva Botanic Garden 92 cases of valuable botany volumes, an accession so large that it took three years to be catalogued and shelved.[116]

The Herbarium (FIGURE 12) grew in tandem with the Library, as the collections of various botanists were combined and purchased by the Garden from estates, acquired through bequests, or secured through official agreement with other institutions.[117] Before J. K. Small had been able to arrange the NYBG Herbarium in the 1890s to conform with the Engler and Prantl taxonomic sequence, Columbia student Britton had had to draw "from hiding places in university closets and offices huge bundles of unclassified and forgotten botanical collections" in order to "organize them into a creditable nucleus for a universal herbarium."[118] Britton thus personally served as a direct link from the nineteenth-century New York botany of John Torrey to the twentieth-century botanical research of NYBG.

The Aesthetic Element

In his ambitious 1896 address to the American Association for the Advancement of Science, Britton acknowledged the role of the botanic garden as a landscaped public space designed to bring orderly natural beauty and civic grandeur to the metropolis. Every aspect of the physical design of a botanic garden, Britton had noted, "should be constructed and arranged with reference to tasteful and decorative landscape effect."

113 "Got Rare Botanical Library," *The New York Times*, September 21, 1902.

114 *Bulletin of the New York Botanical Garden* 5 (1905).

115 Ibid.

116 Callery, "Collecting Collections," 48, see note 110.

117 Ibid., 45.

118 Ibid.

With the exception of formal horticultural displays near architecturally imposing structures, decorative plantings, he claimed, should follow a "natural treatment," one that was in accordance with Olmstedian ideals of pastoral scenery and green urban oases. This ideal, Britton opined, should be extended even to systematic plantings whenever possible, since "the rectilinear arrangement of plant beds found in most older gardens [had] become abhorrent to landscape lovers." Acknowledging the degree to which his native city had been overshadowed, horticulturally, by Boston and Philadelphia, Britton made "the aesthetic element" a priority at the new botanic garden. "The cultivation of decorative plants, and especially the fostering of a taste for them," he proclaimed, were among the important functions of the botanical garden.[119] In addition to taking a leading role in the creation of a citywide horticultural group, the New-York Horticultural Society, under the institutional sponsorship of NYBG, Britton made every effort to create a Garden environment that both rivaled that of Kew and made profitable use of its own idiosyncratic landscape features.

The Garden Plantations

Initial planting of the Garden's parcel of Bronx Park land commenced in the 1890s, once the site was approved and the controversies over layout and organizational matters were settled. Real changes in the new Garden landscape took time to appear, however, because many of the plantings were grown from seed. George Nash, then head gardener, along with a staff of hired horticultural laborers, began laying out the pinetum (FIGURE 13), a collection of cone-bearing trees, in 1901. Covering 30 acres, the pinetum's species were grouped according to genera. Firs, spruces, and hemlocks were well represented; species imported from the conifer-rich southern US, Pacific Northwest and Japan predominated.[120]

Within two years, Nash had begun planting other trees for decorative purposes and to provide shade to Garden pathways between the Conservatory and Museum. In 1903, the great lines of tulip trees that today lead to the Museum building were planted. Nash set out 26 of them, although four eventually succumbed to drought. Seventy-three other trees were also added for ornament. Along the southern plaza of the great glasshouse, Britton instructed Nash to plant the trees in regular, formal rows, "the formality of the surroundings requiring this treatment." The remaining trees were placed "at irregular intervals," so as to avoid such formality.[121]

119 Britton, "Botanical Gardens: Origins and Development," see note 5.

120 New York Botanical Garden. *Descriptive Guide to the Grounds, Buildings, and Collections* (1907): 39. Nash was the second head gardener of the NYBG; he had been preceded briefly by Samuel Henshaw, an accomplished professional horticulturist and a member of the New-York Horticultural Society that had been briefly revived in the 1880s.

121 G. V. Nash, "Report on the Plantations," *Bulletin of the New York Botanical Garden* 3 (1903): 246.

FIGURE 12 Main Herbarium room on the third floor of the Museum building, 1908.

East of the site reserved for the large Conservatory, Nash began planting the various gardens of herbaceous plants.[122] On an eight-acre plot, Nash's men established several gardens of hardy plants arranged in ways to illustrate elementary botanical principles: one or more of each bed of plants was devoted to a single plant family, each of which was accompanied by labels bearing both a common and a scientific name. They included a systematic garden designed to convey a basic understanding of taxonomic groupings, a morphological garden to showcase certain plant organs and leaf forms, and an economic garden of plants with pharmacological and industrial uses. Though these gardens were designed in such a way as to convey principles of botanical systematizing, the layout eschewed formal landscape symmetry in order to "preserve the natural beauties of the glade."[123]

Northeast of the Museum building, the horticultural staff planted a 15-acre "fruticetum," or collection of hardy shrubs also arranged in systematic order.[124] In this plot were represented 330 species of the exotic ornamentals of East Asian origin that had been secured by donation, exchange, or exploration, as well North American temperate-zone plants. In addition, an arbor for the cultivation of vines was under construction. Along the entire eastern length of the Garden, running from north to

122 Herbaceous plants, unlike shrubs or trees, do not have any permanent structures above ground. "Hardiness" refers to a plant's ability to withstand a given location's lowest winter temperatures.

123 New York Botanical Garden, *Descriptive Guide* (1907): 60–61, see note 120; "The Plantations," Journal of the New York Botanical Garden 1 (April 1900): 48.

124 *Fruticetum* is derived from the Latin "frutex," or shrub. See Medbury, "Taxonomy and Arboretum Design," 18, note 47.

FIGURE 13 Newly planted pinetum with Conservatory in the background, 1902–1903.

south, Britton planned a deciduous arboretum. To this end, he had Nash and his men yearly add specimens of aspen, cottonwood, beech, oaks, elm, apple, plum, plane trees, magnolias, flowering cherry, locusts, maples, buckeyes, lindens, ash, willows, walnut, birch, and poplar. This arboretum was supplemented by the 50-acre hemlock forest that straddled the Bronx River in the Garden's eastern half. In total, the deciduous arboretum comprised 90 acres of Garden land and included 40 species of native trees and, eventually, 120 species introduced by the horticultural staff.[125] Finally, Nash planted a 40-foot-wide screen of ornamental herbaceous flowering plants as well as trees along the borders of the Garden that were not already forested. It was Britton's hope that such borders would elicit admiration from the local public and provide an attractive incentive to Garden members and potential benefactors during the annual spring members' tour.[126]

The Conservatory: A Garden under Glass

An integral part of Britton's idea of aesthetic development for the Garden, the great Conservatory had been designed in 1897,[127] and Britton broke ground in 1899 (FIGURE 14). Among the structures built on the site, only this building managed to achieve architectural distinction. This was due in large measure to the choice of the Lord and

125 "The Plantations," 48, see note 123.

126 New York Botanical Garden, *Descriptive Guide* (1907): 65–85, see note 120; "The Plantations," see note 123.

127 N. L. Britton, "Report of the Director-in-Chief," *Bulletin of the New York Botanical Garden* 1 (February 15, 1898): 164–165.

FIGURE 14 Partially completed Conservatory showing the steel and iron superstructure of the central palm-house dome, 1899–1901.

Burnham firm, then the most prominent of North American glasshouse builders, to create the display space needed for the Garden's planned exhibits of tropical plant specimens. Designed with an exterior style to match the Renaissance motifs of the Museum building, the Conservatory consisted of 15 interconnected houses that in total enclosed an area of 45,000 square feet, the first horticultural glasshouse to do so.

Frederick Lord, a carpenter and amateur glazier, had established the firm in the late 1850s, and by the 1870s, he and his son-in-law William Burnham had earned a reputation for their work in the Hudson Valley, creating heated conservatories for New York estate owners. Two decades prior to receiving the NYBG commission, Lord merged his firm with another glasshouse contractor in New Jersey. Together, the united companies became the most prolific manufacturers of greenhouses in North America. In addition to conducting a bustling business in prefabricated glasshouse designs throughout the US and Canada (customers chose plans from a published catalogue), Frederick Lord's firm designed the 1879 Palm House in San Francisco's Golden Gate Park. In 1881, Lord and Burnham constructed railroad tycoon Jay Gould's massive conservatory at his Tarrytown, New York estate. Three hundred seventy-six feet in length, with a central palm-house dome that housed over 320 varieties of plants, Gould's glasshouse likely served as a model for the Botanical Garden's later structure, built from the same innovative materials. Later, Lord and Burnham also built a conservatory for the Brooklyn Botanic Garden. The consolidated firm's success brought still more business, because design innovations in prefabrication had lowered costs, and various municipalities and private enterprises borrowed from garden architecture to commission glass arcades that enclosed restaurants, markets, winter gardens, and other public spaces.[128]

The nineteenth century popularity of the glasshouse, aside from its utilitarian botanical purposes in an institutional setting, stemmed from city dwellers' continued fascination with nature and represented the middle- and upper-class obsessions with the Wardian case on a grand scale. The practical desire to master and display nature could be combined in a piece of striking architecture to enshrine a set of exotic specimens representative of tropical locales contained within American colonial realms.[129] These desires, as the successes of Lord and Burnham in North America and W&D Bailey in 1820s England attest, constituted a trans-Atlantic phenomenon.[130] Such concerns also reflected the nineteenth-century Anglo-American desire to restore nature to the urban landscape and might be considered part of the larger project of the parks movement that had resulted in both Central Park and NYBG.[131] Britton readily exploited American horticultural Anglophilia as he sought approval of his chosen Lord and Burnham design and funding for its construction from the philanthropists he had carefully cultivated to sit on the Garden's Board of Managers.

Two wings of the connected series of glasshouses were constructed on an east-west axis, extending from a central palm house topped by a 90-foot dome. By the turn of the twentieth century, such domes were considered standard issues. Decades earlier, the extraordinarily influential British horticultural writer J. C. Loudon had experimented with the innovative use of a curved glass roof. This configuration admitted the sun's rays at the optimal angle of 90 degrees for at least part of the day to the plants contained within, thus, maximizing the amount of solar energy available for photosynthesis in northerly latitudes. Loudon's other mid nineteenth century innovations included the use of wrought iron bars to form curved skeletal frames, permitting architects to construct ever-larger domes.[132]

The wings of the Conservatory were then extended southward, forming an enclosed court, which Britton reserved for aquatic plantings and other "ornamental

128 John Hix, *The Glasshouse* (London: Phaidon, 1996): 54, 127, 135–136.

129 Georg Kohlmaier, *Houses of Glass: A Nineteenth-Century Building Type* (Cambridge, MA: MIT Press, 1996): 7.

130 John Hix, *The Glass House* (Cambridge, MA: MIT Press, 1974): 21–22. See also Christine Doell, *Gardens of the Gilded Age: Nineteenth-Century Gardens and Homegrounds of New York State* (Syracuse, NY: Syracuse University Press, 1986): 76.

131 Kohlmaier, *Houses of Glass*, 14–18, see note 129.

132 Peter Marston, *The Book of the Conservatory* (London: Weidenfeld and Nicolson, 1992); Hix, The *Glasshouse*, see note 128; Kohlmaier, *Houses of Glass*, see note 129; S. Koppelkamm, *Glasshouses and Winter Gardens of the Nineteenth Century*, trans. Kathrine Talbot (New York: Rizzoli, 1981); and George H. M. Lawrence, ed. *America's Garden Legacy: A Taste for Pleasure* (Philadelphia: Pennsylvania Horticultural Society, 1978): 44–79.

features," including the famous South American floating lilies, giant *Victoria regia*, which held such fascination for Victorians.[133] Overall, the style of the great Conservatory would be described in the same terms as the Museum architecture: "modernized Italian Renaissance."[134] This entire edifice was designed to be essentially fireproof. It was hoped that by constructing the glass houses with a steel and iron superstructure and a bluestone facade for the foundation and by limiting the use of wood to expansion-permitting casings that held the thousands of individual panes of greenhouse glass, the Garden's Conservatory would escape the fate of its New York predecessor, the "Crystal Palace" of Manhattan, which had been destroyed by fire in the 1850s.

A heating system designed to maintain a minimum temperature of 70 degrees Fahrenheit was composed of heated mains submerged in trenches beneath the floor of the Conservatory. These trenches extended to the powerhouse, which, to avoid an unsightly juxtaposition of the utilitarian structure and the delicate architecture of the glass building, was situated some distance away. Here, the influence of Decimus Burton's famous Palm House at Kew on Lord and Burnham's design was readily apparent: Burton had done the same in constructing Kew's conservatory heating system and had erected an "Italianate service tower" 550 feet from the Palm House dome.[135] Within the individual glass houses, "commodious" paths were laid to admit the curious, who would be encouraged to learn about the exotic plant life nurtured within the structure's transparent walls while strolling through the exhibits. Britton planned to include rooms devoted to cacti, ferns, tropical fruits, and aquatic plants.[136]

John R. Sheehan's company commenced excavation of the site during the spring of 1899.[137] By the summer of 1900, the first eight houses of the range were complete and were receiving visitors at least seven hours daily, while construction continued (FIGURE 15). The visitors initially had to dodge leaks from a porous roof and escaping steam from faulty heating mains. Remember that the site Britton had originally chosen in 1896 was a plateau approximately 900 feet to the southeast of the commanding hilltop position upon which the Museum was erected. This arrangement

133 New York Botanical Garden, *Descriptive Guide* (1907): 17–18, see note 120.

134 *Bulletin of the New York Botanical Garden* 1 (February 15, 1898): 164–165.

135 Hix, *The Glasshouse*, see note 128.

136 "Report of the Director-in-Chief" (1898), see note 127.

137 This may have been the same John Sheehan who was an influential Tammany political figure in 1897. David Hammack, in *Power and Society: Greater New York at the Turn of the Century* (New York, Columbia University Press, 1987): 117, identified that person as contractor "John R. Sheehan." But later accounts of Britton's dealings with Parks Department personnel and his garden superintendent, "Tammany figure" Arthur Corbett, attest to the Director-in-Chief's facility in navigating Tammany channels to secure money for NYBG purposes. Thus, it would not be surprising if Britton had managed to award the contract for the construction of the Conservatory to a politically well-connected builder. See Gleason, "Thumbnail Sketches," note 46.

FIGURE 15 Breaking ground for the Conservatory, January 3, 1899. Nathaniel and Elizabeth Britton are visible at the right.

would have enabled the Museum and Conservatory to share a powerhouse for heat. Instead, what is now known as the Conservatory was constructed at its present position on Southern Boulevard, on a plot reserved in Britton's original 1896 plan for a *second* glasshouse. This change was part of the compromise reached with C. S. Sargent's committee of landscape experts hired by the New York City Parks Department in 1897 to review Britton's original site plan. The Enid A. Haupt Conservatory now stands where it is because of Sargent's considerable influence in the 1890s as the editor of *Garden and Forest*, author of a multivolume taxonomic survey of the trees of North America, and chief administrator of Harvard's arboretum.[138]

As the Garden received gifts of plants, exchanged specimens with other botanical institutions, and began sending collecting expeditions throughout the hemisphere, the various glasshouses began to fill. In 1901 alone, 1,600 plants were added to the living collections. Gifts were sent by W. R. Smith, curator of the US Government's botanic garden on the Mall in Washington, DC, and from William Thistleton-Dyer's collections at Kew. In addition, Britton inaugurated a "Conservatory fund" with which to keep the houses stocked with live specimens, donating $100 of his own money and convincing J. P. Morgan to do the same. Andrew Carnegie donated $250 for this purpose that year. Within a year of its completion, the Conservatory would house 12,000 plants, representing 5,800 species and varieties, all arranged (as far as horticultural exigencies allowed) to illustrate familial relationships.[139]

138 Lincoln Pierson to N .L. Britton, May 22, 1900, N. L. Britton Administrative Correspondence, New York Botanical Garden Archives and *Bulletin of the New York Botanical Garden* 2.

139 G. V. Nash, "Report of the Curator of the Plantations, 1902 " *Bulletin of the New York Botanical Garden* 2 (1903): 409–510. Though Britton was careful to laud the Conservatory's design in his

FIGURE 16 Rustic bridge over lily ponds inside the aquatic house of the Conservatory, 1906.

The full set of Conservatory glasshouses was completed by January of 1902, at a cost of over $180,000.[140] Within four years, nine of the 15 houses had been devoted to tropical plants, and the horticultural staff grew specimens originating in Southern Asia, the West Indies, South America, Australia, Java, and China as well as the southern US.[141] Of the remaining six glasshouses, three were devoted to temperate species, arranged according to the Engler and Prantl plan; two contained cacti and other xerophytic plants; and one held a collection of aquatics, a visitor favorite, where a rustic footbridge (FIGURE 16) was erected over the tank to enable the curious to peer down upon large water lilies.[142]

This steady influx of plants meant that Britton's outdoor staff struggled to maintain the educational and horticultural integrity of the exhibits while faced with a dwindling amount of greenhouse space. Even after culling over 1,700 specimens from the various houses, the growth of the remaining 11,200 plants meant that space was extremely scarce. Even the propagating houses, which were located at a distance

---

official communications with the managers, he also conceded the need for modifications to the interior to accommodate horticultural needs of the newly acquired plant exhibits. It had been found necessary to alter the degree of light that entered certain houses; a combination of stippled lead paint and retractable roller shades was used to help control temperature and humidity levels. See N. L. Britton, "Report of the Secretary and Director-in-Chief for 1901," *Bulletin of the New York Botanical Garden* 2 (1903): 246.

140 *Journal of the New York Botanical Garden* 3 (February 1902): 23.

141 *Journal of the New York Botanical Garden* 7 (March 1906): 51.

142 Ibid. Xerophytic plants are those that thrive under dry conditions.

from the main Conservatory and the public areas of the Garden, were overcrowded. Britton believed that Garden-sponsored expeditions remained the best means of increasing the various Museum, Herbarium, Conservatory, and plantation collections. As the spoils of his expeditionary program continued to arrive, and the contingent of researchers in residence needing space for their experimental work with live plants increased, room for basic horticultural tasks such as propagation of ornamentals became increasingly rare.[143] This dynamic was an indication of institutional priorities in the coming years: though Britton greatly valued the physical appearance of the Garden, the demands of research consistently outweighed those of ornamental horticulture.

The Garden's Publishing Mission: Scholarly and Popular Imprints

Even as the Museum stood unfinished and Conservatory construction was just getting underway, Britton established several publishing venues for the work of the Garden's scientific and horticultural staffs. At a September 1899 meeting of the Board of Scientific Directors, he first proposed issuing a monthly journal aimed at a general readership in the hopes of stimulating public interest and increasing the Garden's membership rolls and endowment. Designed to treat subjects of botanical interest in lay terminology and to report to philanthropic patrons on the work of the NYBG staff, the first issue of the *Journal of the New York Botanical Garden* appeared in January of 1900. In this inaugural issue, Britton and various members of the curatorial staff published brief descriptions of Garden departments then undergoing organization and construction. Accompanying diagrams provided readers with a sense of the progress being made as contractors labored to complete the Museum building, the Conservatory, and the system of Garden driveways. In addition, Britton instructed each staff member returning from an expedition to publish an account of his travels here. Photographs of the new Library reading room and other Garden departments, along with monthly reports on accessions to the Herbarium, the Museum, the Conservatory, and the Library, were included in subsequent issues of the *Journal*, which was placed under the editorial direction of Daniel T. MacDougal, the first director of the NYBG laboratories and Britton's first Assistant Director.

In addition, Britton and the Scientific Directors planned a Garden *Bulletin*, as well as a set of Garden *Memoirs*. Mailed directly to members, the *Bulletin of the New York Botanical Garden* contained more detailed accounts of NYBG events; the annual reports of the scientific, curatorial and horticultural staff; and official communications from Britton himself to the Boards of Managers and Scientific Directors.[144] In the

143 *Bulletin of the New York Botanical Garden* 4 (1904): 76–77.

144 The initial volume of the *Bulletin* was funded with an appropriation of $100 from the Garden treasury. Minutes of the Scientific Directors, March 4, 1896, New York Botanical Garden Archives.

pages of the *Memoirs of the New York Botanical Garden*, Garden researchers, explorers, students, and visiting fellows published articles aimed at the community of botanical experts. With these venues for the dissemination of knowledge derived from Garden-sponsored research, Britton intended to gain national and international attention. To ensure a measure of success in this effort, he persuaded the Scientific Directors to pass a resolution expressly requiring any Garden employee or investigator associated with NYBG to secure the director-in-chief's permission before publishing results in periodicals not issued by the Garden's Press.

All of these periodical ventures served as a prelude to the truly monumental publishing task undertaken by NYBG beginning in the early 1900s and continuing through the twentieth century: the North American Flora. In April of 1901, with a sizable bequest from the estate of Garden incorporator C. P. Daly, Britton established a "David Lydig Fund," which the Scientific Directors earmarked for the expense of compiling this systematic reconnaissance of the flora of the US, Canada, and Mexico. In initiating the ambitious project, Britton invited William Trelease, then director of the Missouri Botanical Garden, Byron Halstead of Rutgers University's agricultural college, Frederick Coville of the USDA, E. L. Greene of Catholic University, J. M. Coulter of the University of Chicago, and numerous other botanists from various government agencies and universities to contribute monographs in their subjects of expertise. In addition, Britton continued to collaborate with other individuals and institutions during the 1910s and 1920s on two other major publishing projects, *The Cactaceae*, a four-volume portrait of the Cactus family, and *The Scientific Survey of Porto Rico and the Virgin Islands*, a massive reconnaissance of the natural history of US Caribbean possessions orchestrated by Britton and conducted under the auspices of the New York Academy of Sciences.

The Public Makes a Space for Itself

With the flurry of press coverage of the various controversies surrounding the Garden's founding and the local Bronx papers' habit of sending a reporter to NYBG on a regular basis in order to secure fodder for a column, Britton's new institution maintained a relatively prominent profile. This publicity, along with the hundreds of acres of gardens and woods open daily, did indeed attract the sizable crowds (FIGURE 17) that Britton and Addison Brown had predicted when planning the Garden layout. However, they could not predict some of the ways in which the Garden's popular constituency made use of the city's newest public landscape. Britton sought to impose as much control over "his" institution as he could, through sponsored programs of lectures and tours, horticultural shows, membership drives, and regular communication with both the city police and the local press. And yet local residents as well as Garden visitors from various parts of the city commonly turned the new Bronx Park into a landscape that suited a diversity of needs.

FIGURE 17 Visitors approaching the Conservatory, 1910.

One of Britton's earliest efforts to use this landscape to educate a broader audience than professional botanists was undertaken as part of a larger project of his wife's, the protection of native wildflowers. Elizabeth Britton was one of the founding members of the Wildflower Preservation Society of America, organized in 1902. In her later years, involvement with this organization would actually demand more attention than her own botanical research. Using $3,200 donated by a pair of concerned New York sisters, Caroline and Olivia Phelps Stokes, the Brittons inaugurated an essay contest in which three prizes were awarded annually to the best pieces submitted for publication in the Garden *Journal* that advocated protection of native plants of the Northeast. The cash prizes were derived from the income of the invested gift of the Stokes sisters. He then distributed copies of these essays to various other botanical periodicals and newspapers. By doing so, Britton claimed, the victorious writers reached a nationwide audience of "many thousands" and provided valuable pedagogic material to American schoolteachers (many of whom had also received copies).[145] Among the earliest winners were students conducting research at the Garden, including MA candidate Mary P. Anderson, who took her message of plant conservation directly to her students in the Bronx public schools.[146]

145 N. L. Britton, "Results of the Use of the Stokes Fund for the Preservation of Native Plants" *Journal of the New York Botanical Garden* 3 (1902): 179–180.

146 See *Journal of the New York Botanical Garden* 3 (January 1902), 5 (1904), and 10 (1909): 160–63.

If public behavior on the grounds of the Garden was any indication, however, Anderson's message seemed to be lost in translation. The May 5, 1901, edition of the *New-York Tribune* illustrated supplement carried the headline, "Vanishing Wild Flowers," in which a reporter related the tale of the depredations upon the park lands by various picnicking parties. With a "misdirected enthusiasm for nature," these bands of Bronx residents had managed to "eliminate" from the borough the arbutus plant, even though, the writer asserted, it was likely that "the flowers so carelessly torn from their homes probably give little or no satisfaction to those who make away with them." In language freighted with class-based assumptions about the popular taste of the mass of New York's heavily immigrant population, the *Tribune* writer decried the commonplace practice of flower picking in public parks.

> Is there a flower lover who has gone out into the country with a party of young people unaccustomed to find themselves surrounded with green who had not noted with something very like hopeless rage the immediate rush on every growing thing in the neighborhood, its instant uprooting and subsequent careless tossing aside?... A few dejected blossoms are all that is left when home is reached; nothing of any value remains out of all the lives butchered to make an East Side holiday.... Mothers... encourage their children to squeeze flowers in their hot little hands, and reflect on the poetic beauty of this union of the innocents of nature and the innocents of humanity. The flower's right to existence nobody takes into account or the harm done to the children by allowing them to think that they may destroy life as they choose.[147]

Though it is likely that such use of open and wooded lands had been common during the many years that the Bronx was a rural ward of the city, the reporter promulgated the Brittons' view that such behavior was singularly inappropriate.[148] The writer professed bewilderment at the locals' enthusiasm for the arbutus in particular. "Is there a more unsatisfactory flower to make attractive in a vase?" The same must be said of other "delicate wood plants," the reporter continued. "They depend upon the charm of their surroundings. A sunset painted on the ceiling of a Harlem flat would be hardly more out of place than these transplanted blossoms."[149]

Ironically, much of the blame for these misdeeds was placed by horticultural writers on "the ill-judged and misdirected energies of certain so-called nature study movements."[150] It is tempting to characterize such sentiment as a sort of backlash

147 *New-York Tribune Illustrated Supplement* May 5, 1901.

148 Accounts of local "Italians" collecting firewood from the Garden forest can be found in contemporary newspapers, e.g., "Gathering Fuel in Bronx Woods," *The New York Times*, September 20, 1903.

149 "Vanishing Wild Flowers," *The New-York Herald Tribune*, May 5, 1901.

150 "A Fund for the Preservation of Native Wild Flowers," *American Gardening* October 12, 1901.

against a form of botanical endeavor accessible to women; yet women were among the most active of critics of "misdirected" nature study. Rather, Elizabeth Britton's battle with flower-picking Garden visitors reflects her own class status and educational background, attributes shared with her female allies in this struggle. The disappearance of certain species of attractive native plants and the destruction of flower beds in the Garden itself, Mrs. Britton believed, were the result of a misbehaving public bent not on acquiring knowledge of the natural world through urban "fieldwork," but on gathering decorative objects for their homes. Her sentiments were echoed by writers for *American Gardening*, who asserted that the "precious heritage of wild flowers . . . is sometimes too lightly regarded by the thoughtless multitude, and wanton destruction is the result. The demolition of what is beautiful," *American Gardening* concluded, "is a crime."[151] Concerns over Bronx Park's wildflower population even extended southward into the zoo, as William Hornaday, first director of the newly founded Zoological Gardens, decried the "wonton treatment of flowers by visitors."[152]

In an attempt to check such behavior, the director-in-chief posted signs prominently throughout the Garden grounds that established rules of conduct. Such signs had been used earlier to discourage littering to no avail; the amount of plant theft and cutting, however, had been unanticipated. Britton had even optimistically reported to the Managers in the earliest issues of the Garden *Bulletin* that flower picking had *not* been a problem. Yet, by the fall of 1902, Britton and John Eustis, Bronx Borough Park Commissioner, had drawn up an extensive list of such rules. First among the proscribed behaviors was "the picking of flowers, leaves, fruits, nuts, or the breaking of branches of any plants, either wild or cultivated, the uprooting of plants of any kind, the defacing of trees, and the carrying of flowers, fruits, or plants into or from the grounds of the Garden." Also excluded were ball playing, hunting, littering, or using the Bronx River for boating or fishing. As with other contemporary American public parks or gardens, NYBG, Britton felt, deserved a public that was "aware of the aesthetic features of the design, in search of a contact with nature, and socially disciplined."[153] Such restrictions were in keeping with a contemporary upper-class aesthetic eager to ban folk entertainments. (Sports such as baseball and cycling had been allowed in Central Park only after widespread popular pressure. C. S. Sargent, in particular, had been a vocal opponent of such innovation, arguing in the pages of

151 "The Preservation of Native Plants" *American Gardening*, December 28, 1901.

152 "Destroy Plants in Bronx Park: Wanton Treatment of the Flowers by Visitors," *The Mail and Express,* May 10, 1902.

153 J. B. Jackson, *Discovering the Vernacular Landscape* (New Haven: Yale University Press, 1984). See also Galen Cranz, *The Politics of Park Design: A History of Urban Parks in America* (Cambridg. MA: MIT Press, 1982); Peter J. Schmitt, *Back to Nature: The Arcadian Myth in Urban America* (New York: Oxford University Press, 1969); and Norman Newton, *Design on the Land: The Development of Landscape Architecture* (Cambridge, MA: Belknap, 1971).

Garden and Forest that Olmsted's traditional, Romantic conception of the Park as a refreshing urban oasis was of paramount importance, outweighing lower-class requests for athletic space and room for public pageants, including parades.[154])

Signs at NYBG proved relatively ineffective, however, in the absence of a fence that completely enclosed the grounds and a sizable contingent of police officers stationed in the area. Though several Garden employees were detailed to walk the grounds (and more employees were added on weekends), flower picking and less innocuous forms of vandalism continued from year to year. Garden employees confiscated armloads of picked flowers from exiting patrons, who professed ignorance of the rules. Local boys even stole the lead identifying labels affixed to outdoor plants, melting them down and selling them to scrap metal dealers.[155] Other garden visitors had to be kept from straying from the designated paths through the hemlock forest, because Britton feared the damage energetic hikers and picnickers would do to the trees' delicate root systems.[156] Littering remained a persistent problem, and Britton had to assign a keeper to the forest to gather the debris left by weekend crowds.

In 1909, Garden alumna Mary Anderson found herself not only lamenting the imminent demise of the region's wild flora, but also despairing for the physical health of the entire set of Garden plantations. As a nature study instructor and grade school teacher, Anderson felt a particular responsibility to curb these public enthusiasms for violating posted NYBG rules. In the winning essay for that year's Stokes Fund contest, Anderson reported the results of a study she conducted of her students' comprehension of those same rules. What she purported to discover, in perhaps an unintentionally comic manner, was the total lack of influence that Britton's code of conduct had among the general public.

She began with an account of a day spent among the "multitudes" of Lower East Siders who "pour[ed] in an almost continuous stream through the [Third Avenue Elevated Railway] entrance" of the Garden on a spring Sunday. Though she was quick to note "with humble heart, and open mind" that "all [were] welcome to whatever the Botanical Garden [offered] in the way of education, of beauty, of sunshine and fresh air," nevertheless, she continued, "unfortunately where so much [was] offered, still more [was] taken." Over the course of her day spent in informal anthropological fieldwork, Anderson tallied at least 30 instances in which she confronted a Garden visitor, child or adult, engaged in some form of officially prohibited behavior. Among

154 Cranz, *The Politics of Park Design*, 15–22, see note 153.

155 Minutes of the Scientific Directors of the New York Botanical Garden, April 13, 1912, New York Botanical Garden Archives.

156 In 1915, the city's Board of Estimate arranged with Britton for 10 acres of an additional parcel of land granted to the Garden to be reserved specifically for picnicking parties. *Journal of the New York Botanical Garden* 19 (1918): 79.

these "multitudes" were a man and wife hauling a "great armful" of flowering dogwood branches from the Garden. Accosting them, Anderson informed them that "every effort was being made to keep the wild flowers and plants of all kinds, so that people might come and see what the country is like in springtime or at any other time." Her listeners, she reported, were slow to comprehend the import of her admonishment. Only gradually, she asserted, was the couple enlightened. "At first," Anderson wrote, "the man stared uncomprehendingly, but I gave him time, and he slowly worked it out: 'Gosh! If everybody tuk some, thar wouldn't be none!"[157] Next, she came upon a pair of rowdy siblings breaking branches and picking lilacs. Their parents' only response to Anderson's protests, she wrote, was a "drawled" remark delivered with a "vacant stare," that they couldn't "shtop 'em."[158] Still others, "Italians," "children from the Lower East Side," and "shop girls," were caught by Anderson or Garden employees in various acts of horticultural piracy.

Anderson's impromptu police work extended, she professed, to "the better class" as well. One such woman, whose status was attested to by her dress, informed Anderson that she knew how to smuggle botanical spoils past guards at the entrances. Still another was accompanied by her husband, who, to Anderson's dismay, proudly proclaimed his refusal to read any posted notices within the park grounds. Only upon further discussion with this latter group (who Anderson judged by their "good English, good clothes, and above all, good nature" to be "considerably above the average people") did she discover the source of much of the trouble: the posted signs, her respectable informants said, were simply too long. "I can't imagine myself stopping to read any such thing when there is so much that is beautiful to look at," the woman exclaimed. Anderson concluded her informal study with an admission that "all sorts and conditions, all ages and nationalities of men, women and children must be dealt with." Simpler signs posted more frequently would gain the attention of more visitors.[159] Nevertheless, within two years of Anderson's prize-winning jeremiad against Garden vandalism, Elizabeth Britton ensured that yet another set of signs would be posted after she confiscated 410 jack-in-the-pulpit plants from a class of

157 M. P. Anderson, "The Passing of the Wild Flowers: An Investigation of the Problem of New York City" *Journal of the New York Botanical Garden* 10 (1909): 147–163.

158 Ibid., 150.

159 She hoped that such signs would be readily comprehensible to children, who, she discovered, had been almost universally oblivious to the content of the old notices. In a survey she conducted of first through fifth grade students at the Horace Mann School (children from "homes of culture and comfort"), she found to her dismay that a substantial majority of them found the signs unintelligible. "What is true of the ability of the child under twelve years of age to read and understand English," she concluded, "is also true of the majority of foreigners who come to the Garden."

schoolchildren on a field trip. Despite the Brittons' best efforts, the Botanical Garden often served as New Yorkers' cutting garden in the early twentieth century.[160]

A Cultivated Public: The Revived Horticultural Society

While attempting to control the various uses to which the local population put the Garden landscape, Britton established another public link with the city by helping to reinvigorate its long-standing, if dormant, horticultural traditions. Despite its wealthy and influential membership, composed of some of New York's financial and cultural elite, the reincarnated version of the old New-York Horticultural Society that had surfaced in the 1880s had not lasted into the succeeding decade. Ironically, then, just as the Botanical Garden, the goal of the city's professional and amateur horticulturists for most of the nineteenth century, was being developed, the local horticultural society once again faded from the scene. Though florists' groups were active around the turn of the twentieth century and the American Institute continued its annual exhibitions (which included some horticultural displays), the Garden itself had been essentially the culmination of the efforts not of horticulturists but of the city's botanists. Members of the Torrey Botanical Club and professional scientists associated with Columbia College, backed by the institutional prestige and financial resources of that school's philanthropic network, had been the driving forces behind the Garden's creation, and the scientific priorities of the Garden (exploration, research, and advanced as well as public education) were direct results of these origins.

Though a scientific botanist, Britton refused to ignore what he had characterized as the "aesthetic" and horticultural obligations of the Garden. Once NYBG was established, and its physical plant was nearly complete, he seized the opportunity to continue the horticultural traditions of his New York forebears by offering the Garden as a venue for horticultural exhibits and meetings of a newly established Horticultural Society of New York (HSNY). It had been formed in 1901 and within a year had held its first horticultural exhibition, attracting 500 visitors. During ensuing shows, Britton with top hat perched on his diminutive head, made a point of greeting the Society's members at the Garden's entrance and then escorted them through the grounds to the Museum building. There, in conjunction with the floral displays, lectures were held, and the Garden's Library and Museum exhibits were open for members' inspection. In addition, the Society and the Garden had extended formal invitations to participate in the proceedings to members of the New York Florists' Club, the Farmer's Club of the American Institute, and the American Rose Society. At the close of the festivities, the Garden awarded premiums of its own to the best displays of

160 Elizabeth Britton, "Thoughtless Destruction of Jack-in-the-Pulpit," *Journal of the New York Botanical Garden* 13 (1912): 68–69.

ferns, cut flowers, horticultural novelties, succulents (cacti), orchids, and vegetables.[161] *American Gardening* pronounced the HSNY's first summer exhibit, in June of 1901, "an unqualified success," and the city's *Commercial Advertiser* seconded the praise in a breathless account of a "long, white, cool [exhibit] room" in which rows of brilliant displays of peonies and roses provided an air of "simplicity and naive beauty."[162]

Assessing Britton's Vision

By following a set of principles of botanical garden design that he articulated in 1896, Nathaniel Lord Britton and a professional staff of curators, horticulturists, and architects created an important new educational institution and a productive venue for the prosecution of botanical research in the newly consolidated city's northernmost borough. Possessed of an acute historical consciousness of his local botanical predecessors, Britton wished for his native city a Garden to equal those he had admired in the Old World. A product of the nineteenth century's natural history tradition, Britton succeeded in creating a twentieth century botanic garden complete with the requisite vehicles for scholarly research and publication as well as tools for less specialized public education. An integral part of New York's new botanic garden, of course, were its grand Museum and public Conservatory, structures that culminated the city's century-long series of attempts at building a civic landscape suitable for the nation's commercial capital. These efforts had begun decades earlier and had become manifest in the innovations in architecture (skyscrapers) and park building (Central Park) sponsored by the city's commercial elite throughout the last half of the century. The Garden's creation in the 1890s by a coalition of New York philanthropists and professional scientists might be seen as the final attempt of nineteenth century elites to appropriate for their city a portion of public space for the validation of their city's civic culture.[163]

Of course, all of these new Garden activities took place within the confines of a public space whose meaning was still to be negotiated over the course of the Garden's first decade by various city classes and Garden constituencies. In general, Britton understood the overall mandate of botanic gardens at the end of the nineteenth century to be a difficult one: the maintenance of a public museum within the landscaped boundaries of an accessible park that, in turn, served as the setting for professional research and university education. Such garden-museum-parks were intended to promote national economic and scientific aims as well as provide genteel moral uplift,

161 "The Horticultural Society of New York," *Journal of the New York Botanical Garden* 4 (1903): 79. Note that Britton's link to organized horticulture included his service in the early 1900s as botanist of the Society of American Florists and Ornamental Horticulturists. See ibid., 87.

162 *American Gardening*, June 15, 1901; *Commercial Advertiser*, June 13, 1901.

163 Domosh, Invented Cities, 75–76, see note 9.

through the literal example of cultivation, to a visiting republican citizenry. These citizens, in turn, would be impressed by the constructive and philanthropic powers of American capital. Contemporaries in the American botanical community, especially eminent University of Chicago botanist J. M. Coulter and Brooklyn Botanic Garden director C. S. Gager, shared Britton's view of the expansive educational mandate of botanic gardens.[164] As Britton discovered, however, the response of the public to this call to cultivation was never predictable. All of the new Garden activities took place within the confines of a very specific public space, one newly annexed to the city from a former periphery, whose meaning was still to be negotiated over the course of the Garden's first decade by various city classes and Garden constituencies. Thus, even as Britton and his philanthropic allies fashioned a formal Italianate setting for science and civic display from a previously agricultural landscape, visitors from surrounding neighborhoods continued to work the land as they had for decades. Foragers collected firewood for kitchens, while small boys discovered valuable new sources of lucrative metal and women and children collected flowers for home and classroom. Dedicated attempts to control such behavior were met with either resistance or indifference.

Despite the apparent discrepancies between the public's perception of the Garden landscape and his own vision of its varied purposes, Britton remained genuinely dedicated to maintaining a public space for the education of laymen that simultaneously served as a setting for genteel, passive, outdoor recreation. Through the regular adult lectures in the Museum's auditorium and the nature study tours aimed at city schoolchildren, the Garden managed to perform the aesthetic and the philanthropic functions he had enumerated in 1896. Yet, as a scientist and university educator committed to the training of professional botanists, Britton considered the *primary* aims of NYBG to be scientific and economic in scope, and he adopted as his model the imperial British botanical network headquartered at Kew.[165] That the top floor of the Museum,

164 For a discussion of the cultural dimensions of the practice of natural history and the prescriptive elements inherent in the practices of the American life sciences, see N. Jardine, J. A. Secord, and E. C. Spary, eds., *Cultures of Natural History* (Cambridge, UK: Cambridge University Press, 1996): 8 and Philip Pauly, *Biologists and the Promise of American Life: From Meriwether Lewis to Alfred Kinsey* (Princeton, NJ: Princeton University Press, 2000). For contemporary assessments of the educational value of botanic gardens and the "scientific" habits of mind these institutions inculcated into a receptive public, see J. M. Coulter, "The Social, Educational, and Scientific Value of Botanic Gardens," address given April 19, 1917 at the dedication of the Brooklyn Botanical Garden Laboratories and C. S. Gager, address given the same day, both reprinted in Science 45 (June 29, 1917): 643–652. Gager, Britton's former employee while director of the NYBG laboratories, asserted that the "greatest educational need" was the training of US citizens in the scientific habit of critical thought, i.e., the abilities to "observe accurately, to record faithfully, to reason logically, to keep an open mind, to welcome truth regardless of consequences, quickly to recognize error [and] to recognize charlatanism."

165 It should be noted that his colonial British counterparts were themselves often under pressure to relegate scientific botany to the background and create merely an open-air setting for colonists'

blessed with the best lighting of the entire building, was reserved for the Library, Herbarium, offices, and laboratories was perhaps not accidental.

Graduates of NYBG's doctoral programs, attendees at Saturday afternoon lectures, and visitors to the Museum's popular microscopic exhibits did provide the Garden with an audience at least potentially receptive to its larger message of education and cultural uplift. Ultimately, by marshaling a remarkable set of philanthropic resources Nathaniel Lord Britton succeeded in attaining for New York a stature as a center for botanical knowledge commensurate with the city's rank as *the* American metropolis.[166]

---

picnics and games, as noted by Donal McCracken in *Gardens of Empire: Botanical Institutions of the Victorian British Empire* (London: Leicester University Press, 1997).

166 Thomas Bender, *New York Intellect: A History of Intellectual Life in New York City, from 1750 to the Beginnings of Our Own Time* (Baltimore, MD: Johns Hopkins University Press), 198

Chapter 5
A Botanical Monroe Doctrine

In pursuit of the ambitious goal of creating an "American Kew," Britton sought the establishment of a specifically *American* botany, one in which American botanists, directed from the Bronx, would gather, examine, classify, and name the flora of the Western Hemisphere. These efforts were intended to complete a taxonomic portrait of the United States' natural domain of influence and to identify plants that Britton deemed "truly American" and, thus, worthy of display in the Garden's public Conservatory. With the assistance of his philanthropic connections and the rise of a modern funding apparatus for American science (specifically the Carnegie Institution of Washington), Britton was able to initiate a dedicated program of botanical reconnaissance of the Western Hemisphere. Each year from 1898–1929, the New York Botanical Garden (NYBG) sponsored botanical expeditions in the Caribbean, Central and South America, and western North America.

In order to place these activities in a larger social and political context, this chapter opens with a brief consideration of Britton's British model, the Royal Botanic Gardens at Kew, which during the 1890s sat at the center of a global network of imperial botanical research. Britton's efforts to pursue his agenda of botanical nationalism were eagerly chronicled in the New York and national press, and stories highlighted both the exotic exploits of the city's intrepid botanists and the role of the Garden in creating landscaped public space in the Bronx. Such efforts were afforded attention because of their perceived value to the city and to the nation's scientific reputation. In part, Britton believed, that reputation would be secured if the system of botanical nomenclature used to identify the earth's flora could be made to suit American needs. Thus, this chapter also briefly delineates Britton's role in the acrimonious nomenclature debates that had begun in the 1890s and continued throughout the rest of his tenure at the Botanical Garden.

Finally, the chapter is primarily a portrait of botanical collaboration typical of the NYBG's research activity during Britton's tenure as director-in-chief. As a case study, the story of the writing of *The Cactaceae* (a four-volume taxonomic treatment of the cactus family, comprised of notable Western Hemisphere natives) serves as an illustration of the ways in which such botanical research was conceived, funded, and accomplished. As a long-term project sponsored in part by the newly established

Carnegie Institution of Washington and carried out by both New York and US government botanists, production of *The Cactaceae* underscores Nathaniel Britton's scientific aims and his botanical nationalism. Britton's collaboration with the Smithsonian's Joseph Nelson Rose also highlights the difficulties inherent in setting such an ambitious agenda at a time when the modern system of scientific philanthropy was just beginning to take shape.

A British Model

As the Museum and the Conservatory were under construction at the turn of the twentieth century, Nathaniel Britton began rallying support from the philanthropists who constituted his Board of Managers to underwrite his "American Kew." Britton's model, the Royal Botanic Gardens, situated northwest of London, had been for the previous half century "the botanical center of the world, carrying on all the economic and scientific work of the Empire." Under the sequential directorships of William Jackson Hooker, his son Joseph Dalton Hooker, and subsequently William Thistleton-Dyer (the younger Hooker's son-in-law), Kew's administrators had expanded upon the imperial legacy of Joseph Banks. These men took Kew beyond its origins as a royal and ornamental pleasure ground to place it (and its far-flung network of botanical stations) in the service of Victoria's empire. In this capacity, Kew flourished. Since the 1880s, Thistleton-Dyer had helped expand British botany beyond its traditional task of nineteenth century taxonomy to include Continental botanical concerns with comparative morphology, life cycles, heredity, physiology, and pathology, all carried out with the aid of laboratory research. This acceptance of the mantle of a "new botany" by a mature British botanical network served the needs of imperial economic horticulture quite effectively, and British botanists' work with live plants was stimulated by the opportunities for fieldwork afforded by colonial settings. This fieldwork was in turn encouraged by politicians and imperial administrators eager for expert scientific advice with economic import. By the late nineteenth century, Thistleton-Dyer's botanical network comprised a vast system of scientific patronage, as men trained at Kew filled colonial posts in agriculture, horticulture, forestry, and botany.[1]

For British botanists, John Lindley's 1841 description of Kew's mandate to provide "the mother country [with] everything that is useful in the vegetable kingdom" remained valid into the twentieth century. In the 1870s, for instance, Joseph Hooker's

1 Richard Drayton, "Imperial Science and a Scientific Empire: Kew Gardens and the Uses of Nature, 1772–1903," (PhD Dissertation, Yale University, 1993): 386–393 and D. R. Headrick, *The Tentacles of Progress: Technology Transfer in the Age of Imperialism, 1850–1940* (New York: Oxford University Press, 1988): 213–214. See also Drayton's *Nature's Government: Science, Imperial Britain, and the "Improvement" of the World* (New Haven, CT: Yale University Press, 2000). The emergence of a "new" botany is discussed in Chapter 7.

collectors worked to transfer rubber from Brazilian forests to the British colony at Ceylon, and in 1879, Hooker established a museum of Tropical Economic Botany. By 1896, Kew stood at the center of a network of over 30 gardens within the British Empire, each serving as a government source of information on botanical questions of economic importance.[2] Parliament and colonial administrators, "in the hopes of discovering new green El Dorados," continued the eighteenth-century tradition of sending botanizing expeditions throughout the tropics in search of the organic raw materials of Western commerce and industrialization: cotton, indigo, palm oil, gutta-percha, rubber, cinchona, tea, coffee, sisal, and sugar cane.[3]

As tropical colonies became enmeshed in the Western marketplace, they supplied raw materials for the products of nineteenth-century industrialization. Increases in demand for these colonial resources led to the development, historian D. R. Headrick writes, of an "imperialism of free trade," in which non-Western regions could be brought into a world economy even without direct military conquest and control.[4] It was in this context that the US of the 1890s sought to enter the wider international arena of colonial politics and to create its own network of profitable dependencies. As noted in Chapter 3 , Britton sought the resources to establish NYBG at a less-than-propitious time. The economic contraction in the wake of the Panic of 1893 continued through 1897, as the nation's industrial production outpaced demand, leading to widespread social unrest (a trend that had begun in the 1870s and 1880s), a concomitant sense of malaise on the part of American elites, and the worst economic depression up to that time in US history.[5] Faced with such pressure, the American governmental, social, and business elites began supplementing classical economists' traditional reliance upon monetary policies to ameliorate the business cycle by developing a theory of overproduction and stagnation. These American capitalists, including the NYBG's Vice-President Andrew Carnegie, advocated an updated mercantilist strategy, one designed to mitigate the negative economic effects of a narrowing domestic marketplace.[6]

In this unsettled context, the creation of an urban garden modeled after Kew, one that could potentially provide the uplifting educational effects of museum botany in a genteel outdoor setting, took on even greater importance. In an era marked by dire pronouncements about the closing of frontiers and subsequent skepticism toward the myth of a western American Eden, a new public space had the potential, in the

2 Headrick, *Tentacles of Progress*, 213–214, see note 1.

3 Ibid., 6, 211.

4 Ibid., 6.

5 Walter LaFeber, *The New Empire: An Interpretation of American Expansion, 1860–1898* (Ithaca, NY: Cornell University Press, 1963): 8–16 and Timothy McCormick, *China Market: America's Quest for Informal Empire, 1893–1901* (Chicago: Quadrangle Books, 1967): 29–30.

6 McCormick, *China Market*, 19–29, see note 5.

minds of many plutocratic New Yorkers, to pacify a frightening array of disruptive social forces. In addition, through a dedicated program of economically oriented exploration, an American Kew would open up commercial possibilities for a nation run by a newly powerful "economic man."[7] As the US began, in historian Walter LaFeber's memorable phrase, "measuring itself for Britain's shoes," influential Americans, including historian Frederick Jackson Turner and naval power theorist Alfred T. Mahan, urged adoption of certain British imperial features. Even as "twisting the lion's tail" remained politically popular among members of Congress, these men came to value the growing American military alliance with British sea power.[8] Upon such foundations were built a "new empire" of trade, one that represented the culmination of a half-century of American foreign policy.[9]

Britton's institutional Anglophilia was of a piece with such sentiments, as was his desire to promulgate a "botanical Monroe Doctrine" in which US botanists would dominate the practice of botany in the Western Hemisphere. Although apparently contradictory, these impulses were, in fact, complementary facets of an overall plan. Britton remained staunchly committed to fostering an independent American botany and scientific sphere of influence within the Western Hemisphere, one analogous to Kew's reach within its own botanical empire. This aim was also apparent in his stubborn insistence upon an "American Code" of botanical nomenclature. And yet, for the NYBG director-in-chief, Kew's crucial characteristics were its expansive reach and its ties to nationalistic concerns. Of necessity Britton's first priorities remained the floristic tasks that British botanists were beginning to supplement with attention to other types of botanical research. A comparatively immature American national botany still needed to complete the North American survey projects initiated by Britton's US predecessors, Asa Gray and John Torrey.[10] Beyond the continental US, work such as

7 LaFeber, *The New Empire*, 65–67, 95, see note 5.

8 Ibid., 9, 100.

9 Ibid., 67, 95–100. It is important to note, however, that even as certain facets of British imperialism began to seem worthy of emulation, American policymakers in the 1890s and 1900s would forge their own colonial outlook, one informed by an overriding idea of an American imperial exceptionalism, discussed in Chapter 6. In addition, it is equally crucial to recognize that although New York City plutocrats certainly invested in the new botanic garden in a hopeful effort to ameliorate the effects of rapid immigration and economic upheaval, the Garden's history, though necessarily understood within the context of a city's elite searching for both order and civic prestige, does not consist of a simple narrative of successful efforts at social control.

10 Kew itself had begun assuming its imperial mantle in the 1850s through well-funded and well-documented comprehensive surveys of colonial domains; such work was seen as the logical extension of the concerns of economic botany and led, for example, to the development of the antimalarial drug quinine from the bark of the tropical cinchona tree, a product that was "as vital as the steamship" for the forging of late-nineteenth-century empire. See Drayton, "Imperial Science," 330–334, note 1.

that begun in 1898 in Puerto Rico with a grant from Cornelius Vanderbilt that allowed Britton to inaugurate a three-decade effort to botanize and display the products of this work in the service of the city and the nation.[11]

The crucial nationalistic "British" plant, was, ironically, a Western Hemisphere native, the *Victoria regia*, a massive waterlily whose rapid growth and crowd-pleasing ability to (apparently) support the weight of children on its leaves had made it a sensation at both Kew and the famed Crystal Palace in the 1850s. (Joseph Paxton, the Palace's architect, claimed inspiration for the design of the massive glass structure from the ribbing of the water lily's leaf.) Explorers in British Guiana at mid-century had seized upon the visual power and beauty of this unusual plant to stake a claim for Queen Victoria in northern South America. Its flower, along with the romantic tale of its provenance, eagerly recounted in popular British horticultural journals, made it a potent symbol of the extent of Britain's empire. The flower came to be closely associated with the only British colony in South America and was regularly featured as a symbol on mid-nineteenth-century maps of the region.[12]

The plant captured the attention of the American middle- and upper-class gardening public as well, although in the US, it lacked the nationalist associations that attracted British crowds to Kew. Recognizing the plant's drawing power, Britton had erected an aquatic garden in a court outside the main conservatory in which *V. regia* was grown during the summer months. Though identified by its British binomial for the Garden visitors, the plant's provenance in the Western Hemisphere was highlighted in the NYBG's guidebook.[13] That such a magnificent New World native had been appropriated as a symbol of Old World power must have irked the architect of the NYBG's "botanical Monroe Doctrine." Unable to disassociate the world-famous water lily from its imperial British propagators, Britton turned instead to the cactus, whose botanical family was extremely diverse; native to the American domain of botanical influence; and, crucially, equally charismatic. He sought to appropriate the visual majesty of the giant saguaro cactus and had renamed it for one of America's own regal patrons of science, Andrew Carnegie. Installed in the recreated desert climate of the cactus house (FIGURE 18) in the Conservatory, *Carnegiea gigantea* and its prickly cousins drew crowds of urban New Yorkers, who could be made newly appreciative of the power and scope of their civic garden specifically and American science in

11 See Brian Boom, "Botanical Expeditions of the New York Botanical Garden," *Brittonia* 48 (July–September 1996): 297–307.

12 D. Graham Burnett, *Masters of All They Surveyed: Exploration, Geography, and a British El Dorado* (Chicago: University of Chicago Press, 2000): 148–155.

13 The water lily was too large to store with other tender plants indoors over the winter and was grown from seed annually. See New York Botanical Garden, *Descriptive Guide to the Grounds, Buildings, and Collections* (1907): 17–18.

FIGURE 18 Part of the collection in the cactus house of the Conservatory, 1902.

general. With such plants on display, Britton planned a massive taxonomic portrait of the Cactaceae designed to secure priority for his brand of nomenclature and to establish a claim for the Garden in New World botany.

The Press Celebrates New York's Botanical Nationalists

As a result of Britton's efforts to draw crowds to the Bronx, NYBG became a frequent object of admiring attention for national gardening periodicals and the local newspapers of the early twentieth century. The director-in-chief regularly welcomed reporters from New York papers into the Garden and kept close tabs on the coverage his institution received in the popular press. Neither he nor his wife was hesitant to pen letters to the editors of the various periodicals that portrayed NYBG. Britton also published articles in the popular press in a dedicated effort to attract notice and increase the ranks of dues-paying members. He also courted potential sponsors through the Garden *Journal*, which was filled with reports on the progress of construction and information on researchers and students in residence.

In general, the publicity afforded the Garden by the New York papers and national magazines was quite positive. After the 1890s, when Charles Dana's *New York Sun* had sided with Charles Sargent in the controversies surrounding the Garden's founding (see Chapter 3), the city's established periodicals had begun championing New York's newest civic institution. In 1901, *Metropolitan Magazine* devoted an eight-page, illustrated spread to a profile of NYBG.[14] Readers of this piece were treated to large photos of the exterior and interior of the new Conservatory, as well as the façade of the Museum and some of the herbaceous grounds. The author, Cromwell Childe, devoted considerable space to recounting the Garden's genesis in a movement of public-spirited citizens and touted its rapid ascent to the ranks of leading scientific institutions. "It is not often," he wrote, "that an institution of this sort bursts into being completely equipped, with funds that are sufficient at least to carry on the work without worry, to provide for reasonable expansion. Yet that is precisely the situation at Bronx Park." The entire article reads as a promotional brochure; it might have been written by either of the Brittons themselves. Childe's civic pride was on display when he concluded that, although "the public in New York and elsewhere [had] thought too lightly" of the new Garden up to that point, it was nevertheless destined to become "both a showplace and a scientific institution . . . in the highest degree." In fact, Childe was a friendly acquaintance of the Brittons, and his assertion that "a city or a nation can have no feature of more exceptional interest and charm than a botanic garden of the highest type," along with his description of the Garden as "a people's breathing space, carried to its greatest point of perfection," reflect the bias and interests of a member of his class.[15]

If Childe's characterization of the Garden's finances was entirely too sanguine (a major dynamic at work during the Britton years was the constant search for funds), his overall enthusiasm was attributable to a widespread journalistic enchantment with the exotic overseas expeditions of NYBG botanists. Indeed Childe's article had begun not with a description of the beauty of the Conservatory or the spaciousness of the new Museum, but instead with an account of the whereabouts of several Garden scientists. "This month," Childe reported, "in five different quarters of the globe, five men are working in the cause of American science.... They are of the scientific staff of The New York Botanical Garden, representing the five expeditions that institution has sent out to far distant fields for the collection of specimens and for research." Childe enthusiastically highlighted the work of Britton's assistant Percy Wilson in Sumatra; Garden pteridologist (fern expert) Lucien Underwood in Puerto Rico; D .T. MacDougal in Montana; R. S. Williams in Bolivia; and Marshall Howe in the Canadian Maritime provinces.[16]

In the ensuing decades, other popular periodicals followed suit, sending writers to interview Britton and his staff or publishing full reports of the Garden's activities

14 Cromwell Childe, "The Great New Work of the New York Botanical Garden: What This Picturesque Institution, the Most Representative in America, Is Planning to Accomplish in the Near Future" *Metropolitan Magazine* (1901): 263–271.

15 Childe's allusion to public indifference (those who thought "too lightly" of the Garden) may have been a reference to the controversies surrounding the NYBG's organization in the 1890s.

16 Childe, The Great New Work, 263, see note 14.

penned by the botanists and horticulturists themselves. As might be expected of journalism at the turn of the twentieth century, these accounts are notable for their sensationalist portraits, designed to underscore the heroism of New York's intrepid botanists. These narratives often portrayed, in contemporary colonialist fashion, a dedicated white researcher confronted with an array of hardships, which often included encounters with members of local populations, who were depicted in a racist popular idiom typical of the time.

Each of these Garden-sponsored expeditions so spectacularly chronicled in the popular press was a part of Britton's broader plan to promote his vision of an independent American botany. The Garden's field workers would collect live and dried (herbarium) specimens, including cacti, that would be shipped to New York for study; naming; classification; and, when desirable, propagation and display. The process of naming plant specimens, however, was fraught with controversy, as the international botanical community struggled to adapt a universal standard to which all working taxonomists could refer, thus reducing the degree of confusion that resulted from the continual discovery of new flora. Britton was at the center of this debate, which raged for decades beginning in the late nineteenth century, and provided a charged context in which he promoted American botanical nationalism.

Britton's Role in the Nomenclature Controversy

The knowledge gained through the floral studies that proliferated throughout the age of European discovery was meant to be shared among an increasingly international set of scientific practitioners and government and colonial administrators. To this end, "floras" were compiled as accounts of plants of a given geographic area; "manuals" were prepared to aid in identification of an area's plant life; and "monographs" were published to delineate the extent and members of particular groups of plants. All of these projects were undertaken in part to provide a clear portrait of natural resources that, in turn, might be harnessed for human needs.[17]

Discussions regarding the proper conventions for assigning names to living creatures were not new to nineteenth century science. Because the traditional goal of natural history had been the cataloguing and ordering of all living creatures (and the non-sentient "products of nature" such as minerals), it is unsurprising that differences of opinion emerged among those conducting such surveys. In 1842, naturalists in the British Association for the Advancement of Science had agreed to reduce confusion in the international zoological community by establishing a set of rules that included the use of the first name given a particular specimen (the "law of priority"), the use of Linnaean binomials, and a standardized spelling. In a similar meeting in Paris in

17 George H. M. Lawrence, *Taxonomy of Vascular Plants* (New York: Macmillan Company, 1952): 6–7; 30–34.

1867, botanists attempted to follow suit.[18] The result was completely satisfactory to no one, and the attempt set in motion a late-nineteenth and early-twentieth-century botanical debate that resulted in broken friendships among botanists, estrangement among scientific institutions, and a series of internecine attacks on professional reputations.[19]

Worldwide, virtually every botanist, amateur or professional, agreed on the utility of a universally agreed-upon standard. Nevertheless, dissent over the nature of that convention was inevitable. Though the Paris Code of 1867 represented a crucial step toward international nomenclatural agreement, it never achieved the universal standard sought after by botanists throughout the Atlantic world. By the late 1880s, American botanists were openly questioning the Code's utility. E. E. Stearns, a leading member of New York's Torrey Botanical Club, noted the inherent confusion in the attempt to classify and assign names to the myriad species of plants explorers discovered in North America:

> Ingenious botanists have species at their mercy, and are free to combine, separate and recombine, and to christen the successive groups very much at their own sweet will. No conceivable law can govern the chaos of genera, for a genus is not so much a reality as it is a botanist's idea. A species, however or a variety even, is a tangible something, with at least a measure of positive identity and fixity, quite independent of botanists' fantasies, and the part of the name which belongs to it as a species is the one element of nomenclature which may have and should have positive permanence. Whether we like it or not, the fact remains that the *original trivial name* is the only foundation on which we can hope to build anything approaching a fixed nomenclature. To ascertain this as exactly as possible for all our North American plants, to confirm it when already in vogue, and to reestablish it promptly and firmly wherever necessary, is the self appointed task of an increasing number of American botanists.[20]

Overall, Stearns's goal was adherence to a rule of strict priority in botanical nomenclature: i.e., whichever name had been published first would remain the recognized scientific name for a given plant. Britton's own entrance into the nomenclature wars had come in the 1880s, when, as a member of the Torrey Club, he had begun work on a local flora of New York. This work, published in 1888, followed strict priority, which itself generated a considerable amount of controversy, much of it from English botanist

18 Paul L. Farber, *Finding Order in Nature: The Naturalist Tradition from Linnaeus to E. O. Wilson* (Baltimore, MD: Johns Hopkins University Press, 2000): 47.

19 H. A. Gleason, "The Scientific Work of Nathaniel Lord Britton," unpublished manuscript, New York Botanical Garden Archives.

20 Stearns, E. E. "The Nomenclature Question and How To Settle It" *Bulletin of the Torrey Botanical Club* 15 (1888): 230–235.

James Britten.[21] (Such adherence meant at times abandoning a name long in use to return to an earlier published epithet that was less commonly known; this was the crux of many objections to Britton's stance.)

By the early 1890s, Stearns and Britton decided to challenge the Paris Code's convention openly. Although they agreed that the 1867 standards had been a step in the right direction, they sought to modify them. Strict adherence to the rule of nomenclatural priority was to be of paramount importance in the revised convention, audaciously dubbed the American Code at the 1892 meeting in Rochester, New York of the Botanical Club of the American Association for the Advancement of Science (AAAS).[22] (Subsequent modifications in 1893 at the Madison, Wisconsin meeting of the AAAS Botanical Club and in 1907, in response to a new attempt in 1904 at an international code, retained Britton's ironclad commitment to priority.) Among the original signatories of the 1892 American Code, besides Nathaniel Britton, were John Coulter, H. H. Rusby (later the NYBG's curator of economic collections), USDA botanist Frederic Coville, Lester Ward, Frederic Clements, George Moore, and J. C. Arthur. Other key supporters of the proposed reform included Charles E. Bessey of the University of Nebraska.[23]

John Coulter, then president of Indiana University, was happy to support Britton in his effort to reform nomenclature, but hoped that the "community of botanists who publish" would abide by the American Code because he was "heartily

21 Nathaniel Britton gave a good deal of thought and took great care in countering James Britten's vocal opposition to the American Code, writing and rewriting a response and reading drafts of this response aloud to Mrs. Britton while on an extended research trip to England in 1888. Britten was a botanist at Kew (1846–1924; assistant at Kew herbarium, 1869–71; assistant at British Museum-Natural History, Botany Department, 1871–1909; and author of *Dictionary of English Plant Names*, 1878–86), When Britton finally met him in person, Elizabeth Britton recalled, "Nat told James Britten that he had not come to talk nomenclature—that it was a matter of principle with him and not of opinion and that until some argument was brought forward that would cause him to change his opinion—he intended to keep on." See Diary of Elizabeth Britton, manuscript in the Elizabeth Britton records, Personal Papers, New York Botanical Garden Archives.

22 "Proceedings of the Botanical Club of the Forty-First Meeting of the American Association for the Advancement of Science, Rochester, NY, August 18–24, 1892" *Bulletin of the Torrey Botanical Club* 19 (1892): 281–297.

23 "American Code of Botanical Nomenclature, 1907" *Bulletin of the Torrey Botanical Club* 34: 167–78 and C. E. Bessey to N. L. Britton, August 14, 1895 in Correspondence and Reports, Committee on Nomenclature, Botanical Club of AAAS, 1895–1896, New York Botanical Garden Archives. See also Dan H. Nicolson, "A History of Botanical Nomenclature" *Annals of the Missouri Botanical Garden* 78 (1991): 33–56 for a complete overview of the nomenclature wars written by a professional botanist. Definitive accounts of these nomenclature debates have already been written; see Sharon Kingsland, *The Evolution of American Ecology, 1890–2000* (Baltimore, MD: Johns Hopkins University Press, 2005): 40–61; and Richard Overfield, *Science with Practice: Charles E. Bessey and the Maturing of American Botany* (Davenport: Iowa State University Press, 1993): 101–107, 111–116, 187–88.

tired of the discussion, and hope[d] that this will settle it, [so] that we can all turn our attention to more important and harmonious things."[24] Coulter's desire to move on was largely a function of the degree of detail involved with nomenclatural reform. A century before e-mail, he, Britton, and other key members of the AAAS found themselves exchanging letter after letter, trying to compromise on the sort of important minutiae such as punctuation and capitalization that characterized the work of "publishing botanists."[25]

Despite Coulter's fatigue with continued disagreement, other botanists wrote Britton to offer their sincere congratulations. S. M. Tracy, director of the Mississippi Agricultural Station, commended the work of Britton's AAAS committee at Rochester, adding that he thought the international community would find common ground in the Americans' compromise and thus finally set "definite rules for work."[26] Other Britton allies, including Lucien Underwood, who would become a member of the NYBG staff and was instrumental in drawing up the new code, shared his belief that a special opportunity existed for American botanists to affect the conventions of the science for the better. "I believe more than ever," he confided to Britton, "in the power of American botany to swing the whole business as we think best if we go a little slowly and remember that ideas come to most Europeans by freight and have to be absorbed by capillary action in homeopathic doses." The reformers, Underwood contended, were moving in the right direction, but would have to advance their ideas diplomatically.

Frederic Coville of the USDA agreed, acknowledging that reformers would require the support of the international (European) botanical community in order to succeed.[27] As Underwood surmised, the issue of nomenclatural reform was a sensitive one, rife with potential disagreement. In the US, sharp differences of opinion were expressed and debated within the leading botanical journals. The editors of the *Journal of Botany, Botanical Gazette*, and *Science* engaged in heated debates over the wisdom of the reforms proposed by Britton's AAAS committee, going so far as to accuse one another of attempting to stifle discussion.[28]

24 J. M. Coulter to N. L. Britton, August 13, 1892, Nomenclature Correspondence, Nathaniel Lord Britton Records, Series One, New York Botanical Garden Archives (cited subsequently as Nomenclature Correspondence, New York Botanical Garden Archives).

25 Charles Sprague Sargent of Harvard's Arnold Arboretum, for example, wrote asking Britton about the possibility of keeping duplicate binomials (*Catalpa catalpa*, for example).

26 S. M. Tracy to N. L. Britton, September 21, 1892, Nomenclature Correspondence, New York Botanical Garden Archives.

27 F. V. Coville to N. L. Britton, September 3, 1895, Nomenclature Correspondence, New York Botanical Garden Archives.

28 See, e.g., the letter by Charles Reid Barnes, of the University of Wisconsin–Madison and an editor of the *Botanical Gazette*, to the editor of the *Journal of Botany*, August, 1895; and the *Journal of Botany* 33 (July 1895): 212.

By 1895, however, the reformers led by Britton were encountering resistance among not European botanists but from Americans working at Harvard. A Harvard publication carried an article in direct response to a reformist piece published by Lester Ward in the *Bulletin of the Torrey Botanical Club*. This "Harvard Circular," written mainly by Harvard professor Benjamin L. Robinson and signed by 74 botanists, contended that the rules of the American Code proposed at Rochester were not binding. They noted that there was no foreign support for those rules, and that key international authorities considered them to be in direct conflict with accepted rules from the Paris Code of 1867. The circular's signatories considered the Rochester rules too radical and likely to produce a great deal of confusion, because their strict adherence to the rule of priority would mean overturning many familiar plant names.[29]

Though Britton wanted to respond immediately in print to this challenge, other members of the AAAS committee were not certain what that response should be. The USDA's Coville counseled patience, specifically urging his allies to ignore the Harvard challenge, while reiterating his belief that European botanists were their most important audience. He predicted that their proposed reforms would sway international opinion within a decade, and he believed that the Harvard objections would be eliminated by acceptance of the new principles overseas. "It is evident," he told Britton, "that a great many of the signers of the Harvard circular are going to take the stand that they do not oppose the new code, but that they do oppose an *American* code. We can do much toward unifying American botanists by preparing openly for international acceptance of the principles."[30] Nevertheless, Coville considered the circular a personal affront to Nathaniel Britton.[31]

Of course, a key element behind the Harvard botanists' protest was Britton's continued determination to undermine Harvard's authority. His attempts to reform the rules of botanical nomenclature constituted part of his quest to assume the mantle of Asa Gray in American botany. Lucien Underwood recognized the degree to which these opponents "so greatly admired Asa Gray that they saw the entire reform movement as an attempt to discredit him personally." As long as "all systematic botany was the idea of one man [Gray]," Underwood noted, "such a system would work." But, by the 1890s, with "half dozen centers for systematic work in America instead of one," he contended, "there must be some system that is uniform and based on principles not chance. . . ."[32] E. L. Greene, another key Britton ally in this movement, vociferously challenged Gray's institutional descendants at Harvard, who con-

29 Overfield, *Science with Practice*, 111–112, see note 23.

30 F. V. Coville to N. L. Britton, September 3, 1895, Nomenclature Correspondence, New York Botanical Garden Archives.

31 Overfield, *Science with Practice*, 112, see note 23.

32 Ibid., 115, Underwood is quoted.

tinued to follow the lead of Kew botanists. Greene and Britton, as adherents to the rule of strict priority, adamantly refused to do so. As historian Richard Overfield noted, "to Gray the practice of strict priority [had been] letting rules dictate to good sense," whereas for Greene it meant "placing nomenclature under the 'rule of laws' rather than under the arbitrary control of a few persons such as Gray and Hooker." Gray's power over nomenclature questions in the US had been virtually unchallenged in his lifetime. Britton initiated this serious challenge to Gray's methods only after the latter's death in 1888.[33]

Nebraska's Charles Bessey penned a public response to the Harvard protest.[34] In a July 1895 article in *The American Naturalist*, he criticized opponents of the American Code for leaving too much open to the individual botanist, thus, undermining the concept of a uniform standard. Bessey, privately angered by the circular, went so far as to ridicule the Harvard botanists as unscientific, linking them caustically to opponents of Darwinism: "There are still some people who believe in the fixity and original independence of species, and hence of varieties, also, and for whom the facts of development and evolution have no significance. For such, [a proposed Harvard modification of the American Code] is a logical necessity."[35] To members of the Torrey Club, Bessey maintained that "there could be no middle ground," in the battle between the Cambridge botanists' viewpoint and that of Britton and his American Code collaborators. "Under the action of the simple and just law of priority [the Harvard names] must give way to earlier ones. No list of exceptions to the law of priority can remain stable."[36]

By the time the Botanical Club of the AAAS met again in August, 1895, Britton and Liberty Hyde Bailey of Cornell University were actively seeking to win over the Harvard botanists by including them more directly in their deliberations. They invited both C. S. Sargent and B. L. Robinson to serve with them on their committee; Sargent apparently agreed, but Robinson refused. The Buffalo, New York, meeting solved nothing, however, and the debate over nomenclature continued through the 1890s into the twentieth century, and throughout the formative years of NYBG. Crucial among the tenets of the American Code was the recognition that, as Underwood put it, "for the stability of nomenclature it is essential that each generic name be inseparably connected with a particular species as its type."[37] Type species of a given

33 Ibid., 101–107.

34 Ibid., 114.

35 Ibid. and C. E. Bessey, *The American Naturalist* (July 1895): 666–668.

36 C. E. Bessey to N. L. Britton, August 16, 1895, Nomenclature Correspondence, New York Botanical Garden Archives.

37 Minutes of the Nomenclature Committee of the Botanical Club of the AAAS, December 29, 1900 at Cosmos Club, Washington, DC, Nomenclature Correspondence, New York Botanical Garden Archives.

genus, Britton held, should be so designated by the botanical author by the use of specific epithets such as "officinalis, communis, vulgaris, verus, germinus or typicus."[38] This so-called "type method," along with the principle of priority being given to earliest published names, characterized the American Code as Britton promulgated it from 1892 throughout his career. A crucial facet of Britton's nomenclature system, this requirement for names to refer to actual "types" (which were herbarium specimens of plants to which a name was first given by the author of a new species), distinguished it from traditional European practice. Types were not part of European nomenclature because long European familiarity had created a sort of folk knowledge regarding plant names, thus, in the opinion of the majority of European botanists, obviating the need for type specimens.[39] Use of a "type method," one Britton contemporary noted, was a developing concept over the course of Britton's career and throughout the 40-year period of dispute over botanical nomenclature. Understanding of the type method was not universal in the botanical community. In nomenclature, this rule of using "types" was applied as follows:

> A species includes a large number of individuals. Exact study of these individuals sometimes indicates that they represent two or more species, rather than one only. In the division of this collective group into two or more species, which one shall bear the original specific name, and to which shall new names be given? The type method require[d] that the name shall be retained for the particular plant upon which the original author based his description of the species, and for all other plants associated with it in the same species. Similarly, in the division of a genus into two or more genera, the type-method require[d] that the original generic name be retained for one of the species known to the author of the genus, so that the name is used essentially in the sense of the author.[40]

By 1902, with agreement on proposed reforms still far from unanimous, Britton had urged the AAAS to form a larger commission, with permanent standing committees, each of which would consider nomenclature for certain families of plants. The duties of this expanded committee would include serving as a "court of appeal for questions of interpretation which may be submitted to it by the special committees," and the special committee of each subject [spermatophytes, algae, etc.] "shall be charged with the duty of typifying all generic names and synonyms under the code and shall submit to the General Committee a detailed report of such work in form for publication."

38 Meeting of the Nomenclature Committee at AAAS meetings in Pittsburgh, June 30, 1902, Nomenclature Correspondence, New York Botanical Garden Archives.

39 Personal interview, William Buck, PhD, Curator of Bryophytes, New York Botanical Garden, February 15, 2000.

40 H. A. Gleason, "The Scientific Work of Nathaniel Lord Britton" *Proceedings of the American Philosophical Society* 104 (1960): 213.

Members of these proposed standing committees included the usual roster of botanists, among them Elizabeth Britton.[41]

By early 1903, these committees were to have begun their first nomenclature-related tasks, selecting type species for every genus of North American plants (a job Coulter noted, that entailed "no end of drudgery"), yet progress was slow in coming.[42] By 1904, a compromise code hammered out at a Philadelphia meeting was ready for international consideration, and Britton sent out the newest set of proposed rules to John Briquet, secretary of the International Botanical Congress at Vienna for 1905. He also suggested that a permanent international nomenclature committee, similar to that set up by the AAAS, be established. Meanwhile, Harvard botanists also submitted nomenclatural rules that they hoped would be taken up by the assembled international community at Vienna.[43]

The Vienna meeting, held in June and attended by over 500 botanists, however, proved profoundly disappointing to Britton and his botanical allies. "The Nomenclature Congress," Britton reported to his cactus collaborator, Smithsonian botanist Joseph Nelson Rose, "rather wiped the floor with us."[44] Although the international gathering accepted the importance of the principle of priority, the Vienna Congress allowed for exceptions to Britton's vision that meant that the American Code remained essentially unreconciled with international standards; the "upper hand," wrote the NYBG's Henry Gleason (an American Code partisan) went to the Boston faction of US botanists.[45] Britton, angered, claimed that his potential allies had been underrepresented and effectively deprived of equitable voting power.[46] F. S. Earle, director of the Cuban Agricultural Experiment Station, wondered what possible tack the reformers might take in light of the Vienna Congress's failure to consider the Americans' proposed nomenclature rules. In a letter to *Science*, Earle expressed outrage at the reformers' perceived treatment at Vienna: "the absurd recommendation to make exceptions from

41 Included among these committee members were the names of many usual Britton collaborators, including David White, F. H. Knowlton, Arthur Hollick, Marshall Howe, George Moore, Tracy Hazen, F. S. Earle, Frederic E. Clements, O. F. Cook , C. L. Shear, J. C. Arthur, L. M. Underwood, Alex Evans, John Coulter, William Trelease, F. Coville, and H. H. Rusby. N. L. Britton, Nomenclature Correspondence files, 1902, New York Botanical Garden Archives.

42 J. Coulter to N. L. Britton, February 23, 1903, Nomenclature Correspondence, New York Botanical Garden Archives.

43 N. L. Britton to J. Briquet, May 9, 1904, Nomenclature Correspondence, New York Botanical Garden Archives.

44 N. L. Britton to J. N. Rose, July 26, 1905, J. N. Rose Correspondence, Smithsonian Institution Archives.

45 Gleason, "The Scientific Work of Nathaniel Lord Britton," 22, see note 19. Britton's response to the Vienna meeting can be found in his summary of the proceedings published as "Nomenclature at the Vienna International Botanical Congress" *Science* 22 (August 18, 1905): 217–219.

46 Ibid. (Britton)

the rules adopted in the case of over 400 generic names makes it morally certain that these rules will not be final and will not settle the vexed question of nomenclature. It also seems morally certain that these rules will not be even temporarily accepted by the majority of American systematic botanists." The American solution, Earle urged Britton, would be simply to "go ahead and do as we please." He boasted that "the Europeans will be falling into line all right twenty-five years hence." The USDA's Coville merely professed his "disgust" and hoped privately that "the Vienna Code would break its own back."[47] Given the continued work on such mammoth taxonomic undertakings as the *Flora of North America*, fellow reformer Frederic Clements was able to temper his disappointment with confidence, asserting that they should continue using the American Code as it had been most recently amended. It would then seem "inevitable," he contended, "that even a Harvard man will ultimately have to give way and become a little modern in spite of himself."[48]

Through the next five years leading up to the 1910 Congress at Brussels, where the AAAS was granted a single vote in the nomenclature discussions, Britton and his cohort had essentially done just what Clements and Earle suggested, publishing their voluminous taxonomic work using the American Code and tweaking the reformed rules as they went along. Britton had begun effectively promulgating the Code years earlier when, in 1898, he had published the last of his three-volume *Illustrated Flora*, a manual of botany aimed at a wide audience of both professional and amateur plant hunters in the Northeast. Britton's *Flora* enjoyed a key advantage over Asa Gray's rival *Manual* (then in its sixth printing): the fact that it was illustrated made it even more useful to avocational botanizers, who were potentially more likely to accept Britton's nomenclatural reforms. Britton had even introduced this flora with an explicit plea for his vision of reform:

> The necessity for these rules of nomenclature arose from the great confusion that has existed through the many different botanical names for the same species or genera. Some species have had from ten to twenty different names, and, worse still, different plants have often had the same name. For about 200,000 known species of plants, there are not fewer than 700,000 recorded names. Such a chaotic condition of nomenclature is not only unscientific, burdensome and confusing . . . but the difficulty . . . of identification which it causes . . . must make it . . . a serious and constant obstruction in the path of botanical inquiry.[49]

47 F. Coville to N. L. Britton, July 19, 1905 and F. Earle to N. L. Britton, August 8, 1905, Nomenclature Correspondence, New York Botanical Garden Archives.

48 F. Clements to N. L. Britton, October 25, 1906, Nomenclature Correspondence, New York Botanical Garden Archives.

49 N. L. Britton and Addison Brown, *An Illustrated Flora of the Northern United States and Canada, from Newfoundland to the Parallel of the Southern Boundary of Virginia, and from the Atlantic Ocean Westward*

Meanwhile, the Harvard botanists, including Sargent, continued the schism in American botany by adhering to the rules enumerated at Vienna. Sargent's dedication to avoiding Britton's reforms continued throughout his career. His Harvard colleagues followed suit. As late as 1917, a USDA forester complained to the director of NYBG that Sargent's "personal authority" and "eminence" in dendrology meant that the Vienna Code would by default gain adherents in the forestry community.[50] The First World War only strengthened Britton's conviction that an American Code was a necessity. He asserted to A. S. Hitchcock, a USDA grass specialist, that he would never refer to a "Teutonic-made code" as "international."[51] At war's end, however, Hitchcock professed a strong desire to heal the schism among American botanists:

> Those bitter-enders who might be prejudiced against the American Code, if this were adopted entire, will, I think, be ready to accept a code which represents the result of a careful consideration of all codes. The code as we now have it does not bear superficially much resemblance to the American Code, but fundamentally it is in entire agreement with that. I am very desirous that we should get the backing of the Botanical Society without stirring up old prejudices.[52]

Hitchcock subsequently managed to establish a permanent standing committee of the Botanical Society of America (BSA) to adjudicate questions of nomenclature. Such a committee, he felt, would enable American botanists to "investigate the subject of nomenclature de novo" and to consult all existing codes in an effort to arrive at a satisfactory system for all. Harmony and unification in the postwar world, Hitchcock felt, were of paramount importance, more so even than a strictly "logical" code that followed "rigid" rules.[53]

---

to the 102d Meridian 2nd edition (New York: Dover Publications, 1970). Nevertheless, at the same time that the Code enjoyed a boost from the popularity of the *Illustrated Flora*, it did not catch on with a large proportion of readers interested in forestry. Britton's 1908 manual on North American trees never achieved the popularity of Sargent's *Silva*, published three years earlier and written using the International Code. Gleason, "The Scientific Work of Nathaniel Lord Britton," see note 19.

50 George B. Sudworth to N. L. Britton, June 22, 1917, Nomenclature Correspondence, New York Botanical Garden Archives.

51 N. L. Britton to A. S. Hitchcock, March 11, 1918, Nomenclature Correspondence, New York Botanical Garden Archives.

52 A. S. Hitchcock to N. L. Britton, December 2, 1918, Nomenclature Correspondence, New York Botanical Garden Archives.

53 See March, 1919 Committee on Nomenclature of the Botanical Society of America, Nomenclature Correspondence, New York Botanical Garden Archives. This particular set of correspondence would be useful to botanists interested in highly specific objections and pleas by individual botanists throughout these controversies; particular personal positions and the evolution of stances by principal actors in the drama can be recreated by recourse to the series of memos circulated

It should be noted that despite the decades worth of work completed by Britton et al., the American Code as he envisioned it was never officially adopted as nomenclatural policy by the AAAS's Botanical Club. Given the very clear schism among North American botanists, such an endorsement was improbable. Even after the Great War, when Hitchcock attempted to heal the divide, the American Code remained highly controversial. A poll of 119 US botanists conducted by Hitchcock in 1919 revealed that 81 of them favored the International Code (with modifications), 23 preferred the American Code, and 15 were in favor of reconciling the two nomenclatural systems in some way. "From the above," Hitchcock concluded,

> it seems clear that it would be inadvisable for our committee to attempt to commit the Society in favor of either code. While the Society as a whole would undoubtedly give a majority vote in favor of the International Code, it could not carry with it the strong minority which is in favor of the American Code, and there would result a schism tending to disturb the harmonious relations of the members.

Hitchcock was being overly diplomatic; such a schism had been a fact of life for decades by the time he wrote those words.[54] Sympathetic to the logic of both sides, Hitchcock made clear the laudable motives behind each camp's adherence to its decades-old stance. Those in favor of the International Code preferred it precisely *because* of its international character, i.e., "an agreement among botanists is of greater importance [to them] than an ideal code to which botanists will not agree with a fair amount of unanimity." Followers of the American Code, he asserted, "feel that it is better to formulate a local or ideal code and trust to its merits finally to secure the adherence of all botanists." His personal sympathy, however, is perhaps in evidence in his use of biological metaphor as he characterized these respective positions. Whereas the International Code's adherents remained committed to "a compromise based upon tradition, historic evolutions, and custom," the idealists of the American Code camp were ready to make "a quick jump or mutation toward a result based upon the logical development of principles." True logic, he seemed to imply, dictated adoption of an ideal code, whose value would in the end be persuasive. Hitchcock's argument for conciliation was optimistic. Almost 30 years after the Rochester meeting, agreement among US botanists had yet to be reached.[55]

---

among the principals: L. R. Abrams (Stanford); Britton; Edward A. Burt; Alexander Evans (Yale); J. M. Greenman (Missouri Botanic Garden); Hitchcock (Bureau of Plant Industry, USDA); M. A. Howe (NYBG); C. L. Shear; and Witmer Stone (Academy of Natural Sciences, Philadelphia). Conspicuously absent were any Bostonians.

54 "Committee on Nomenclature, Botanical Society of America, Circular 2," July 7, 1919, Nomenclature Correspondence, New York Botanical Garden Archives.

55 Ibid. Note that Hitchcock, although sympathetic with Britton's ideals, could nevertheless see that a strict adherence to the rule of priority of publication in naming plants might in some instances require modification. "The principle of priority of publications is not fundamental (see principle 3

Yet again, the solution was to form committees: one to consider adaptations of the International Code, and one to consider modification of the American Code. Both committees, Hitchcock hoped, would arrive at ways to adopt the respective nomenclatural systems that would lead toward reconciliation of the two into a unified whole. Such changes could then be proposed at the next International Congress. The standing BSA committee on nomenclature would also, Hitchcock proposed, "introduce greater flexibility [into the American Code] but attain precision and uniformity."[56] In order to win over the greatest number of overseas adherents, Hitchcock proposed abandoning the name "American Code" in favor of "Type Basis Code," a name, he hoped that would

> make it plain that our code is not a mere variant of the Vienna Code, put forward by a small group of dissenting botanists, but is based upon principles fundamentally different from those underlying the Vienna Code, and is supported by approximately half the taxonomists of America. The concept of types being the most important difference, it seems desirable to have this indicated by the title. We hope to attract adherents among foreign botanists who shall appreciate the advantages of the type-basis concept and thus render the code international in scope.[57]

By the early 1920s, Hitchcock was confident that the time had arrived for compromise and incorporation of the key tenets of the American Code of nomenclature, if Britton and his allied botanists were willing to concede ground on somewhat more minor points, e.g., the retention of a list of certain names whose derivation did not strictly adhere to the American Code's rule of priority of publication. Based on this conviction, Hitchcock forged ahead, despite the misgivings of a stubborn Britton, who remained wedded to the principles as they had been set out in the 1890s.[58] Despite, or perhaps because of, such signs of international compromise and the apparent gaining of

in the Code) in the sense that it is absolute and indispensable, but rather is a convenient principle upon which the great majority of botanists may hope to agree. Strict priority is contrary to the usual evolution in terminology and may tend to obstruct healthy development." See "Committee on Nomenclature of the BSA," March 15, 1920, Nomenclature Correspondence, New York Botanical Garden Archives.

56 A. S. Hitchcock, "Committee on Nomenclature, Botanical Society of America, Circular 3," April 28, 1920, Nomenclature Correspondence, New York Botanical Garden Archives.

57 "Committee on Nomenclature, Botanical Society of America, Circular 4," 1920, Nomenclature Correspondence, New York Botanical Garden Archives.

58 A. S. Hitchcock to N. L. Britton, July 3, 1922; N. L. Britton to A. S. Hitchcock, July 14, 1922. He was able to issue Circular 7, April 24, 1924, under the aegis of Committee on Botanical Nomenclature, BSA. See also N. L. Britton to A. S. Hitchcock, May 16, 1924, Nomenclature Correspondence, New York Botanical Garden Archives. Britton reiterated his firm belief in a strict rule of priority in naming in an October 1, 1924 letter to T. A. Sprague of the Royal Botanical Gardens at Kew: "I have no faith whatever in an arbitrary list of generic names excluded from the operation of the rule of priority having any permanent stability, and I have always been looking for some means to

ground by the long-struggling American Code, some Harvard botanists remained firmly committed to thwarting Britton's notion of nomenclatural reform. In December of 1924, M. L. Fernald and C. L. Weatherby prepared a strongly worded six-page rebuke of Hitchcock's attempt at compromise:

> It certainly is not worthwhile for the majority of botanists to make undesirable changes in nomenclature in order to attract the adhesion of those who, by their past action, have shown unwillingness to accept majority decisions which do not wholly coincide with their wishes. By accepting the International Rules [i.e., Vienna code] those who had followed the Kew Rule gave up much more than the followers of the American Code would have had to give up had they also accepted them. *The International Rules themselves are a compromise,* [emphasis in original] reached by sacrifice on many sides. . . .

"Nevertheless," Fernald concluded in a conciliatory, if high-handed manner: "if the proposals are adopted by a truly representative and duly constituted international congress, we will, of course, accept them, as before." In this way, the decades-old New York-Boston botany argument continued into the 1920s with Harvard botanists maintaining their willingness to compromise, and attempting to draw a contrast between their own, principled democratic stance and that of Britton, who, they maintained, remained unwilling to negotiate even in the last years of his career.[59] For his part, Britton gradually withdrew from the fray, ceding the ground to Hitchcock in advance of his retirement. He remained pessimistic through the 1920s that agreement among botanists would be reached, at least along the lines he had proposed. "I do not believe now that any comprehensive and elaborate system of rules is going to have any permanent effect," he wrote British botanist G. Claridge Druce. "There is too much independence of thought to permit any general or continuous adhesion."[60]

Britton's persistence was to be partially rewarded four decades after he helped precipitate the sharp divide over plant names. At the International Congress at Cambridge, England in 1930, a "spirited debate" ensued, the result of which was the creation of a new International Code essentially "in near harmony," one Britton partisan later wrote, with the American Code.[61] Writing in 1991, Smithsonian botanist Dan Nicolson characterized Hitchcock, not Britton, as the key to these eventual

get rid of it." Nomenclature Correspondence, New York Botanical Garden Archives.

59 Nomenclature Correspondence, New York Botanical Garden Archives.

60 N. L. Britton to G. Claridge Druce, October 7, 1927, Nomenclature Correspondence, New York Botanical Garden Archives.

61 Britton himself played no direct role in these proceedings, as he was officially retired from the Garden. See Gleason, "The Scientific Work of Nathaniel Lord Britton," 212–213, note 40.

compromises forged in nomenclature starting in the late 1920s. After the 1930 Cambridge meeting, the schism in US botany was ended, with Brittonian ideas to some extent degree vindicated.[62]

Ironically, by war's end, Britton's objection to an international method of nomenclature and his botanical nationalism had adopted a different target. Whereas his traditional antagonism had been toward English principles of naming, by 1920 his particular "enemy" was now the "Teutonic" principles laid down at Vienna, though this bit of nationalism in naming may have been to some degree an attempt to gain allies through a bit of convenient, postwar posturing.[63] In response to Hitchcock's poll of US botanists, he wrote: "My dear Hitchcock, I expect that if you had indicated to members of the Botanical Society that the so-called International Code was a German-made proposition adopted at Vienna, you might have had a different series of replies."[64] Later twentieth-century systematists have characterized Britton's stance on nomenclature as "pretentious and jingoistic," and indicative of a destructive American "insularity" that by the 1920s had resulted in a diminution of US botanical credibility.[65] It is interesting to compare this characterization of a Brittonian "conservatism," with the "leftist radicalism" ascribed to NYBG director by George H. M. Lawrence:

> The leftist American faction [the American Code adherents] was headed by NL Britton, a scientist . . . not imbued with idolatry of botanical masters nor steeped in botanical lore and heritage. He entered the field of taxonomic botany unencumbered by sentimental prejudice or bias. Britton's ideas of absolute priority were logical, if neither sound nor practical. He refused to accept the more widespread view that nomenclature has a practical function, that of giving a concise reference to a particular concept (i.e., a name to a taxon). Priority is only a means to an end, and as others have pointed out, 'the happy idea of *nomina generica conservanda* [the idea of conserving well-known, if not prior, generic names] was the saving of a situation.

62 Nicolson, "History of Botanical Nomenclature," 39–40, see note 23.

63 Britton's jab against "Teutonic" botany might also have been a veiled criticism of his old nemesis, Harvard botany, many of whose late-nineteenth-century practitioners had had German training. Ironically, given this crude jingoism, Britton had been among the earliest of American naturalists to adopt German taxonomic reforms in the 1890s. Here again, Harvard played a role: B .L. Robinson and M. L. Fernald, editorial custodians of Gray's *Manual* after Asa Gray's death, did not revise that work using the Engler and Prantl classificatory system until 1908. This was over a decade after New York's botanists had begun publishing their own work based upon the German system, in a conscious attempt to undermine Gray's authority. See Overfield, *Science with Practice*, 187–189, note 23.

64 N. L. Britton to Hitchcock, July 12, 1919, Nomenclature Correspondence, New York Botanical Garden Archives.

65 David Frodin, *Guide to Standard Floras of the World* (Cambridge, UK: Cambridge University Press, 1984): 33–34.

Britton's Harvard opponents, Sargent first among them, Lawrence correspondingly characterized as a "rightist element… entrenched and steeped to a high degree with the botanical traditions of revered preceptors, of botanically endowed heritage, and of near-sentimental conservatisms; all laudable sentiments but characteristics that sometimes obscured the objectivity that makes for science."[66] However Britton's role in nomenclature reform is characterized (using either extreme of politicized labels), it is clear that his efforts were of a piece with his overall desire to promote New York botany in general, and, through the development of a nationalistic conception of taxonomic science, NYBG in particular. His decades-long attempts to establish his vision of botanical names were bolstered by his success in publishing literally thousands of those names in Garden publications and elsewhere. One of the most significant of these naming projects was *The Cactaceae*.

A Case Study: Botanical Collaborators and an American Monograph

As a scientist, Nathaniel Britton served as a transitional figure in New York City history. Trained in geology by J. S. Newberry, a scientific generalist and institution-builder, he came to the study of botany in a roundabout manner somewhat more typical of the early nineteenth century than the academically oriented 1880s and 1890s, when he forged a public career. As a young boy in Staten Island, Britton had maintained his own herbarium specimens, pressed and dried under the supervision of a helpful older neighbor. Later, as a young graduate of Columbia's School of Mines working on his PhD dissertation under the auspices of the New Jersey Geologic Survey, Britton was in some respects one of the last traditional "natural historians" to attain prominence in the city's academic and scientific history. He became adept at negotiating the developing bureaucratic world of modern American scientific philanthropy. And his commitment to the institutional success of NYBG led him to encourage the pursuit by younger staff members of the latest forms of botanical investigation. Nevertheless, by temperament and in practice, Britton remained a traditional taxonomist, concerned not with the teleology of natural theology, but instead with the construction of a naturalistic (botanical) nationalism. He sought to use his pulpit as director of the American Kew, in order to declare, in the words of a close colleague, a "botanical Monroe Doctrine."[67]

Britton augmented the free publicity afforded the Garden in various periodicals with regular coverage of botanical expeditions in the Garden *Journal*. Aimed at a wide audience of potential members, philanthropists, and visitors, these pieces were also an excellent forum to highlight his vision for a New York-centered American botany. Yet, these publications, as popular as they were, did not further these nationalistic claims within professional taxonomy. To do so, Britton identified plants he characterized

66 Lawrence, *Taxonomy of Vascular Plants*, 218–219, see note 17.

67 Henry A. Gleason apparently coined the phrase.

as "truly American," the cactus family, or Cactaceae, and set out to use the Garden's resources to write a definitive monograph that would unambiguously establish US botanical hegemony in the New World. Once collected and named in print, live specimens of these American plants would be grown in a spectacular conservatory setting, impressing urban visitors with the scope of New York science.

In the 1900s, Britton presented his particular brand of botany, taxonomic surveys of tropical and western American regions, as a useful scientific tool for the expansion of American knowledge and influence. This, in turn, would diminish US dependence upon Old World producers, repositories, and disseminators of natural historical knowledge. The long-term study of the cactus family that occupied a substantial proportion of Britton's professional time from the early 1900s until the 1920s exemplifies the ways in which Britton pursued this ambitious agenda for NYBG. "The Cactus Project," as it came to be known among the participants, was prosecuted by a pair of botanists convinced of the necessity of having American scientists and explorers collect and classify this vast family of plants native to the Western Hemisphere. Britton's clever use of available philanthropic resources subsequently furthered his aims for New York's botanical garden, whose scientific reputation and geographic influence were bolstered by the success of the four-volume monograph that resulted from Britton's collaboration with Joseph Nelson Rose.

Britton's relationship with Rose dated to the 1880s, when the Columbia botany professor and editor of the *Bulletin of the Torrey Botanical Club* sent Rose a herbarium sample and asked him to submit a contribution for publication in the *Bulletin*. Again in 1894, Britton wrote Rose in Washington, asking him to contribute a paper on the Liliaceae for a nascent project he was coediting with John Coulter, the *Systematic Botany of North America*. The Brittons' contacts in the capital city were numerous, and these types of exchanges of specimens and of information were typical of the network of correspondence that connected American botanists at the end of the nineteenth century. Both Brittons corresponded with a number of government botanists and agricultural scientists in the course of their taxonomic work at Columbia in the 1880s and 1890s. Botanists connected to the US National Herbarium at the Smithsonian and the Department of Agriculture were particularly helpful to both Elizabeth and Nathaniel Britton. Nathaniel corresponded regularly, for instance, with George Vasey of the USDA, while Elizabeth traded specimens with Edwin Willet, an assistant Secretary of the Smithsonian. Elizabeth considered her collaboration with the DC botanists particularly satisfying on a patriotic level: "I consider it a duty of every American student," she wrote Willet in 1893, "to place a type in the National Herbarium."[68] Meanwhile, her husband found his friendship with government scientists

68 E. Britton to E. Willet, February 13, 1893, Record Group 220, Box 3, folder 6, Archives of the Smithsonian Institution.

especially helpful as he catalogued and arranged John Torrey's old herbarium for Columbia's Botany Department in the late 1880s.[69]

The particular relationship that Britton forged with Rose, however, proved especially significant to the careers of both men. During their decades-long collaboration, the two men greatly enriched the herbaria of their respective institutions, while establishing well-deserved reputations as the foremost experts on what Britton dubbed a "truly American" plant, the cactus.[70] In 1904, Britton enlisted Rose and the resources of the National Museum for his cause: the enlargement of the living cactus collection of NYBG. Britton hoped to augment his collections in order to overtake the considerable cactus resources of France and Germany and, thus, to rank as the world's finest. The cactus, Britton argued, was native to the Western Hemisphere, and, thus, the proper object of sustained study by American botanists. "If he could have done so," the NYBG's Assistant Director Henry Gleason recalled, Britton "would have extended the Monroe Doctrine to botanical research. He lost no opportunity to secure to this country any advantages which might arise from priority."[71] This desire to secure priority both in scientific study and publication on American plants (and the most comprehensive and important herbarium and live collections of those plants in the world) became the driving force behind Britton's career.

By 1904, the Garden's cactus collection was the world's third-largest, mostly as a result of the NYBG's expeditions to the US Southwest. (NYBG's Assistant Director Daniel MacDougal's sojourns to Arizona, northern Mexico, and Baja California had been undertaken with the primary purpose of cactus collection.) And yet the ultimate end of cactus research, Britton proclaimed at the start of the project, was not simply the expansion of the NYBG's holdings or even the compilation of a definitive monograph on this particular plant family. Rather, he hoped that knowledge of the Cactaceae would enable botanists to aid in the reclamation of desert lands for agriculture.[72] The Carnegie Institution of Washington (CIW), in fact, had awarded a grant to famous California horticultural innovator Luther Burbank to pursue this

69 N. Britton to G. Vasey, February 6, 1886, Record Group 220, Box 3, Folder 7, Archives of the Smithsonian Institution.

70 In the case of the cactus, of course, the term "American" would have to include the entire Western Hemisphere, as cacti are native to North and South America.

71 Gleason, "The Scientific Work of Nathaniel Lord Britton," 54, see note 19. Gleason noted Britton's ability to get botanical diagnoses in print rapidly. In discussing the work Britton conducted in Puerto Rico (see Chapter 6), which overlapped with similar work by a German botanist, Gleason noted: "On some occasions duplicate collections [from Puerto Rico] were sent to New York and to Berlin. Before [Ignatz Urban] could complete his ponderous descriptions, and his erudite discussions, Britton had his brief diagnoses actually in print and the type specimens, christened with a Brittonian name, safely deposited in the herbarium of the New York Botanical Garden."

72 "A Plant Truly American," *The New York Sun*, Sunday, May 22, 1904.

FIGURE 19
Joseph Nelson Rose, date unknown.

goal. In the early 1900s, Burbank was involved in a 12-year hybridizing effort to cultivate a spineless cactus to be used as cattle forage in the desert Southwest. He never succeeded in developing an entirely spineless variety, although his work was chronicled by the press, and government experiment stations were inundated with inquiries about the potential of cacti to transform desert into productive range land.[73] Such grandiose aims, however, were most likely indicative of Britton's astute use of the press to further his own institutional goals, rather than a specific plan of action. Though undoubtedly the completed four-volume cactus monograph proved useful to settlers and planners in the arid southwest, Britton's primary concerns were priority of publication and the scientific prestige and renown that subsequently would accrue to NYBG.

In Joseph Nelson Rose (FIGURE 19), Britton found an accommodating junior partner. Where Britton remained both aloof and domineering, Rose, a "tall, good-looking man," maintained a "very quiet disposition" that enabled him easily to fill the role of dedicated scientific subordinate to the NYBG's director-in-chief. Britton served as the animating force behind his ambitious cactus plans, marshaling resources in a competitive funding climate, forging alliances among the nation's botanical institutions, and directing Rose in the writing of the taxonomic treatments that would become *The Cactaceae*.[74] Quiet, dutiful Rose, from his government post at the US

73 Peter Dreyer, *A Gardener Touched with Genius: The Life of Luther Burbank* (Berkeley: University of California Press, 1985): 139–143.

74 Henry Allan Gleason, "Thumbnail Sketches of Botanists" 1961, unpublished typescript, New York

National Museum (USNM) of Natural History and in a temporary capacity as a Carnegie Institution research associate, undertook the vast majority of the field work. Britton remained tied to his myriad administrative duties in the Bronx.

Born in 1862, Rose was only three years Britton's junior. During the late 1880s, he had studied under John Merle Coulter, taking his doctorate with a thesis on the Umbelliferae, a family of plants that includes such prosaic plants as Queen Anne's lace, carrots, and celery. Over the course of his active career in taxonomy, he established a solid reputation as yet another intrepid American field worker, journeying to Central and South America, the Caribbean, Mexico, and the American Southwest in pursuit of specimens.[75] By the turn of the twentieth century, Rose's position at the USNM made him an attractive candidate for a collaborative effort on a monographic study of these succulent plants. Britton had continued his correspondence with Rose through the 1900s, conferring with his government colleague on taxonomic matters. Occasionally, as his demanding schedule as Garden administrator and New York scientific entrepreneur permitted, Britton would journey south or west for field trips with Rose in the vicinity of the Potomac or the Delaware Rivers.[76]

Though the CIW did not start funding Britton and Rose's work on an official "Cactus Investigation" until 1911, the two botanists began their taxonomic efforts on these succulents in earnest in 1904. On his annual late-summer trips to the Caribbean, Britton had been gathering specimens of cacti; he also charged Garden explorers sent to the region with the same task. Simultaneously, Britton received shipments of cactus plants from various correspondents, including some professional botanists, who knew of his desire to supplement the public displays in the Conservatory's succulent house. As these collections grew, Rose proposed that they undertake an intensive effort to collect and study the Cactaceae and to publish the results in a definitive monograph. Though saddled with numerous responsibilities, Britton agreed. "I am inclined to think that [this project] is the most important thing that we can take up during the next two or three years," Britton said, and he promised Rose he would "do everything in [his] power to make it a success."[77] His estimate of the time it would take to complete the work proved profoundly optimistic: the final volume of the Carnegie-financed *Cactaceae* was not published until the early 1920s. Yet Britton

---

Botanical Garden Archives; Gleason, "The Scientific Work of Nathaniel lord Britton," 64, see note 19. *The Cactaceae* ultimately comprised four volumes, published from 1919–1924 by the Carnegie Institution of Washington. It remains in print as the definitive monographic treatment of the subject.

75 William Trelease, "Joseph Nelson Rose, 1862–1928," Science 62: 598.

76 See N. L. Britton to J. N. Rose, June 3, 1904, J. N. Rose Correspondence, Smithsonian Institution Archives (cited subsequently as Rose Correspondence, Smithsonian).

77 N. L. Britton to J. N. Rose, September 27, 1904, Rose Correspondence, Smithsonian.

readily committed himself to the effort, in the process ambitiously extending the reach of the Garden's floristic agenda in the Western Hemisphere. He had declared his botanical Monroe Doctrine.

Nathaniel Britton's eagerness to establish the Garden's scientific reputation did not entail a reckless disregard of financial costs. When faced with the opportunity to obtain more specimens, Britton consistently devised the most economical means of doing so. He allied the Garden with joint expeditions initiated by other institutions, so long as all financial obligations were clear, and the Garden was ensured of receiving its fair share of living and dried specimens in good condition. As he noted to Rose before a Baja California Expedition in 1904, "I think it will be necessary to restrict an agreement to pay for specimens, to the conditions that they be received here in good order, as I am not willing to waste money."[78] When, due to prior funding allocations, he was unable to contribute directly to the financing of such exploring trips, Britton agreed to exchange specimens from the NYBG's growing herbarium for new collections. A careful auditor, Britton also insisted upon receiving receipts for expenses incurred by collectors whose work he sponsored. Ever frugal with Garden funds (and consistently forced throughout his tenure as director-in-chief to raise money for various Garden needs as they arose), Britton insisted on getting what he paid for.

During the summers of 1905 and 1906, Rose embarked for Mexico, where he began collecting cactus specimens in earnest. In anticipation of a growing collection, Britton began simultaneously reordering the sets of collections in the Garden's Conservatory and propagating houses. Throughout his expeditions south of the Rio Grande, Rose sent shipments of collected plants north to New York, where they arrived by steamer and were met by Garden employees at dockside. Considering the logistics involved and the distance covered by these shipments, it is worth noting that Britton was pleased with the results of Rose's work; new plants continued arriving at the Bronx into the autumn of 1906.[79]

As Rose made his summer forays into Mexico, returning to his USNM curatorial duties in the winter while corresponding with New York on the sorting and naming of specimens he obtained, Britton continued sending Garden botanists into other regions with the same end in mind: increasing the NYBG cactus collections of living specimens for the conservatory displays and pressed specimens for the herbarium (which would enable the Garden to accrue prestige as the major repository of information on this American plant family). The Garden horticultural supervisor, G. V. Nash, spent a good deal of time in Haiti, whereas the Brittons made regular trips to other islands in the West Indies. Throughout these early expeditions, Rose

78 N. L. Britton to J. N. Rose, December 7, 1904, Rose Correspondence, Smithsonian.

79 N. L. Britton to J. N. Rose, October 3, 1906, Rose Correspondence, Smithsonian.

continually appealed to Britton for financial support, and the New Yorker responded with funds when he could. In light of the continual expense involved with a dedicated program of exploration and plant collection, it is unsurprising that Rose would choose to honor Andrew Carnegie, a major benefactor of the Garden and a patron of the Britton-Rose cactus efforts, with the commemorative name, *Carnegiea*, given with Carnegie's permission to a genus of giant cacti in 1908.[80]

By 1909, these collecting efforts had begun paying off in print. As a result of several years' work, Britton and Rose were prepared to publish a series of short taxonomic papers. These manuscripts, intended for the pages of *Contributions from the National Museum*, comprised a "complete reclassification of the old composite genus, Cereus," and contained several proposed novel genera.[81] When, due to a backlog in the Government Printing Office, the National Museum could not publish them immediately that summer, Rose requested Britton rush them into print in one of the Garden's journals. Timeliness, of course, was of paramount importance, because according to Britton's own staunch belief in nomenclatural priority, their new generic names would be worthless if unpublished. In addition, until such priority was established, Britton was prevented from publishing further material under the Garden's own imprint that was based upon this earlier work.[82] After a direct appeal to the Assistant Secretary of the Smithsonian, Britton was able to ensure the *Cereus* papers' publication by the Smithsonian, and the priority of his new taxonomic arrangements was assured. By the time Rose returned to Mexico in late winter of 1910, these botanical collaborators had established reputations as the foremost authorities on the Cactaceae.[83]

Rose's personal wealth shrank in direct proportion to the amount of time he spent in the field bolstering his reputation as a cactus expert. Repeated collecting trips to Mexico and South America had left him personally poorer, despite overall financial support of these trips from outside sources and a continuing salary from the National Museum. In the fall of 1910, even as he attempted to organize another ambitious sojourn to cactus-rich regions in Baja California, Rose confided to Britton that he

80 N. L. Britton to J. N. Rose, July 30, August 5, August 12, 1908, Rose Correspondence, Smithsonian. See also *New York Sun*, December 27, 1908.

81 F. V. Coville to J. N. Rose, response written on letter of J. N. Rose to F. V. Coville, May 14, 1909, Rose Correspondence, Smithsonian,

82 N. L. Britton to Richard Rathbun, Assistant Secretary of the Smithsonian Institution, May 26, 1909.

83 Although Rose spent a good deal of time in fieldwork during the years 1904–1910, he was not solely responsible for augmenting the cactus collections of the NYBG and the USNM. Both Britton and Rose were eager correspondents of others in the field whose collecting acumen was beyond reproach, accepting shipments of specimens in exchange or for a fee. In this way, Britton was able to keep the cost of his program of botanical reconnaissance to a minimum. See J. N. Rose to N. L. Britton, August 16, 1910, Rose Correspondence, Smithsonian.

would find it difficult to take part in further expeditions.[84] Unlike Britton, whose personal wealth (though modest by the standards of other NYBG founders) enabled him to make annual trips to the Caribbean or Europe throughout his career, Rose was entirely dependent upon his middle-class salary as a government scientist.[85] His ability to continue research and fieldwork was contingent upon the discovery of a more reliable funding source than either the Garden or the USNM. In early 1911, with help from an old NYBG colleague, Britton was able to ensure Rose's continued success. In doing so, of course, he also secured further renown for his own institution.

The New Politics of Funding: The Carnegie Institution of Washington

On January 24, 1911, Britton met with Daniel MacDougal, the former assistant NYBG director who, capitalizing on early fieldwork in the Southwest, had resigned in 1906 to establish the Carnegie Institution of Washington's (CIW) Desert Botanical Laboratory (DBL) in Arizona.[86] MacDougal agreed that Britton and Rose, with CIW support, should expand their efforts into a comprehensive study of the entire Cactaceae, engaging in a thorough survey of the Western Hemisphere and ultimately publishing a full systematic treatment in a series of well-illustrated volumes. Britton was nearly giddy at the prospect. This "would enable us to work together as much as we wanted," he wrote Rose. He predicted confidently that the completed work "would be the greatest thing of the kind ever done, no doubt, by the time we get through with it."[87] While Rose considered ways to accept MacDougal's inviting proposition without resigning his position at the Smithsonian, he accompanied another major expedition to Baja California under the auspices of the American Museum of Natural History. During the two months he spent on the US Fisheries Commission ship *Albatross* that spring, he collected still more cactus specimens for both the USNM and NYBG. Upon his return, Britton continued pressing him to find a way to accept the Carnegie Institution's offer.

During the summer of 1911, Britton and MacDougal conferred on a specific outline for the Cactus Investigation. From his summer laboratory at Carmel, California,

84 J. N. Rose to N. L. Britton, October 13, 1910, Rose Correspondence, Smithsonian.

85 Henry Gleason estimated Britton's personal wealth before the Depression at approximately $7 million. Britton, Gleason asserted, had been fortunate to inherit "considerable money, most of which was invested in high-class stocks." The Brittons lived "simply" at home (though Nathaniel enjoyed a reputation as an epicure), but when traveling in Puerto Rico, they stayed at a luxury hotel where, according to Gleason, they lived lavishly and Nathaniel was known as the "millionaire botanist." See Chapter 6 and Gleason's *Thumbnail Sketches*, 39, note 74.

86 On MacDougal, see Sharon Kingsland, "The Battling Botanist: Daniel T. MacDougal, Mutation Theory, and the Rise of Experimental Evolutionary Biology in America, 1900-1912" *Isis* 82 (1991): 479–509, and *The Evolution of American Ecology*, 70–77, note 23.

87 N. L. Britton to J. N. Rose, January 25, 1911, Rose Correspondence, Smithsonian.

MacDougal wrote Britton of his efforts to secure a salary and a paid assistant for Rose during the project. Rose would retain his position and salary at the USNM, but would take regular leaves from his duties to work as a Carnegie "research associate." Under pressure from the Carnegie board of trustees, MacDougal had predicted that Britton and Rose would be ready to publish their first volume within three years. Britton suggested they should begin the series with a treatment of the four genera *Pereskia*, *Pereskiopsis*, *Opuntia*, and *Nopalea*, arguing that this was the most feasible way to meet the promised deadline. Special consideration of further work on the economic importance of the genus *Opuntia*, they agreed, would have to await later volumes. Britton too would be offered the official title of Research Associate, though he would remain unpaid.[88]

The particular logistics of this ambitious undertaking were not as trivial as they might seem. By 1911, the Carnegie Institution had been in existence for only 10 years and had been successful in funding a series of scientific projects. But decisions to allocate support for certain research, while refusing assistance in other situations, were controversial ones within the organization. These decisions, in turn, had a significant effect upon the ways in which science was practiced in the US in the early twentieth century.

Upon its establishment by Andrew Carnegie in 1901 with an endowment of $10 million, the CIW had become the most important source of research funds for American scientists in a wide array of disciplines. When Daniel Coit Gilman, former president of Johns Hopkins University, entered office as the Carnegie Institution's first chief executive, he received a host of both solicited and unsolicited advice on how to disburse its largesse among the nation's researchers. In an academic landscape devoid of modern governmental and foundation support, Carnegie's endowment was a highly significant act of academic patronage.[89] Between 1903 and 1920, the Institution distributed over $100,000 annually in grants to researchers in various fields. With this font of funding in mind, Columbia University anthropologist Franz Boas was moved to describe the CIW as nothing less than an opportunity "to discover the principles and laws of human development with the view of utilizing them to regulate the present and to mold the future of the race."[90] Carnegie Board members and administrators, however, contended among themselves over the setting of funding priorities, making recourse to what they each perceived Carnegie's intentions to be.

88 D. T. MacDougal to N. L. Britton, June 16, 1911, Rose Correspondence, Smithsonian.

89 Robert Kohler, *Partners in Science: Foundations and Natural Scientists, 1900–1945* (Chicago: University of Chicago Press, 1991): 15.

90 ibid.; Nathan Reingold, "National Science Policy in a Private Foundation: The Carnegie Institute of Washington," in Alexandra Oleson and John Voss, eds. *The Organization of Knowledge in Modern America, 1860–1920* (Baltimore, MD: Johns Hopkins University Press, 1979): 313–341.

Carnegie's motives in establishing the Institution were relatively unclear, however. Although he made the emphatic claim that he intended to stimulate basic research, once the foundation was established, he left its day-to-day administration to the trustees and employees. The specific fields of inquiry to be funded, then, seemed open to debate, and the founder's wishes were not necessarily available guides to setting priorities.[91] Lacking a clear mandate, Gilman's tenure at CIW was marked by persistent conflict with two key trustees: C. D. Walcott, successor to the legendary John Wesley Powell as head of the US Geological Survey, and John Shaw Billings, founder of the New York Public Library.[92] With this dissension among a set of adversaries each powerful in his own right, Gilman remained unable in his brief presidency to enumerate a clear standard for funding decisions or to set institutional priorities.

His successor, Robert Woodward, who presided over the Carnegie Institution from 1904–1920, was able to achieve a degree of administrative coherence and executive leadership that had eluded Gilman. Woodward, a geologist and mathematician, had served as Columbia University's Dean of the Faculty of Pure Science. He had been chosen Gilman's successor with the aid of Billings, who presumed he had found in his fellow New Yorker a ready ally in administrative struggles. Over the course of the ensuing decades, Billings grew disappointed, as Woodward managed to consolidate his power and consequently to steer the Institution in a direction that was to some extent his own.[93] During Woodward's presidency, CIW administrators and trustees still competed to set funding priorities. Some of Woodward's advisors argued that the quest to discover those individual investigators that Carnegie had dubbed "exceptional men" worthy of financial support would lead naturally to the universities. Others contended that these exceptional individuals would likely be found outside what, by the early 1900s, had become the familiar contours of modern academe. For their part, many university-based researchers and administrators, echoing Columbia's Boas, saw the CIW as a bonanza. Eager to take advantage of this new resource, these men resented what they perceived to be outside meddling by CIW grant administrators. By the end of Gilman's tenure, the Institution had been granting a small number of fellowships to "promising" individuals and a larger number to "established" investigators.[94] Woodward, however, firmly believed that the best way to advance research would be to fund proven people, *either* at universities or in-house research depart-

91 Reingold, "National Science Policy," 314, see note 90.

92 Walcott later became the Secretary of the Smithsonian Institution as well as President of the National Academy of Sciences. See Reingold, "National Science Policy," 315, note 90.

93 Ibid., 323–324.

94 Ibid., 319–321. Kohler notes that ìacademics used to the laissez-faire of grants committees run by friends and colleagues were unprepared for a granting agency that had its own agenda and its own ideas about how academic science should be practiced." *See Partners in Science*, 16, note 89.

ments and laboratories.[95] The DBL, as the CIW's own botanical branch, would subsequently serve as sponsor and midwife to the Britton–Rose Cactus Investigation.

Prior to Woodward's arrival in Washington, USDA botanist Frederick V. Coville and US Forest Service head Gifford Pinchot had joined Britton as part of a CIW advisory committee on botany. Coville, an expert on desert flora, was convinced of a coming population boom in the American Southwest and had long been an advocate of a Southwestern botanical station similar to the tropical laboratory Britton established in 1903 on Jamaica.[96] To this end, Coville had enlisted the aid of the NYBG's Assistant Director MacDougal in a reconnaissance of Arizona in 1902. By the end of the year, Coville had persuaded the CIW trustees to appropriate $8,000 for the purchase of 40 acres of desert land outside the University of Arizona's Agricultural Experiment Station at Tucson and the establishment of research facilities.[97] In 1906, Britton's former assistant MacDougal was appointed Director of the new Carnegie division, the first of its kind.

MacDougal was a likely candidate for the position, for in addition to his knowledge of desert flora and the administrative skills he had honed under Britton in the Bronx, he had gained a degree of international renown in botanical circles for both his experimental work and his ability to invent and build useful pieces of laboratory equipment. During his tenure as head of the NYBG laboratories, MacDougal had published a paper suggesting that he had found evidence to support Dutch botanist Hugo deVries's mutation theory of evolution, making him, at the time of his appointment to head the DBL, a prototypical "exceptional man."[98] Soon after taking the new post, however, MacDougal found himself at odds with President Woodward concerning expenditures by the DBL for equipment and salaries; it was thus significant that he had been able to secure Carnegie funds for the planned publication of *The Cactaceae*.[99]

95 Reingold, "National Science Policy," 323–324, see note 90.

96 Ray Bowers, *Mr. Carnegie's Plant Biologists: The Ancestry of Carnegie Institution's Division of Plant Biology* (Washington, DC: Carnegie Institution of Washington, 1992): 3–4.

97 Ibid., 4. See also William G. Maginnies, *Discovering the Desert: Legacy of the Carnegie Desert Botanical Laboratory* (Tucson: University of Arizona Press, 1981) and Carnegie Institution of Washington, *Carnegie Institution of Washington: Founded by Andrew Carnegie* (Washington, DC: The Carnegie Institution, 1929).

98 Bowers, *Mr. Carnegie's Plant Biologists*, 5, see note 96. DeVries's mutation theory held that large abrupt changes in the genetic material of plants accounted for the relatively rapid appearance of new species. See the *New York Times*, October 4, 1908 piece triumphantly titled, "Dr. MacDougal's Remarkable Discovery Tends to Shake Evolution Theories—Man Able to Change Form and Colors of Flowers at Will," for an idea of MacDougal's contemporary reputation.

99 Bowers, *Mr. Carnegie's Plant Biologists*, 8, see note 96.

Despite MacDougal's tempestuous relationship with Woodward, his championing of the Britton-Rose monographic efforts was not necessarily the long shot it seemed. As noted above, Woodward, the twentieth century's "first modern manager of science,"[100] was convinced that the proper recipients of Carnegie support would be "proven specialists within an established discipline."[101] By 1911, with almost a decade of work on the Cactus family complete, Britton and Rose, like Luther Burbank, clearly met this standard and thus, by Woodward's own standards, merited CIW assistance. Whereas Burbank's work was conceived and funded with a specific agricultural goal in mind, the Britton-Rose project was understood to be a piece of necessary "pure" research into the taxonomy of this potentially valuable (and inarguably fascinating and showy) plant family. Funds for this ambitious publishing project would be devoted directly to the collaborators' work on the monograph's volumes. In keeping with Woodward's policy, financial assistance would *not* be granted directly to NYBG and could not be used to underwrite the Garden's other institutional aims. Of course, Britton, as a member of Columbia's Faculty of Pure Science and a Vice-President of the American Association for the Advancement of Science, was well known to the CIW president in 1911.[102] Thus, his appointment as a Carnegie "research associate" was in keeping with Woodward's overall agenda: the establishment of a special set of external researchers directly responsible to the CIW.[103]

Woodward's plan, however, was still at odds with the views of proper grant administration held by many of his trustees. These men contended that the Institution should search for relatively *unknown* "exceptional men" for whom Carnegie funds would make a crucial difference in their ability to conduct research.[104] This difference

100 Kohler, *Partners in Science*, 16, see note 89.

101 Reingold, "National Science Policy," 324, see note 90.

102 Besides his relationship with MacDougal and the DBL, Britton's personal connections to the CIW dated to the Institution's founding. In 1902, citing the "valuable economic features of [such an] investigation" and the "many new applications [of tropical plants] and their products in the arts and manufactures," Britton had applied for and apparently received $5,000 for a six-month botanical survey of the proposed "Nicaraguan Canal" region, the "Danish West Indies," and Puerto Rico, although there is no record at the Garden to confirm that these particular investigations took place. See N. L. Britton to Trustees of the Carnegie Institution of Washington, March 26, 1902, Carnegie Institution of Washington Archives. Britton also had been unsuccessfully proposed as a candidate to fill a 1903 vacancy on the CIW Board of Trustees by New York philanthropists C. F. Cox and W. E. Dodge. Carnegie Institution President Gilman politely demurred, asserting that he thought it best that "the Board should be made up chiefly of business men" not scientific men. See C. F. Cox to W. E. Dodge, January 29, 1903, W. E. Dodge to D. C. Gilman, January 29, 1903, and D. C. Gilman to W. E. Dodge, January 31, 1903, "Britton, Nathaniel L." folder, Carnegie Institution of Washington Archives.

103 Kohler, *Partners in Science*, 20, see note 89.

104 Ibid., 23.

of opinion had led to a decline in the number of overall grants from 1905–1909. Although the CIW Board continued to appropriate funds aimed at discovering obscure "geniuses," Woodward chose to award fewer and fewer of these grants, while increasing support (and the size of awards) to a relatively small set of accomplished researchers. In effect, Woodward was putting his own vision of a network of accomplished extramural CIW researchers, his research associates, into action.[105] In 1911, Carnegie donated another $10 million to the Institution, effectively doubling its initial endowment. Most of these new funds went directly to the new in-house CIW departments, including the DBL.[106]

In addition to its direct relevance for the success of Britton and Rose's cactus endeavor, these internal machinations at the Carnegie Institution are significant for the degree to which they illustrate the uncertainty of research funding in the early twentieth century. Woodward's struggle with his trustees also illuminates the role that financial support of professional activities played in the shaping of academic and institutional mores at organizations reliant upon CIW for support. Within these contexts, Britton and Rose appeared to be ideal research associates for Woodward's vision of the Carnegie Institution. As MacDougal prepared to secure Carnegie funds for the Cactus Project, Woodward continued to struggle with both his board and the wider university culture committed to its own standard practices of research and training of graduate students. Robert Kohler notes:

> Woodward, as head of a research institution, was bound to see grants as an extension of his institution's mission. Obsessed with the mechanics of the selection process, Woodward failed to understand that a workable large-scale system of grants would require accommodation with the realities of academic work.[107]

NYBG, with titular Columbia professor Britton at its head, was in fact in many ways an academic institution. Britton and his staff trained PhD candidates and conducted original research for publication. Yet, despite this role in postgraduate education, in other ways, the Garden's mission was more closely aligned with the priorities of a research institute run by business-oriented foundation executives. Aside from the obvious institutional similarities (e.g., NYBG and CIW shared a high profile wealthy founder, Andrew Carnegie), the Garden under Britton's management entertained priorities that extended beyond the training of new investigators. In addition, by funding Britton et al., Woodward would be following his stated aim of supporting those who "directed research, not those who were directed" and would

105 Ibid., 25.
106 Ibid., 26.
107 Ibid., 32.

be financing only one research project whose parameters seemed clearly defined and strictly delineated.[108] The success of these accomplished taxonomists and explorers in the decade leading up to their winning of CIW funds only increased their attractiveness as candidates for support. Finally, a project directed by Britton was likely to gain Woodward's support because of the extent to which Britton shared the CIW president's outlook. If Woodward's long tenure at the head of the CIW was marked by his attempts "through the extramural grants [the "research associateships"] to teach business ideals to academic scientists," in some sense, so also was Britton's reign at New York's own botanical research institution. Britton himself was a rigorous taskmaster with an auditor's keen eye for the logistical and financial details of expeditions and investigations undertaken by Garden scientists. He also shared with Woodward, on a smaller scale, the responsibilities of allocating a limited amount of funds among numerous, worthy, research priorities. Overall, then, Britton and Rose seemed poised to become, with MacDougal's sponsorship and Woodward's acquiescence, a Carnegie success story.

Starting the Cactus Project

Designation as a Carnegie "research associate," a semantic sleight-of-hand practiced by President Woodward in an effort to institutionalize his own funding priorities, did not carry with it the security most researchers associated with academic tenure.[109] Instead, as Joseph Rose discovered, his precarious situation continued despite this new source of income. In spite of the seeming fit between the respective outlooks of Britton and CIW president Woodward as the Cactus Project got underway, Rose discovered firsthand the difficulties attendant with serving as a CIW research associate. At the root of his troubles was, as always, money. As planning for the project began in earnest during 1911, MacDougal, under pressure from Woodward to minimize expenses, suggested to Britton that the Garden and the Smithsonian take responsibility for all expenses related to field work and traveling during their compilation of the first volume on *Pereskia*. Conferring with Rose, Britton acknowledged the impossibility of this suggestion. He suggested as a solution that Rose plan on living in New York during the summer, while they collaborated on the determination of specimens collected in the field. Britton attempted to cheer Rose in the face of this setback, assuring him that the "Bronx climate is better than that of Washington!"[110] With a family in Washington, however, Rose could not relocate to New York for any significant length of time; instead he and Britton resolved to begin their efforts while trying to keep to the initial annual budget of $7,000 that MacDougal was able to give them.

108 Ibid., 30.

109 Kohler, *Partners in Science*, 32–36, see note 89.

110 N. L. Britton to J. N. Rose, June 22, 1911, Rose Correspondence, Smithsonian.

With much of the internal jockeying and bureaucratic struggles within the Carnegie Institution apparently behind them, Britton and Rose could turn to the logistical tasks at hand. Accustomed during their earlier cactus work to cobbling together material support from as many institutions as possible, they continued to exploit whatever resources were at hand, even with the appropriation from Woodward at their disposal. Early in the spring of 1912, Britton contacted Charles Walcott, a Carnegie trustee and Secretary of the Smithsonian, asking him to help secure for the cactus project the use of US naval vessels in the Caribbean. With an ambitious scope for their taxonomic revisions, Britton informed Walcott that a thorough reconnaissance of the Caribbean basin was necessary. "One of the first things to be accomplished," he wrote, "is to obtain accurate knowledge of the earlier published species [of cacti], commencing with those of Linnaeus and Philip Miller." Hoping to begin explorations at Cuba and then proceed to islands off the northern coast of South America and the Dutch West Indies, Britton requested Walcott to inquire if naval vessels were available to land his men on the shores of these islands. Such expeditions, Britton assured the Secretary, "would make extensive collections of plants and botanical specimens which would greatly enrich the collections of the United States National Museum, and would undoubtedly add greatly to our knowledge of the West Indian flora and its distribution." He proceeded to outline for Walcott an ambitious 76-day itinerary that would involve landings at no fewer than 28 islands over the course of the following winter (of 1913). The party was to consist of five explorers, including Rose, Britton, two assistants, and a botanical photographer.[111] At MacDougal's urging, Walcott agreed to forward Britton's request to George von L. Meyer, Secretary of the Navy. However, Secretary Meyer demurred, postponing a definite answer until he could determine the availability of smaller Navy vessels of the type required.[112]

While waiting to see if naval support would be available, Rose traveled to Europe in 1912 (the first year of CIW support) to examine several key collections of living and herbarium specimens of cactus. During a three-month tour, he examined the extensive holdings at the Royal Botanic Gardens at Kew, where he was able to inspect several type specimens of American species. He then moved on to Berlin and subsequently to La Mortola (Hanbury Botanic Garden) in Italy, which at that time Britton considered "undoubtedly the best in the world." In this latter locale, curator Alwin Berger pledged complete cooperation with the CIW investigation and began shipping live specimens to the USNM. In addition to detailing Rose to inspect major European collections, Britton during 1912 plotted out a long-term strategy for the project. After the planned exploration of the Caribbean he had outlined to Walcott, he and Rose envisioned a "thorough" examination of the deserts of Argentina.

111 N. L. Britton to C. D. Walcott, April 24, 1912, Rose Correspondence, Smithsonian.
112 George Von L. Meyer to C. D. Walcott, May 3 1912, Rose Correspondence, Smithsonian.

Simultaneously, through a concerted effort to publicize the undertaking via methods such as the February 1912 announcement in *Science*, they sought to take advantage of a far-flung network of available botanical collectors, both professionals and reliable amateurs, in order to cover as much ground as possible. In 1912, Rose reported with satisfaction that "much material is being sent in by volunteer collectors and botanists throughout the southwest." In addition, the Carnegie cactus collaborators sought to make efficient use of government contacts by encouraging federal scientists to collect and send specimens to Rose in Washington or Britton in the Bronx. In that first year, Woodward's new "research associates" coordinated the efforts of approximately 20 US government explorers, who provided them with material whenever possible. Optimistic regarding their prospects of success, Rose nevertheless tried to be realistic about the effort required to complete a monographic undertaking of this scope. The following "two or three years," he predicted, would necessarily be devoted solely to collecting and organizing cactus specimens. Overall, however, he felt that a "fairly good beginning [had] been made," even though he and Britton had yet to venture into the field together under Carnegie sponsorship.[113]

By late fall, 1912, arrangements for the Caribbean fieldwork had yet to be finalized; the Navy had failed to commit resources to the project.[114] In planning their "gunboat scheme," the men carefully considered their potential crew members. Britton rejected the idea of including J. C. Arthur because of his poor health, instead proposing John Cowell, head of the Buffalo Botanic Garden and a frequent Britton collaborator; photographer Stewardson Brown; and William R. Fitch, Rose's USNM assistant.[115] By the end of the month, however, events in the Caribbean had become too unsettled for American foreign policymakers and the military for Britton to entertain serious hopes of securing Navy personnel for the expedition. Consequently, the NYBG director proposed traveling to the region on a commercial liner, a plan involving the expenditure of roughly $2,000 more than if they had been assisted by the Navy.[116] The extra funds, Britton promised, could be secured from the Garden budget.[117] In the meantime, Rose continued to receive specimens for description, photography or illustration, and taxonomic assignment.

Finally, in 1913 , Britton and Rose, accompanied by Elizabeth Britton, J. A. Shafer, William Fitch, Paul Russell, and Delia Marble, made it into the field. Leaving Assistant Director William Murrill behind to deal with the day-to-day administration of the

113 "Memorandum on Cactus Investigation," J. N. Rose, August 1912, Rose Correspondence, Smithsonian.

114 J. N. Rose to N. L. Britton, October 12, 1912, Rose Correspondence, Smithsonian.

115 N. L. Britton to J. N. Rose, October 18, 1912, Rose Correspondence, Smithsonian.

116 N. L. Britton to J. N. Rose, October 31, 1912, Rose Correspondence, Smithsonian.

117 N. L. Britton to J. N. Rose, November 6, 1912, Rose Correspondence, Smithsonian.

Garden, Britton and his party sailed for the Virgin Islands and Puerto Rico. On the overall itinerary were the islands of St. Thomas, St. John, Tortola, Virgin Gorda, St. Croix, St. Kitts, Antigua, Hispaniola, and Curaçao, in addition to Puerto Rico.[118] They arrived with 3,000 driers; six collecting boxes; approximately 9,000 sheets of paper upon which to mount new herbarium specimens; dozens of preserving jars; and several pounds of the preservative formalin.[119] In addition, they carried several letters of introduction to local dignitaries and to representatives of the US Navy in the region.[120] Secretary Walcott provided them permission of the Navy to land on the Puerto Rican island of Culebra, site of a former military base.

Once in the West Indies, the small expedition split into separate parties, with J. A. Shafer and Elizabeth's assistant, Delia Marble, accompanying the Brittons to the Virgin Islands and Puerto Rico, and Rose leading William R. Fitch and Paul G. Russell by local steamer to cover ground on the rest of the planned itinerary, including the islands of Hispaniola, St. Thomas, St. Croix, St. Kitts, and Antigua.[121] Since Puerto Rico had been added to the growing list of American possessions, interest in its tropical vegetation had attracted increasing attention from systematists. German investigators had begun publishing a *Flora Portoricensis*, but, as Britton noted with a mixture of disdain and relief, this 1911 work had been "a critical list of flowering plants and ferns only, but without descriptions... and thus not applicable to general use." This preliminary work on a general Puerto Rican flora had been supplemented by the work of botanists from Chicago's Field Museum of Natural History, Yale University, the US Department of Agriculture (USDA), and the newly established College of Agriculture at the University of Puerto Rico. Britton endorsed these latter efforts, pursued as they were by Americans. The "diffusion of knowledge" and preparation of a comprehensive natural history survey, Britton asserted, "would be

118 J. N. Rose, Memorandum on the Cactus Investigation, 1913, Rose Correspondence, Smithsonian.

119 This was a standard set of equipment for such an expedition. The "driers" were likely sets of felt blotters placed in between layers of plant specimens that themselves were sandwiched between layers of paper. These felt pieces would be removed regularly and dried in the hot tropical sun before being reused. The preserving jars were necessary to pickle specimens of succulent plants. I am indebted to William Buck of the NYBG for these descriptions. For information regarding the apparatus of nineteenth-century botanical field collectors, see *Tyler Whittle, The Plant Hunters: Being an Examination of Collecting with an Account of the Careers and the Methods of a Number of Those Who Have Searched the World for Wild Plants* (Philadelphia: Chilton Book Co., 1970): 109–123.

120 N. L. Britton to J. N. Rose, January 10, 1913; J. N. Rose to N. L. Britton, January 14, 1913, Rose Correspondence, Smithsonian.

121 See Rose, Memorandum, note 118. Once in the field, the Brittons would at times divide their personnel again if it was necessary for an expeditionary party to camp out overnight. Elizabeth is known to have stayed with her female assistant at a hotel, while Nathaniel spent time reconnoitering in the Puerto Rican mountains.

of great value, not alone to Porto Ricans [sic] but as an important contribution to scientific knowledge."[122] This trip, ostensibly part of the cactus investigation, also yielded a wide variety of Puerto Rican plant specimens.

The highlight of the spring 1913 collection season came for Britton on March 14. Upon reaching by automobile an altitude of 2,000 feet in the island's Cordillera Central, his party gathered specimens of what he later categorized as 213 species, a result he proudly asserted was "a record-breaker in botanical field work." Waxing nearly poetic in his account published in the NYBG *Journal*, Britton described the mountain setting as "an enchanting forest [where] orchids festooned and covered the tree trunks" and ferns grew in "great variety and luxuriance everywhere," while "mosses carpeted the banks" and "flowering shrubs and climbers were unexpectedly large." Caught in a passing tropical downpour, the Brittons and their companions were saturated, but dried off in "the subsequent sunshine" and returned to the town of Ponce after this "joyous experience" with an automobile laden with specimens.[123] Then, accompanied by William E. Hess of the Puerto Rican Agricultural Station and Franklin L. Stevens, Director of the College of Agriculture at Mayagüez, Puerto Rico, the Britton party continued exploring the island's interior and coastal regions, from mid-March through mid-April, 1913.[124]

Rose was pleased with the work the two exploring parties accomplished and asserted to Carnegie Institution administrators that together they had gathered "undoubtedly the largest collection" of living cacti "ever made in the West Indies."[125] Allowing for a natural level of hyperbole in a bureaucratic memorandum, Rose's assessment of their success was most likely valid. Never before had the resources of several established public and private institutions been devoted to the taxonomic reconnaissance of these islands. These efforts, Rose asserted, allowed the previously "meager descriptions" of earlier botanists to be elaborated, presumably to the benefit of both science and national interest.[126]

After the spring trip to the Caribbean, Rose journeyed in the autumn to both Texas and New Mexico, again hoping to corroborate and build upon the work of earlier generations of botanical travelers. He visited various regions in which type

122 See *Journal of the New York Botanical Garden* 14 (May 1913): 99–104.

123 In "Sastre-D. J. and Eugenio Santiago-Valent," "Botanical Explorations of Puerto Rico by N. L. Britton and E. G. Britton: Their Significance in Plant Conservation, Horticulture, and Education" *Brittonia* 48 (1996): 322–336; and *Journal of the New York Botanical Garden* 14 (May 1913): 99–104.

124 "Sastre-D. J. and Santiago-Valent," "Botanical Explorations," 329, see note 123.

125 Rose, Memorandum, see note 118.

126 Britton's expansive reach entailed, of course, the arrival of enormous quantities of living plant material destined for culture and display within the Garden's facilities. The importation of such material, Britton feared, might involve USDA quarantine procedures, something that by 1913 had to be taken into account when planning an expedition.

specimens of certain cactus species had been collected, hoping to discern with greater accuracy a sense of the range, habitats, and physical forms and variations among the various species he sought.[127]

Logistical considerations also awaited the explorers upon their return to their offices from the field. Rose, seemingly overworked as always, was continually requested to identify specimens sent him from correspondents throughout the country. Such tasks diverted his attention from the necessary postexpedition duties of cataloging, identifying, and describing collections garnered on each trip.[128] And yet these submissions also formed a crucial segment of the Cactus Investigation's field labor force. By 1913, over 100 collectors had voluntarily submitted specimens from around the Western Hemisphere to Rose's Washington office, in the knowledge that he and Britton were undertaking an ambitious floristic task.[129]

Work in succeeding years was meant to involve exploration of the arid regions of southern South America, as well as somewhat briefer journeys by Rose to southwestern Nevada and Utah, Arizona, and California. (Collecting in Mexico, which they acknowledged to be incredibly rich in cactus species, was hampered by political instability associated with the Mexican Revolution.[130]) The Carnegie appropriation for these purposes in 1914 was $8,000, with the proviso that NYBG provide an additional $1,000 for fieldwork expenses. In addition, the USDA agreed to care for the accumulated collections of living cactus specimens during the course of Rose's research. (Additional plants, of course, continued to be sent to the Bronx.) Desert Botanical Laboratory botanists also assisted, collecting succulents from locales throughout the American Southwest.[131] These cooperative arrangements, Britton noted with pride, met with the approval of Andrew Carnegie himself, who arrived at the Garden accompanied by his wife on a spring day in 1914. Britton took the opportunity to show off Mary Eaton's impressive cactus illustrations to the Carnegies, who inspected nearly 100 of her paintings spread across the tables of the Garden's herbarium.Carnegie reportedly expressed his "delight" with the progress of the overall investigation.[132]

127 Rose, Memorandum, see note 118.

128 See, e.g., J. N. Rose to N. L. Britton, June 3, 1913, Rose Correspondence, Smithsonian.

129 Rose, Memorandum, see note 118.

130 *Yearbook of the Carnegie Institution of Washington* 20 (1921): 381–382.

131 D. T. MacDougal to J. N. Rose and N. L. Britton, December 22, 1913, Rose Correspondence, Smithsonian.

132 N. L. Britton to D. T. MacDougal, April 25, 1914, Britton-Rose Cactus Investigation—Other Correspondence, New York Botanical Garden Archives (subsequently cited as Cactus Correspondence). It is interesting to note that Britton's adept courting of philanthropic attention through the use of the illustrated treasures of his library remains a practice in use at the Garden today. New York City foundation executives and potential donors are regularly escorted by Garden administrators on a tour of the library that includes the display of lavishly illustrated floras and finely rendered watercolor portraits of plants by noted botanical artists. These rare volumes

Presumably with a similar Carnegie benediction, Rose, accompanied by his wife who served as expedition photographer, sailed southward in the summer of 1914 toward the arid regions of Peru and Chile.[133] Steaming from New York's Pier 16 on the United Fruit Company's "Santa Marta," the Roses made stops at the botanical garden in Kingston, Jamaica and then at Panama. Landing in Peru almost two weeks later, they discovered as "near an absolute desert" as any place Rose had encountered, locating only three species of plants along the coastline where they had put ashore. Part of this denuded landscape, however, was due to Peruvian husbandry; flocks of sheep and goats "devoured practically everything in their path," he noted.[134]

Upon landing in South America during the Southern Hemisphere's winter, they often found themselves both short of funds and out of contact with either Washington or New York. (When a message did reach Britton in New York, it was a plea for additional funds; Britton sent a personal check to South America for $600.) Despite such uncertainties, the Roses established an initial base of operations in Lima and began collecting in Peruvian mountains up to altitudes of 4,200 meters, even as French and German nationals clogged the steamship liners out of Lima and headed back to their soon-to-be warring homelands.[135] They then made a subsequent journey inland as far as La Paz, Bolivia. After six months engaged in South American fieldwork and travel, the Roses returned home to Washington in late November of 1914, their itinerary extended several weeks by the exigencies of the outbreak of the First World War.[136]

---

are spread on the tables of the NYBG Library's reading room for inspection by the modern philanthropists.

133 As Britton and Rose each worked during the winters of their tenures as Carnegie Institution research associates to describe specimens, additional paintings and drawings were prepared by accomplished NYBG artist Mary Eaton, whose salary was partly derived from Rose's Carnegie grant. Eaton's masterful watercolors were intended to serve as illustrated plates in the resulting monograph volumes. Her role in the Cactus project was thus a traditional one, with precedent in Anglo-American and Continental botany dating back to the eighteenth century, when wives and daughters of naturalists were often encouraged to learn the techniques of botanical illustration as an adjunct to the written descriptive and taxonomic efforts of their partners. Such eighteenth century illustrations were intended to convey, according to historian Martin Kemp, "as precise an effect as [possible] of the unadorned presence of a real specimen." Eaton's skillful works filled just such a role. See Martin Kemp, "'Implanted in Our Natures': Humans, Plants, and the Stories of Art," in David Miller, ed., *Visions of Empire: Voyages, Botany, and Representations of Nature* (Cambridge, UK: Cambridge University Press, 1996): 199. Mrs. Rose's work as her husband's field photographer in 1914 might be understood to be a modern continuation of this botanical tradition. In addition, by 1920, Rose's daughter Rebecca served as her father's assistant in his Washington office, helping him with the administrative work involved in compiling taxonomic monographs. See J. N. Rose to N. L. Britton, June 22, 1920, Rose Correspondence, Smithsonian.

134 J. N. Rose, Memorandum, 1914, Rose Correspondence, Smithsonian.

135 J. N. Rose to N. L. Britton, August 8, 1914, Cactus Correspondence, New York Botanical Garden Archives.

136 J. N. Rose, Memorandum, 1914, Rose Correspondence, Smithsonian.

In total, this expedition, one of the most ambitious and productive of the entire Britton-Rose collaboration, netted some 100 jars of preserved specimens of succulents; 200 packets of seeds and bulbs; a collection of algae from the western South American coast; 150 containers of mosses; 500 specimens of parasitic fungi; and specimens of "Senecios," ferns, grasses, and lichens. These botanical spoils were divided among researchers at NYBG, USDA, Royal Botanic Gardens at Kew, University of California, Purdue University, Missouri Botanical Garden, and USNM. Britton's garden received the collections of living cactus plants for propagation and display in his Conservatory and utilitarian glasshouses. (Facilities in Washington for Rose's live collections were limited. The USDA reported a shortage of space in its greenhouses allotted to Rose). In addition, Rose had secured a small collection of animals, including invertebrates, fish, lizards, and even two mammals.[137] Rose returned to South America in 1915 and added plants from the eastern coast to his collections. Overall, these series of efforts resulted in the description of several new genera and species and confirmed Britton's suspicion of affinities between the floras of the Caribbean and the coast of South America. Britton hypothesized that such botanical links might be attributable to a "transoceanic distribution" of the "greater part of the flora" and idly entertained the thought of plotting the course of Caribbean hurricanes as means to identify them as possible agents of botanical transport.[138]

Typically entrepreneurial, Britton used the successes of the project to urge further fieldwork, both to ascertain the accuracy of his theory of biological affinities among regional floras and to collect information on numerous species of cactus that he was convinced had been "poorly described and often wrongly classified" by previous generations.[139] The answer lay in a fuller survey of the northern coast of South America, Britton asserted. "It would be a splendid investigation for the CIW to undertake, in cooperation with other institutions. The amount of novelty and of addition to the knowledge of botanical geography would be stupendous and . . . it would connect the

137 J. N. Rose, Memorandum, 1915, Rose Correspondence, Smithsonian.

138 See J. N. Rose to N. L. Britton from Bahia, Brazil, June 16, 1915 and N. L. Britton to D. T. MacDougal, August 16, 1915, Cactus Correspondence, New York Botanical Garden Archives.

139 Britton wielded a bit more prestige as he attempted to secure further funds, due to his 1914 election to the National Academy of Sciences. MacDougal congratulated him on this honor and commented upon the previous neglect of botany in the Academy's ranks: "We all note your accession to the National Academy with satisfaction. It seems to us high time that that body take some effort to remedy the grotesque aspect of its botanical membership, which is a reproach . . . in many ways." Britton, however, professed nonchalance, noting "the principal activity as regards the National Academy of Sciences since I was elected is paying them a fee of $5, which seems a good deal like belonging to other societies, though perhaps there is something different about it which I do not yet understand." D. T. MacDougal to N. L. Britton, June 7, 1914 and N. L. Britton to D. T. MacDougal, June 12, 1914, Cactus Correspondence, New York Botanical Garden Archives.

southern area covered by North American Flora with the flora of Brazil."[140] Rose agreed, calling the 1915 expedition "the most important one I have ever made."[141]

Meanwhile, the Brittons and a series of colleagues from numerous American botanical institutions made successful spring trips to Puerto Rico in 1914 and 1915. After the end of the First World War, they began making annual trips that continued throughout the 1920s. (These latter trips, however, were not directly associated with the CIW Cactus Project, but instead comprised part of the New York Academy of Science's Survey of Puerto Rico, the subject of Chapter 6.)[142] Continued exploration of South American regions for cactus specimens was hampered both by geography and by the ever-present need for funding. Britton, aware of the importance of optimistic accounts and predictions of research success, edited Rose's report for 1915 by removing the latter's candid assessment of future difficulty in pursuing cactus fieldwork. "We had expected to make large collections in South America," Rose had written, "but we did not appreciate the magnitude of the problem. We knew, of course, that the country [sic] is large, but we did not know that the transportation facilities were so difficult."[143] Britton continued to explore additional possibilities for financing further cactus expeditions. To supplement anticipated Carnegie Institution funds, he and Rose began targeting certain friendly US Senators, seeking direct federal subsidies. Britton addressed New York's Senator O'Gorman and had the NYBG's economic botanist Henry Rusby write the New Jersey US Senate delegation. In his letters, Rusby alluded to the unfulfilled work of Luther Burbank, emphasizing the useful side of the cactus investigation at Britton's urging. Such efforts appear to have been futile, however: the 1916 "Sundry Civil Bill" passed by Congress in June failed to include an appropriation for South American botanical fieldwork, and a similar failure occurred the following year. Funds from the Smithsonian were also scarce, and to complete his autumn 1916 sojourn, Rose was forced to appeal to private collectors willing to underwrite the trip in exchange for specimens. Meanwhile, MacDougal, a steadfast supporter of his former NYBG employer, continued to allocate Carnegie funds to enable Rose to make necessary trips from Washington to meet Britton in New York.[144]

140 N. L. Britton to J. N. Rose, in Rio de Janeiro, July 12, 1915, Cactus Correspondence, New York Botanical Garden Archives.

141 J. N. Rose to N. L. Britton, July 31, 1915, Cactus Correspondence, New York Botanical Garden Archives.

142 Sastre-D. J. and Santiago-Valentin, "Botanical Explorations," 329–334, see note 123. See also N. L. Britton to J. N. Rose, March 30, 1915, Rose Correspondence, Smithsonian.

143 J. N. Rose, "Recent Explorations in the Cactus Deserts of South America," (1915), typescript in Cactus Correspondence, New York Botanical Garden Archives.

144 J. N. Rose to N. L. Britton, June 16, 1916; June 22, 1916; June 30, 1916, September 12, 1916, and December 16, 1916, and N. L. Britton to J. N. Rose, June 19, 1917, Rose Correspondence, Smithsonian. In 1913, MacDougal had attempted to suggest to Britton that Rose relocate to New York for the balance of the project. Though Britton agreed, Rose, beset with personal and financial pressures stemming in part from the poor health of various family members, held fast to his Washington,

The Long Gestation of *The Cactaceae*

Despite Rose's private misgivings about the completeness and scope of their expeditions, MacDougal was confident that the research they had carried out was sufficiently complete for the preparation of publications. As 1916 approached, MacDougal noted to the Carnegie Institution's President Woodward that Britton and Rose had been conducting surveys of the cactus habitats of the Western Hemisphere for four years. Although some areas of South America remained underexplored, he assured Woodward that such gaps in the research would be filled by NYBG expeditions during the coming years. The first volume of *The Cactaceae*, he asserted, would be ready for printing in early 1916.[145] However, the manuscript was not submitted for publication until June of that year. Even as Rose submitted that volume, Britton remained unsatisfied and expressed doubts to Woodward as to the completeness of their fieldwork efforts in Argentina. Noting that the Botanical Garden had contributed "about half" of the expenses for the Cactus Project's fieldwork to date, Britton suggested further cooperation under Carnegie Institution auspices. Perhaps reluctant to continue a four-year project that had yet to yield published results, Woodward apparently never responded to Britton's suggestion directly.[146] Woodward's reticence might also be attributed to the sheer expense of the cactus investigation: by the end of 1916, when the first volume had yet to appear in print, the cost of publication, according to Woodward, had consumed fully one-fifth of the Carnegie Institution's publications budget for the year.

Delays in publication of Volume One were due in large measure to attempts of the CIW's editorial department to control expenses by reducing the number of illustrated plates to be included within the finished work. Carnegie publication funds were particularly scarce that year due to Woodward's efforts to bring out at least two posthumous volumes of the work of C. O. Whitman. As Woodward gently informed Britton at the close of 1916, "it has to be remembered . . . by some of us that the assumption of unlimited resources of the Institution is a myth."[147] Not surprisingly, Britton, Rose, and MacDougal were all disappointed by such cost-cutting measures, which added to their overall discomfort with the unfinished nature of their South American fieldwork. Britton nevertheless gamely replied to Woodward that he would rather see some text cut from the published work than lose crucial illustrations. Engaging somewhat humorously in sympathy with a fellow administrator, Britton commented dryly, "I perceive that even the President of the Carnegie Institution has to confess inability to make the same amount of money accomplish more than one

---

DC, base. See D. T. MacDougal to N. L. Britton, January 5, 1913, Cactus Correspondence, New York Botanical Garden Archives.

145 D. T. MacDougal to R. S. Woodward, December 8, 1915, Rose Correspondence, Smithsonian.

146 N. L. Britton to R. S. Woodward, June 8, 1916, Rose Correspondence. An annotation in Britton's handwriting to a copy of this letter reads, "This letter was not answered."

147 R. S. Woodward to N. L. Britton, December 22, 1916, Rose Correspondence, Smithsonian.

thing at a time. I had hoped that by this time he would have developed some method of removing this bothersome administrative restriction, but I conclude that we must now definitely abandon that hope."[148] Rose's relations with the Carnegie administration were more overtly strained, as Woodward scolded the Smithsonian botanist for questioning the finite nature of the arrangements the CIW had made to support him while he worked up the collections:

> It is impossible for me to understand how the completion of your work with us, in the time originally agreed upon and in accordance with terms originally proposed, should work to your professional disadvantage. It seems to me (and this is purely a guess) that possibly friends, or maybe interested parties, took the attitude and circulated reports that you were to be with the Carnegie Institution permanently, or at least for an indefinite period. I have, in fact, very good reason to believe that this attitude was taken. We did nothing to encourage such a belief.... Meanwhile let me say that I think you are taking a morbid attitude toward this affair, which is in itself doing you an injury.... Affairs in your Institution, or in the National Museum, must be in a very bad way indeed if this work is to your detriment.[149]

If MacDougal had in any way suggested to Rose that Carnegie Institution funds would support him throughout his career, Woodward was taking pains to disabuse Rose of such a notion.

Over the course of the next two years, Britton and Rose completed manuscripts and submitted them well in advance of their scheduled publication by the Carnegie Institution. The problems that plagued the release of Volume One continued to haunt subsequent volumes, as each was delayed by concerns over cost-cutting measures. Though careful not to antagonize Woodward directly, Britton was less judicious in sharing his anger at CIW editorial policies with Rose. *Cactaceae* editor William Barnum particularly annoyed the NYBG director-in-chief by seeking not only to reduce to a minimum the number of Mary Eaton's colored illustrations but also to impose a set of punctuation rules in the botanical citations that were at odds with Britton's favored conventions. As the first volume was prepared for press, Britton angrily informed Barnum that he was "not improving [the citations] at all, [since] the method which we are following in publications had been thought out by many in the past and all of your suggestions and modifications have been considered and rejected."[150] Given

148 N. L. Britton to R. S. Woodward, December 27, 1916, Rose Correspondence, Smithsonian.

149 R. S. Woodward to J. N. Rose, May 30, 1917, New York Botanical Garden Archives.

150 N. L. Britton to William Barnum, April 14, 1917, Rose Correspondence, Smithsonian. Barnum attempted to enforce his editorial viewpoints in his arguments with Britton by asserting that the monograph's volumes were to be targeted at a wider audience than a merely technical one. Britton and Rose prevailed in the controversy over punctuation and contraction of botanical synonyms by citing the precedent of the Kew Index as well as Britton's own work on the North American Flora. William Barnum to N. L. Britton, March 24, 1917 and April 12, 1917, Cactus

Britton's history of antagonizing any botanist who disagreed with him in questions of nomenclature, his chastisement of a Carnegie editor was entirely in character.

Ironically, these publication disputes afforded the final volumes a comprehensiveness they otherwise would have lacked. Delays allowed the cactus collaborators to continue adding new information to otherwise complete manuscripts, as new specimens were received from correspondents in the field. In addition, the collaborators were able to allow time for living specimens in their possession to flower in the greenhouses, thus giving them additional botanical characteristics upon which to base their taxonomic diagnoses.[151] Delaying publication too long, however, meant that the cactus volumes, once issued, risked obsolescence, because other botanists also were in the process of identifying and naming cactus plants, though not on the scale attempted by the Carnegie collaborators. By the start of 1918, Rose confided to Britton, "if the Carnegie Institution holds up this volume much longer, I fear we will have to re-write it."[152]

As they waited anxiously for their work to appear in print, Britton and Rose, ever entrepreneurial fieldworkers, cobbled together another funding package for a trip to Ecuador in 1918. Though Carnegie money was not forthcoming (for the two years 1917 and 1918, the Institution appropriated a total of only $1,600 for the Cactus Investigation), Rose secured funds from Secretary Walcott of the Smithsonian, a promise of USDA support from F. V. Coville, and a pledge of $500 from Harvard's Gray Herbarium.[153] Coville and his USDA colleague David Fairchild were particularly

---

Correspondence, New York Botanical Garden Archives.

151 See, e.g., J. N. Rose to N. L. Britton, July 23, 1920, and May 18, 1921, Cactus Correspondence, New York Botanical Garden Archives.

152 J. N. Rose to N. L. Britton, January 24, 1918, Rose Correspondence, Smithsonian. The cactus investigators were also shadowed by competition from a USDA botanist, David Griffiths, whose work on *Opuntia* was progressing simultaneously. Griffiths was notoriously jealous of the type specimens in his possession and in addition was slow in publishing his work. He was also an outspoken critic of Luther Burbank's efforts to develop a spineless cactus as cattle fodder. Rose appealed several times with varying success to Griffiths' superiors in the Bureau of Plant Industry to gain access to these collections. MacDougal proposed that Griffiths, the author of a 1907 USDA *Bulletin* discussing the use of cacti to feed livestock, be invited to pen a segment of the monograph on the economic potential of the genus. Britton's opinion of Griffiths' taxonomic skills was low, however, and he criticized the USDA cactus specialist for recognizing too many species in small geographic areas, i.e., for being a "splitter." Britton felt Griffiths should be taken "out of *Opuntia* and put into something in which he can do less damage." See N. L. Britton to D. T. MacDougal, June 3, 1911; D. T. MacDougal to N. L. Britton, June 16, 1911; and N. L. Britton to D. T. MacDougal, January 22, 1916, Cactus Correspondence, New York Botanical Garden Archives. See also D. Griffiths and R.F. Hare, "Summary of Recent Investigations of the Value of Cacti as Stock Food," USDA, *Bureau of Plant Industry Bulletin* 102 (1907); and Dreyer, *A Gardener Touched with Genius*, 193, note 73.

153 B. L. Robinson to N. L. Britton, May 7, 1918; J. N. Rose to N. L. Britton, June 14, 1918 and June 29, 1918, Cactus Correspondence, New York Botanical Garden Archives. In 1917, no appropriation was

interested in Rose's ability while in South America to secure seeds and specimens of cinchona and other economically useful plants.[154] Taking his son George with him as expedition photographer, Rose set sail for Ecuador from Havana in late July 1918. Along the way, the Roses collected in the Canal Zone and Haiti as well. Upon arriving in southern Ecuador, they retraced the route taken by Alexander von Humboldt over a century earlier, collecting 160 new species in the process.[155] Meanwhile, Britton continued to receive specimens from his own collectors, including J. K. Small, who scoured the cactus-rich lands of the Florida Keys in the spring of 1919.[156]

When at long last published in late June 1919, Volume One of *The Cactaceae* was well received by reviewers and the botanical community at large, whose attention the authors had courted assiduously.[157] Its appearance was heralded on page one of *Science* in an advertisement purchased by the Carnegie Institution. N. E. Brown, writing in *Nature*, hailed the Britton-Rose achievement as "a splendid work," that was "so excellently planned and so fully illustrated as to be a long way in advance of the very unsatisfactory German works that have hitherto held the field." Allowing for possible postwar anti-German sentiment, Brown's review nevertheless underscored

---

made at all. During the previous five years since the project's inception, Britton and Rose had been granted an average annual sum of approximately $7,500. In 1919 and 1920, the investigators would receive only $2,600 and $2,000, respectively, from the CIW. By 1920, the *Cactaceae* collaborators had been granted over $43,000 in Carnegie funds. This figure did not include the costs of publishing the monograph itself. Britton personally contributed over $2,000 in 1917 to finance a portion of the South American fieldwork. See J. N. Rose, Memorandum regarding the Cactus Investigation, 1920, in Cactus Correspondence, "John C. Merriam" folder, New York Botanical Garden Archives.

154 J. N. Rose to N. L. Britton, June 14, 1918, Rose Correspondence, Smithsonian. Cinchona had been the object of British botanical reconnaissance since the mid-nineteenth century.

155 J. N. Rose to N. L. Britton from Guayaquil, October 20, 1918, Cactus Correspondence, New York Botanical Garden Archives.

156 J. K. Small to N. L. Britton, March 9, 1919, Cactus Correspondence, New York Botanical Garden Archives.

157 N.L. Britton to J. N. Rose, June 7, 1919 and June 12, 1919, Rose Correspondence, Smithsonian. Subsequent to the issuance of Volume One of The *Cactaceae*, Britton and Woodward engaged in an extended exchange of letters regarding the distribution of free copies of the book. The tone of these letters barely masks Woodward's increasing exasperation or Britton's pique at not having enough copies to distribute as gifts to the Garden's Board of Managers. (Britton would get his way, however, as each member of the Garden's Board of Managers was granted a free copy in November, 1919. See Walter Gilbert to N. L. Britton, November 5, 1919, Cactus Correspondence, New York Botanical Garden Archives.) Britton later expressed embarrassment to Rose upon learning that the Royal Botanic Garden at Kew had had to purchase its copy of *The Cactaceae*. "I never supposed for a moment that the botany publications of the Carnegie Institution were not sent gratis to the most important botanical institution in the world." N. L. Britton to J. N. Rose, September 4, 1920. Two years after the publication of 649 copies of Volume One, 441 had been sold or distributed; Rose cited similar figures for Volume Two. See J. N. Rose to N. L. Britton, September 26, 1921, Rose Correspondence, Smithsonian.

the usefulness of the years of fieldwork and taxonomic labor that had gone into *The Cactaceae*. Plant descriptions were in English rather than Latin, and, thus, the monograph would prove accessible to "the multitude rather than to the botanist alone."[158] German botanist Alwin Berger shared Brown's enthusiasm, writing Rose that the volume was "simply marvelous [and] full of wonders." It would remain, he predicted, the standard work on Cactaceae.[159] Henry John Elwes, President of the Royal English Arboricultural Society and founder of the *Quarterly Journal of Forestry*, declared that Volume One was peerless in its field and particularly admirable in its combination of "the researches and knowledge of the botanical traveler, the cultivator, and the herbarium botanist." The genera described in this volume, Elwes noted, were exceedingly difficult, and he commended the authors on their meticulous field work, saying "whatever view future systematists may take of your conclusions, they will not be able to criticize them without access to an amount of material that I think is likely to remain unrivalled."[160] J. M. Coulter, then editing the *Botanical Gazette*, agreed, declaring the final four-volume work to be "undoubtedly the most comprehensive piece of cooperative work that has ever been undertaken in any botanical field. The results have been epoch-making, and will for a long time remain the standard in this group of plants."[161] Though somewhat formulaic in their praise, these reviewers were sincere in recognizing the crucial, if essentially thankless, taxonomic task Britton and Rose had undertaken.[162]

Subsequent volumes were published slowly. The second volume was not issued until September of 1920, and the manuscript of the third was still at the Institution's

158 N. E. Brown, "A New Book on *Cactaceae*" *Nature* (July 7, 1921): 580-581. The use of English in place of Latin, of course, had been a hallmark of Britton's career-long quest to reform nomenclature.

159 Alwin Berger to J. N. Rose, November 29, 1920, Cactus Correspondence, New York Botanical Garden Archives.

160 H. J. Elwes to N. L. Britton, November 11, 1921, Cactus Correspondence, New York Botanical Garden Archives. On Elwes, see Ray Desmond, *Dictionary of British and Irish Horticulturists and Botanists Including Plant Collectors, Flower Painters, and Garden Designers* (Bristol, PA: Taylor & Francis; London: Natural History Museum, 1994).

161 Cited in A. R. Rodgers, *John Merle Coulter: Missionary in Science* (Princeton, NJ: Princeton University Press, 1944): 287.

162 The monograph's appeal was not limited to the community of taxonomic botanists. Australian entomologists seeking a natural insect enemy of an introduced species of *Opuntia* that had overrun Australian wheat fields identified from Britton's work a potential cactus parasite. Though any economic applications of the cactus project would have been intended by Britton and Rose to aid American agriculture and horticulture, nevertheless, the Carnegie Institution hailed this promising application as a "hundredfold" justification of their expenditures. See the *Yearbook of the Carnegie Institution of Washington* 20 (1921): 381–382. Britton's response to the insect infestation of his live *Opuntias* was characteristically acerbic: "Damn that midge!" But he "hoped the Australians got something out of it."

printer a year later. As the editing work progressed, Rose continued to seek funds to complete his fieldwork in South America. Woodward's successor as Carnegie Institution president, John C. Merriam (1869–1945), was not inclined to appropriate funds for further exploration, hoping instead simply to complete the publication of the planned four volumes.[163] Though he assured Britton that the Cactus monograph was "one of our most important enterprises," Merriam wished to reorganize the Institution's botanical departments and was reluctant to commit further resources to a 10-year-old project.[164] By 1921, the Department of Botanical Research had been divided into Frederic Clements's research programs at Santa Barbara, California and Colorado, and MacDougal's labs at Tucson and Carmel. Overall, CIW plant biologists received only 6% of the Institution's budget, and of this, Clements received a third. Unsatisfied, Clements, the pioneer ecologist who developed the famous notion of a "climax" landscape, lobbied for his own, separately funded division. Simultaneously, H. A. Spoehr, hired by MacDougal in 1916, was gaining prominence for his experimental work on the nature of photosynthesis in cacti, and Merriam needed money to keep him in the Carnegie fold.[165]

Within this competitive context, it is unsurprising that by the autumn of 1921, Rose was somewhat disillusioned with his experience as a Carnegie Institution "research associate." He confided to Britton the extent of his disappointment at the CIW's refusal to underwrite further expeditionary work:

> with regard to finishing the Cactus Project I do not believe that Dr. MacDougal is any more anxious to complete this work than we are. Personally, I should be glad to close it up tomorrow and take up some other projects which I have long had in mind, especially the . . . Flora of Ecuador. I think, however, that we owe it to ourselves as well as to the Carnegie Institution to see that this monumental undertaking is brought to a successful conclusion. If we had not taken this stand in the past this whole project would have been a miserable financial and research failure. The delay . . . is due to the fact that we were not able to carry on the necessary field work at the proper time and thus settle the problems by means of specimens which we are vainly trying to solve from descriptions. During the past year I have been in correspondence with

163 J. N. Rose to N. L. Britton, June 8, 1911 and June 11, 1921, Rose Correspondence, Smithsonian.

164 J. C. Merriam to N. L. Britton, June 9, 1921, Cactus Correspondence, New York Botanical Garden Archives. Merriam, a friend of Henry F. Osborn, had spent his career prior to coming to Washington as a professor of vertebrate paleontology at the University of California at Berkeley. A prominent preservationist, he also served as president of the Save the Redwoods League from 1917 until 1942. See Susan R. Schrepfer, *The Fight to Save the Redwoods: A History of Environmental Reform, 1917–1978* (Madison: University of Wisconsin Press, 1983), 5, 45–50 and *The Dictionary of American Biography* (New York: Scribner's, 1946–58).

165 Bowers, *Mr. Carnegie's* Plant Biologists, 14, see note 96.

> more than sixty persons, endeavoring to get information regarding plants which I should have seen and studied long ago in the field.[166]

Rose's determination to conclude the operation was undermined by his belief that they would need continued Carnegie or Smithsonian funds to "bring out Volume Four properly."[167] Nevertheless, with an estimated cost of $1,000 to publish the last volume, Merriam advised Rose to close out the investigation and to consider further studies as part of a future project.[168]

The Significance of *The Cactaceae*

Despite the various administrative obstacles Britton and Rose faced as Carnegie collaborators in a competitive scientific environment, the entire four-volume monograph was ultimately published to general acclaim in the botanical community. By 1924, when the final volume was rushed through the press in two parts in order to head off the publication of several cactus species by a European competitor, Rose had regained some equanimity and was able to reflect on the entire partnership with pride.[169] Although an exasperated Britton asked MacDougal to inform him "whenever you have another little job like this cactus investigation," the published work had become the standard on the subject.[170] In the wake of the project, several pharmaceutical firms began efforts to develop drugs from the hundreds of species described in *The Cactaceae*. "It is beginning to look as though the Cactus monograph was not written in vain," Rose noted dryly to Britton. "Not only is there a large number of people taking up the cultivation of these plants, but an effort is being made to find a real use for them."[171] Rose had been correct in his enthusiastic 19 16 assessment of the project's prospects for success: "nothing like it [had] ever been attempted," and it had indeed become "the standard for future work of its kind."[172]

166 J. N. Rose to N. L. Britton, September 10, 1921, Rose Correspondence, Smithsonian.

167 J. N. Rose to N. L. Britton, November 21, 1921, Rose Correspondence, Smithsonian.

168 J. N. Rose to N. L. Britton, November 29, 1921, and May 25, 1922, Rose Correspondence, Smithsonian. Rose's continuing battle with CIW administrators throughout 1922 did not discourage him from attempting to initiate, with Britton, further large-scale survey projects. He sought to inaugurate a flora of northern South America, under the joint auspices of the US National Museum, the New York Botanical Garden, and the Gray Herbarium of Harvard in late June 1922. See J. N. Rose to N. L. Britton, June 29, 1922, Rose Correspondence.

169 See J. N. Rose to N. L. Britton, September 26, October 4, October 9, and December 14, 1923, Cactus Correspondence, New York Botanical Garden Archives.

170 N. L. Britton to D. T. MacDougal, May 17, 1923, Cactus Correspondence, New York Botanical Garden Archives.

171 J. N. Rose to N. L. Britton, August 15, 1924, Cactus Correspondence, New York Botanical Garden Archives.

172 J. N. Rose to William Barnum, June 29, 1916, Cactus Correspondence, New York Botanical Garden Archives.

Nevertheless, as the somewhat tempestuous relationships among the collaborating institutions and the personalities involved suggest, the entire Cactus Project was a case study of the ways in which extensive scientific projects were pursued and funded in the early twentieth century. As historian Nathan Reingold noted, research institutions and investigators generally resented efforts by the Carnegie Institution, one of the era's major scientific philanthropists, to become more than a disbursing agency.[173] In this sense, Rose's continued difficulties with the CIW's administrative hierarchy was typical of the strained relationships such tensions produced. On the other hand, Rose and Britton had been designated "research associates" within a specific Carnegie Institution subdivision, and thus their activities were subject to a somewhat greater degree of control by CIW administrators than mere grantees would have been. (Britton, of course, as an unpaid associate and an independent administrator of considerable stature, was less subject to pressure than Rose, who remained at all times dependent upon salaried positions to pursue his scientific career.) Such were the vagaries of finding funding for science in the early twentieth century. The most ambitious projects, even those formulated with overtly selfless, nationalistic aims, were contingent upon securing the endorsement of the network of men, including federal bureaucrats, foundation scientists, and private philanthropists, who were in positions to provide substantive financial and logistical assistance.

New World Plants for an American Audience

Nathaniel Lord Britton's conception of a specifically American botany, one managed in part from the offices of his own Museum, had much in common with the grand visions of the previous generation of American naturalists that had included both Asa Gray and Louis Agassiz. He also harbored the expansive ambitions of his contemporaries in the American biological community, especially those federal botanists and biologists with whom he formed working alliances and who sought to promulgate their vision of a biological science that helped to "culture" American life.[174] By the early twentieth century, of course, repeating the centralizing feats of Asa Gray, who had orchestrated American botany from his Cambridge home in the 1850s and 1860s, was impossible.[175] American botany had become far too diverse and was practiced at too disparate a set of institutions for any one figure to recreate Gray's dominance. Yet, even within this context, Britton situated his new institution near the center of American botany. His ability to secure support from a diverse set of individuals and institutions, along with his nationalist's agenda, brought NYBG a degree of renown

173 Reingold, "National Science Policy in a Private Foundation," 335, see note 90.

174 See Philip Pauly, *Biologists and the Promise of American Life* (Princeton, NJ: Princeton University Press, 2000).

175 Ibid., 97–99.

unprecedented for an American botanical garden. Displays of living, growing, and flowering cacti in the Garden's Conservatory would prove, Britton hoped, particularly effective at educating New Yorkers and Americans in general about the striking landscapes they either inhabited, influenced, or ruled.

Britton was the last of nineteenth-century New York's nationalist naturalists, the culmination of a venerable line of Americans intent upon appropriating the nation's biological diversity and splendor for patriotic purposes. Linkages between natural history and "national character," of course, had antedated the nineteenth century; ancient Greeks and Reformation-era theologians alike had sought a correlation, for example, between climate and political systems.[176] Subsequent writers as diverse as seventeenth-century New England clergymen and the Revolutionary-era republican Thomas Paine had appealed to the higher authority of American nature in attempts to discern natural laws of conduct. William Bartram had helped foster a new American wilderness landscape mythology by promulgating images of certain flora and fauna as symbols of national identity.[177] Other eighteenth-century figures in American letters had also yearned for scientific and ecological independence. Congregational minister and cartographer Jedidiah Morse had called for American independence in geographic knowledge. Morse noted that up until the 1780s, New World geography was virtually "unwritten and indeed very imperfectly known to anyone." Previously, Morse added,

> we seldom pretended to write, and hardly to think for ourselves. We humbly received from Great-Britain our laws, our manners, our books, and our modes of thinking; and our youth were educated as British subjects, not as they have since been, as citizens of a free and independent nation. The Revolution has proved favorable to science in general among us; particularly to that of the Geography of our own country.[178]

Others in the early republican era had struggled to establish US cultural independence in art, fiction, philosophy, and language, from Noah Webster's work in creating an American lexicography to Matthew Carey's reworking of post-Revolutionary British maps of North America. Each of these efforts had been acts of intellectual nationalism and provided Britton with strong precedents as he declared botanical independence

176 Clarence Glacken, *Traces on the Rhodian Shore: Nature and Culture in Western Thought from Ancient Times to the End of the Eighteenth Century* (Berkeley: University of California Press, 1967): 621–622.

177 Paul Semonin, "'Nature's Nation': Natural History as Nationalism in the New Republic," *Northwest Review* 30 (1992): 6–9, 12.

178 J. Morse, *Geography Made Easy* 20th edition (Utica, NY: Williams, 1819): iii, cited in John Rennie Short, "A New Mode of Thinking: Creating a National Geography in the Early Republic," in Edward C. Carter, ed., *Surveying the Record: North American Scientific Exploration to 1930* (Philadelphia: American Philosophical Society, 1999): 20–21.

from Old World knowledge. Ironically, he did so by emulating the most successful producers of Old World knowledge.[179]

If Britton's desire for hemispheric hegemony in taxonomy recalled Morse's call for independently derived American knowledge of American landscapes, it also echoed Thomas Jefferson's eighteenth-century rebuttals to the Comte de Buffon. Typically American organisms, including plants, soared to heights unknown in the Old World. Indeed reconstituted displays of the bark of "mammoth trees" had delighted visitors at great exhibitions on both sides of the Atlantic since the mid-nineteenth century. Just as Jefferson had speculated upon the existence of a giant Western woolly mammoth whose bulk would trump the comparatively puny anatomy of the Asian elephant, Britton portrayed the bristling, immense profile of *Carnegiea gigantea* as representative of the special importance of an American landscape, broadly defined.[180] Philadelphia museum entrepreneur Charles Willson Peale had sounded a similar note of naturalistic nationalism, arguing that "America has . . . a conspicuous advantage over all other countries, from the novelty of its vast territories. But a small number is yet known of the amazing variety of animal, vegetable, and mineral productions, in our forests. . . ." From such abundance, Peale contended, American museums could be stocked with specimens to make them among the world's finest. In this way, natural history became national history.[181]

Later, in the nineteenth century, Americans from Thomas Cole to Ralph Waldo Emerson attempted to foster American cultural independence from Europe by underscoring the uniqueness of Americans' relationship to their national landscape. American natural history provided an escape from cultural dependence, "since it allowed the new nation to go beyond its colonial tradition . . . to seek its national identity in the absolute moral authority of nature." Emerson "gave white Americans a new natural national ego by linking their destiny through a single metaphor to the

179 Short, "A New Mode of Thinking," 26–27, see note 178.

180 Jefferson's effort to rebut European degeneracy theory is found in his *Notes on Virginia*. See especially "Query VI: A Notice of the mines and other subterraneous riches; its trees, plants, fruits, etc.," in Adrienne Koch and William Peden, eds. *The Life* and *Selected Writings of Thomas Jefferson* (New York: Modern Library, 1972): 198–215. Jefferson's direct riposte to Buffon's controversial contention begins on page 206. See also Pauly, *Biologists and the Promise of American Life*, 18–19, note 174. In 1908, in honor of Andrew Carnegie, the joint patron of the Botanical Garden and the Carnegie Institution of Washington, Britton and Rose placed the saguaro into a newly created genus, *Carnegiea*, renaming the plant *Carnegiea gigantea*. See Daniel MacDougal to Robert Woodward, November 16, 1908 and Woodward to MacDougal, November 20, 1908, Daniel MacDougal files, Archives of the Carnegie Institution of Washington. See also Gleason, "The Scientific Work of Nathaniel Lord Britton," 54, note 19.

181 Peale cited in Christoph Irmscher, *The Poetics of Natural History: From John Bartram to William James* (New Brunswick, NJ: Rutgers University Press, 1999): 57.

powerful energy of geological time emerging with new authority in American society where the naturalistic assumptions of racial superiority were married to the evangelical conceits of a chosen people."[182]

There is insufficient evidence to trace such a coherent ideological position in Britton's thought. Yet his conception of American botany can be understood within the context of this series of cultural and historic precedents, begun in the earliest years of the new republic. Decades before Britton's career began in earnest, Harvard's Louis Agassiz had conceived a subscription venture to finance an ambitious series of *Contributions to the Natural History of the United States of America*, a publication with which he planned to "exalt the character of America," while raising US science to rival European standards.[183] Britton's efforts to chronicle the natural history of the vast, charismatic cactus family of the New World, coupled with his declaration of nomenclatural independence, comprise the last in this series of nineteenth century efforts to create and nurture a scientific nationalism.

In a private note to Joseph Rose in 1920, Britton acknowledged that Kew remained the most important of the world's botanic gardens, a notably candid admission from an administrator prone to institutional boosterism. In this narrow sense, his goal of creating a new Kew remained unfulfilled; his American garden had failed to supplant entirely the preeminent European one. Yet Britton's botanical nationalism nevertheless succeeded: he had managed to build up such a significant herbarium at NYBG that American botanists were freed from the complete dependence upon European collections that had characterized American botany just decades earlier.[184]

During the 1920s, Britton's last decade as director-in-chief, he was plagued by problems with the Garden's infrastructure and financial resources, even as he and his wife pursued distinguished careers through the last years of their lives. Recognition of the incompleteness of Britton's entire agenda, however, should not diminish an appreciation of what he, his staff, and his scientific collaborators managed to accomplish as they set out to build the Garden's international reputation as a preeminent botanical institution.

182 Semonin, "Nature's Nation," 26–28, see note 178.

183 Irmscher, *The Poetics of Natural History*, 234–235, see note 181.

184 Gleason, "The Scientific Work of Nathaniel Lord Britton," 61, see note 19.

Chapter 6
Metropolitan Botany in a Tropical Colony

Between 1895 and 1915, Nathaniel Britton and his staff built the physical garden and established the reputation of the New York Botanical Garden (NYBG) within the international botanical community. An imposing (if architecturally odd) Museum and an impressive Conservatory were erected and began attracting thousands of visitors annually. Projects such as the North American Flora and the *Cactus Investigation* were underway, as Britton collaborated with fellow American botanists in an effort to establish US scientific dominance in New World taxonomic botany. By 1915, when the Garden commemorated the twentieth anniversary of its city land grant, American botany had a well-established and active New York home. During the last decade of Britton's tenure, new structures were erected, and, when funds were available, older buildings were enlarged and modified in accordance with new institutional priorities.

Britton shaped the Garden's priorities during the 1920s to further its institutional prestige. To this end, he continued his entrepreneurial efforts, and organized a scientific survey of Puerto Rico and the Virgin Islands.[1] Organized under the auspices of the New York Academy of Sciences (NYAS), this project entailed coordinating the efforts of the major scientific institutions of the City of New York, including the Garden, the American Museum of Natural History, and Columbia University, in order to compile a comprehensive portrait of the natural history and anthropology of these new American colonies. The survey, along with the writing of *The Cactaceae*, represented the major professional accomplishments of Britton's NYBG career. In the course of rallying various American scientists to the cause of tropical investigation, Britton and the NYAS developed a "metropolitan botany," in which the priorities of an urban science (practiced within New York City and pursued by a team of cosmopolitan

1 The Survey's official title used the contemporary Anglicized spelling, "Porto Rico," to refer to the island. This spelling had been officially sanctioned by US Major General John Brooke, the first peacetime American governor of the island, and remained in regular use on all government documents and maps until 1932. See James Dietz, *Economic History of Puerto Rico: Institutional Change and Capitalist Development* (Princeton, NJ: Princeton University Press, 1986): 84–85. The colonial implications of this orthographic dispute are discussed in Gervasio Luis García, "I Am the Other: Puerto Rico in the Eyes of North Americans, 1898" *Journal of American History* 87 (2000): 49–51.

scientific nationalists) shaped the investigation of tropical US colonies.[2] Puerto Rican plants and geologic samples were collected, classified, and shipped home by the thousands for display in urban museums. In addition, Britton, his staff, and his survey colleagues became directly involved in insular development, lending their expertise to US-led efforts in Puerto Rican forestry, mining, agriculture, and higher education. This survey capped Britton's scientific career and extended through his retirement up until his death in 1934.[3]

In the mid-1920s, both Nathaniel and Elizabeth Britton maintained their own professional identities, pursuing annual tasks organized around an itinerant seasonal schedule that included taxonomic fieldwork and insular development efforts in the Caribbean, administrative duties in New York, and involvement in regional efforts to preserve native North American plants. These activities were complicated by the fact that by 1925, the Garden remained underendowed, with a physical plant showing unmistakable signs of wear. These administrative challenges coincided with, and were linked to, a perceived diminishment of institutional prestige, with a concomitant danger of scientific marginalization and subsequent philanthropic ostracization. Thus, as Britton entered his third and final full decade as director-in-chief in the 1920s, he and the institution he helped create faced obstacles as well as opportunities. It was in this context that his career as one of America's most prominent botanists concluded.

Taking Stock at an Anniversary

The year 1915 was a milestone for NYBG, and Nathaniel Britton was quick to seize upon its symbolic significance to garner attention and donations to his perpetually small operating fund. As Britton continued his cactus work, the Garden hosted a set of commemorative events marking the twentieth anniversary of the grant of the original 250 acres of New York City parkland to the NYBG Corporation. The anniversary celebration, in the form of an informal academic conference punctuated by planned tours of the Garden and other New York metropolitan area sites, filled five days from September 6 through 11, 1915. The festivities were intended to inaugurate an endowment campaign designed to net $1 million for the Garden. Guests were invited from the membership rolls of various regional and national botanical societies. Visitors were directed to exhibitions of Mary Eaton's watercolor illustrations of the Cactaceae and a new collection of Darwiniana purchased for the Garden from the estate of Scientific Director Charles Finney Cox. Planned excursions included day

2 I am indebted to Andrew Mendelsohn of the Centre for the History of Science, Technology, and Medicine, Imperial College, London as well as Jens Lachmund of the Max Planck Institute for the History of Science, Berlin, for their suggestions that helped me to articulate this concept.

3 The Puerto Rico Survey continued after Britton's death and was finally completed in the 1940s.

trips to the coastal dunes of Staten Island, the cedar swamps of Long Island, and the Pine Barrens in southern New Jersey.[4]

The year 1915 also brought approximately 150 additional acres of city parkland within the borders of the Garden.[5] With this parcel came the Lorillard family mansion, built in 1856, which served as additional administrative and exhibition space until it was destroyed by fire in March of 1923. The land, located along the Garden's southern boundary, was transferred to Garden control with the stipulation that a portion would be used for public concerts (a bandstand already stood on the site) and picnics. In addition, a five-acre parcel was set aside for the development of baseball fields, tennis courts, or playgrounds. The city officials making these arrangements knew that New York's growing population was willing to travel for recreation, a situation that, along with the advent of the mass-produced automobile, would lead by the 1920s to epic traffic jams on the east-west arteries of Long Island.[6]

The components of the original Garden plantation had also been augmented considerably over the course of 20 years. The herbaceous plantings in 1915 comprised some 130 beds of plantings, representing over 3,000 species; the fruticetum contained another 2,700-plus specimens. The deciduous arboretum had grown to include 340 species and varieties. Over 1,000 individual trees received the careful attention of Head Gardener George V. Nash's horticultural staff of 50 men. In addition, the Garden's collection of pine trees numbered over 1,500 specimens, representing some 303 species. Inside the Conservatory, Nash's staff cultivated an additional 9,240 species of tender tropical and subtropical plants. The total collection under glass exceeded 17,000 specimens.[7] The NYBG's research facilities had also expanded or had been rearranged to accommodate registered investigators, scholars-in-residence, and staff researchers. Sixteen visiting botanists (including five women) worked in the Garden's laboratories on the uppermost floor of the Museum building, making use of both the ever-grow-

4 See New York Botanical Garden, "Papers Presented at the Celebration of the Twentieth Anniversary of the New York Botanical Garden"; "Report of the Director-in-Chief for 1915," *Bulletin of the New York Botanical Garden* 8 (1912–1917): 85; N. L. Britton, "History of the New York Botanical Garden," address given September 6, 1915 at Anniversary Celebration; "The Twentieth Anniversary Celebration of the New York Botanical Garden," *Journal of the New York Botanical Garden* 16 (October 1915): 203–209, and E. Matzke, "History of Botany at Columbia University," manuscript at New York Botanical Garden Library.

5 "Report of the Secretary and Director-in-Chief for the Year 1915," *Bulletin of the New York Botanical Garden* 8 (1912–1917): 85.

6 Minutes of the New York Botanical Garden Executive Committee, January 15, 1915, New York Botanical Garden Archives. This land would remain part of the Garden until 1938, when it was reclaimed for road building. See also Robert Caro, *The Power Broker: Robert Moses and the Fall of New York* (New York: Vintage, 1975): 149–157.

7 George V. Nash, "Report of the Head Gardener for 1915," *Bulletin of the New York Botanical Garden* 8 (1912–1917): 113–118; *Journal of the New York Botanical Garden* 16 (May 1915).

ing herbarium as well as the expanding library, whose collection had surpassed the 27,000-volume mark.[8]

In 1915 alone, the herbarium absorbed 43,000 specimens. Most of these plants were collected on Garden expeditions throughout the Western Hemisphere. (In its first two decades, the Garden had sponsored over 100 of these research trips.) The amount of material on display had increased as well, as head curator John K. Small's staff added thousands of preserved plant specimens to the rows of museum cases in the synoptic collection. Simultaneously, Rusby's economic museum grew by 8,000 items. In addition, the walls of the museum had been adorned with a set of 200 botanical photographs, designed to illustrate "plant societies" and the "structural characters" of both flowering plants and larger algae and fungi. These educational efforts had been supplemented during the Garden's first two decades by the continuing program of public school nature study lectures; since the program had begun in 1905, approximately 115,000 of the city's schoolchildren had participated, with an average of 450 attending in any given year. Saturday afternoon lectures aimed at lay adults were also popular: a typical day brought 100 New Yorkers to the NYBG's auditorium. In sum, these and other improvements such as new plantings of rhododendron, willows, Japanese holly, conifers, and herbaceous borders had drawn an average attendance of 35,000–45,000 visitors on summer Sundays or holidays, an increase of 5,000–10,000 from the previous year.[9] Helping to draw these sizable crowds was a series of five horticultural exhibits hosted by the Garden in conjunction with the Horticultural Society of New York that Britton had helped revive.

After two decades of growth, NYBG had nearly reached the level of popularity that its incorporators had envisioned, although its dues-paying membership rolls were not as robust as Britton might have wished. Sixty-four annual members had been recruited in 1915, bringing the tally to 837. Members contributing greater sums were designated by various gradations of patronage (such as "fellows"). The total membership roster in 1915 surpassed 1,000. In addition, gifts and bequests from wealthy patrons enabled Britton to begin the construction of a large rose garden designed in part by noted garden architect Beatrix Farrand.[10]

Within less than a decade, however, this apparently successful institution would be scrutinized and found wanting by Frederic S. Lee, the Garden's new President. In conjunction with the Olmsted Brothers firm of landscape architects, Lee, upon

8 *Bulletin of the New York Botanical Garden* 8 (1912–1917).

9 John Kunkel Small, "Report of the Head Curator for 1915," *Bulletin of the New York Botanical Garden* 8 (1912–1917): 118–124; Arthur Corbett, "Report of the Superintendent of Grounds for 1915," ibid.; *Journal of the New York Botanical Garden* 16 (May 1915).

10 *Bulletin of the New York Botanical Garden* 8; Minutes of the Scientific Directors of the New York Botanical Garden, October 9, 1915, New York Botanical Garden Archives.

ascending to the Garden presidency in 1924, began a concerted attempt to remake the Garden in his own image, setting new scientific and public priorities for NYBG in the process. Before discussing Lee and Olmsted's negative assessment of the Garden's stature (the subject of Chapter 7), however, it is necessary to consider the context in which this evaluation occurred. As Olmsted and his landscape designers tramped along the Bronx River in the early 1920s, the Garden's scientists, led by Britton himself, were undertaking the final grand project of Britton's career, one that surpassed the Cactus Investigation in scope and ambition: the "Scientific Survey of Porto Rico and the Virgin Islands," referred to below as the Puerto Rico Survey or the Survey.[11]

Planning the Puerto Rico Survey

The NYAS was the lineal descendant of the nineteenth-century Lyceum of Natural History, where John Torrey and Asa Gray had begun their careers in botany. An Academy member, Britton conceived the Puerto Rico Survey in 1912, when he proposed sending teams of New York's scientists into these new US colonies in the Caribbean. His aims for this ambitious project were several: to compile a comprehensive catalogue of the islands' natural history; to assess potential for economic development; and, once again, to further the scientific prestige of his own institution.[12]

In his 1990 portrait of the NYAS, Simon Baatz discussed the role of the Puerto Rico Survey in the Academy's history as a metropolitan scientific institution.[13] For Baatz, the Academy's strength and its competence to sponsor such an ambitious colonial-scientific project lay in its ability to speak authoritatively for New York's various scientific institutions: from the august American Museum to smaller, specialized societies like the Torrey Botanical Club. Baatz recognized Britton as a crucial figure in this consolidation of New York City area societies. Cooperation brought greater publicity and guaranteed access to the Academy's publications, the NYAS *Annals* and *Proceedings*. Such consolidation, Baatz wrote, illustrated the changed structure of science within New York's urban setting. Britton had established the Botanical Garden on firm political and financial bases in the 1890s, while typically also acting in several other capacities: Columbia professor, librarian of the NYAS, and, eventually, organizer of the Scientific Alliance of New York. With the latter group, Britton sought to harness

11 Following contemporary conventions of American geographic orthography, Britton and his cohort adhered to the Anglicized spelling of the island.

12 In doing so, however, Britton coordinated a large contingent of American scientists and cast his net widely among his many contemporaries and acquaintances in botany and other disciplines in the city and along the Northeastern corridor. As a result, parts of the Survey took on the characteristics of a US government enterprise, but one managed by Britton from New York, under the joint institutional auspices of the Garden and the New York Academy of Sciences.

13 Simon Baatz, *Knowledge, Culture, and Science in the Metropolis: The New York Academy of Sciences, 1817–1970* (New York: The Academy, 1990).

the intellectual power of the increasingly diverse array of specializations that characterized early 1900s New York science. Britton's crucial role in each of these ambitious activities led Baatz to describe the NYBG director, somewhat hyperbolically, as "indubitably the leading individual in New York science."[14]

With intense competition for philanthropic support, Britton decided, Baatz says, to concentrate his efforts on his first priority, the Botanical Garden. Such a decision was a wise one, given contemporary tensions between the older avocational ideal of natural history and the rising class of American professional scientists, who lamented the relative dearth (compared to Europe) of public and private support for 'pure' scientific research. Efforts to secure support for the Scientific Alliance from the Samuel Tilden Trust (whose funds went instead to the New York Public Library) were evidence, Baatz notes, of the mutually supportive but delicate relationship between professional scientific disciplines and the private New York City wealth upon which they were contingent in Britton's day.[15]

Academy Vice-President Henry F. Osborn, Britton's Columbia colleague and eventually his counterpart at the American Museum, also recognized the role New York philanthropy would continue to play in underwriting expansive scientific projects. The New York Academy's success in cultivating such support, Baatz writes, was a function of Osborn and Britton's ability to articulate scientific and institutional concerns in terms of public necessity and civic prestige. Though the Scientific Alliance was short-lived, other efforts of the Academy and the Botanical Garden, including the Puerto Rico Survey, succeeded due to their administrators' ability to sell them to a wider public. Britton promised his patrons a body of knowledge that would advance metropolitan and national interests.[16]

Britton's Scientific Alliance failed but was subsumed within the NYAS, which benefited, Baatz notes, from the influx into its membership of an increasingly professional scientific elite: the Columbia professoriate, curators at the American Museum, and botanists from the Botanical Garden. Of the several active units of the Academy, the biological members surged to prominence under Britton and Osborn's leadership. Osborn forged a particularly strong relationship between the Academy and Columbia's biological scientists, whose research focused upon investigations of evolutionary mechanisms at the cellular level. Baatz writes that the increased prominence in the Academy of men such as T. H. Morgan and Edmund Beecher Wilson had the potentially dangerous effect, however, of alienating the crucial support of the lay members of the biological section, upon whom the Academy was dependent for

14 Ibid., 139–142.

15 Ibid.

16 Ibid., 170–171.

financial support.[17] Although other sections of the NYAS, including James Furman Kemp's geological group, managed to retain popular appeal and public accessibility, the general direction of the institution was toward greater specialization and professionalization.[18]

In 1912, as Academy members considered efforts to appeal to a wider public, Britton presented his ideas for the Puerto Rico Survey. This audacious plan, he claimed, would excite public enthusiasm and, thus, assure it of some measure of financial support. Academy president Emerson McMillan, a wealthy banker and amateur scientist, backed Britton with a sizable contribution of his own money. Britton and McMillan believed that the Survey would be sufficiently broad to appeal to all members of the Academy, including those scientists whose professional interests could be advanced in the new American colonial arena.[19] Britton understood the particular attraction of Puerto Rico for such a study: it was situated in a tropical region of the Western Hemisphere (the "American Mediterranean") that he himself had designated the special province of American scientific efforts, and it had not been surveyed systematically during the first decade and a half of American occupation. In addition, the island was developing rapidly, as the US military and American sugar corporations constructed roads and the American-controlled insular government established agricultural experiment stations and a public land-grant university.[20] In conceiving the survey, Britton devised a clever strategy to retain public support for the Academy (and by extension the Garden), *despite* the context of increasing scientific professionalization in which the Survey itself would be conducted. Overall, as Baatz writes, Britton was "exceptional in his ability not only to pinpoint challenging projects that would capture the imagination and support of his peers in the scientific world, but also to ensure that the requisite organizational and administrative tasks were successfully completed."[21]

The Puerto Rico Survey affords an opportunity to consider the extent to which certain metropolitan values, including middle-class ideals of expertise, leisure, and economic development, were fostered by metropolitan scientists.[22] Britton knew that his surveyors' work would culminate in the publication of a body of expert knowledge that could be used to promote American national interests. The Survey's planned scope,

17 Ibid., 185.

18 Ibid., 194–195.

19 Ibid., 202–205.

20 Ibid., 202–205.

21 Simon Baatz, "Imperial Science and Metropolitan Ambition: The Scientific Survey of Puerto Rico, 1913–1934" in Colon, Julio C. Figueroa, ed. *The Scientific Survey of Puerto Rico and the Virgin Islands: An Eighty-Year Reassessment of the Islands' Natural History* (New York: New York Academy of Sciences, 1996): 2–10.

22 Philip Pauly first noted this missed opportunity in an assessment of Baatz's work, "Review of *Knowledge, Culture, and Science in the Metropolis*," *American Historical Review* 96 (1991): 1607–1608.

a thorough botanical, geological, ecological, zoological, oceanographic, paleontologic, meteorological, and anthropological reconnaissance of the tropical colony, would foster a closer degree of American supervision over Puerto Rican development and would rationalize it in a way that scientifically uninformed capitalists or bureaucrats could not. Britton's organization of the Survey was part of the same nationalistic impulse that had led him to focus his attention on finding, naming, classifying, and displaying his quintessentially "American" plants, the cacti. Similarly, the Survey, orchestrated from Britton's office in the Bronx, the Century Club in Manhattan, or from a luxury hotel room in Puerto Rico, would strengthen NYBG's prestige. During the course of this project, which began in earnest in 1913 and actually outlived him by more than a decade, Britton envisioned a program of rational insular development in forestry, mining, commercial agriculture, and education. It was a development plan founded on his and his compatriots' understanding of American science.

Forging Tropical Ties

The very first NYBG expedition to Puerto Rico occurred in 1898, when A. A. Heller, joined by his wife, sailed south to collect specimens on the island. The Hellers' expenses were paid by NYBG President Cornelius Vanderbilt, who also provided a letter of introduction to local US military authorities from Secretary of War R. A. Alger.[23] This initial foray had been followed in succeeding years by the expeditions of Samuel Henshaw in 1899, L. M. Underwood in 1901, Percy Wilson in 1902, and M. A. Howe in 1903.[24] These early NYBG excursions were typical of the metropolitan botany that characterized the Garden's scientific relationship with the nation's nearest tropical colony. Urban New Yorkers, most educated in American universities, arrived in San Juan, Puerto Rico's capital and largest port city, determined to examine the island's littoral and mountainous interior regions. There they collected, named, and classified what they encountered. Then they shipped their finds back to the US for display and study in urban settings in New York and Washington. Such efforts reflected a cosmopolitan belief in museum collections as signifiers of civic value. Citizens and scientists alike considered the taxonomic comprehensiveness of a city's scientifically catalogued possessions an indicator of civic worth. Britton understood that any products of tropical investigations would, like his cactus displays, attract crowds and instill in American visitors a sense of the value of scientific as well as colonial endeavors.[25]

23 A copy of this letter, dated November 21, 1898, is in the Nathaniel Lord Britton Administrative Correspondence, New York Botanical Garden Archives.

24 Brian Boom, "Botanical Expeditions of the New York Botanical Garden" Brittonia 48 (1996): 297–307.

25 The idea of a civic significance for local museum collections is explored by Lynn K. Nyhart in "Civic and Economic Zoology in Nineteenth Century: The 'Living Communities' of Karl Mobius" *Isis* 89 (1998): 605–630.

These encounters with the tropical colony could be described in the familiar vernacular of the frontier, a vocabulary that maintained a particular resonance in the public imagination long after Frederick Jackson Turner's 1893 jeremiad decrying the loss of opportunities for continental American expansion.[26] This newer frontier opportunity, however, was recognizably smaller, and in effect more suitable for a thoroughly systematic, scientific expropriation and examination than the western US had ever been. The nineteenth-century surveys of the West conducted by the US Geological Survey had been undertaken by men no less determined than the Garden's tropical explorers to subjugate and exploit a particular landscape. These expeditions, like their early-twentieth-century descendants, took months or even years to complete. Yet, even the legendary tenacity of John Wesley Powell (an early director of the US Geological Survey and noted explorer of the Grand Canyon) could in reality lead to the successful reconnaissance of only a miniscule portion of the Western landscape. By contrast, NYBG botanists and their New York Academy of Sciences cohort (which, given Britton's close ties to Washington botanists, *did* include workers from the federal scientific bureaucracies) could reasonably expect to examine a considerable proportion of "Porto Rico" in some detail and in a thoroughly scientific manner.[27] *The New York Times* wrote of Britton's Puerto Rican exploits that the island was both "so moderate of size that it can be covered readily," and "so extensive as to furnish 2,000 species" for study and importation into the metropolis.[28] Garden administrator Henry Gleason, conducting the ecological portion of the Puerto Rico Survey, made the analogy with earlier Western exploration explicit. Nineteenth-century work, he asserted, had been undertaken without the benefit of ecological methods, and, consequently, "billions of dollars were thrown away." "Attempts were made," Gleason said, "to use land for purposes for which it was not adapted." Workers in Puerto Rico, in contrast, were employing scientific techniques that would prove, he maintained, "vital to agriculture," and would aid insular authorities in determining which "lands should be used for agriculture, what for grazing and what for forests."[29]

Approximately two-thirds the size of Connecticut, Puerto Rico was a fairly convenient object of scrutiny for the battery of metropolitan scientists who descended

26 Harold P. Simonson, ed. *F. J. Turner, The Significance of the Frontier in American History* (New York: Frederick Unger, 1963). In 1893, Turner argued that the key to American history had been found at its expanding frontier, where Europeans became Americanized. The closing of that western frontier marked the end of the first period of American history. See Patricia N. Limerick, *The Legacy of Conquest: The Unbroken Past of the American West* (New York: Norton, 1987): 20–21.

27 Included, for example, in the Survey's work in Puerto Rico was federal botanist and explorer William Ralph Maxon, whose collecting expeditions to South America had been on behalf of the Smithsonian as well as the NYBG.

28 Editorial, "Plant Ecology," *New York Times*, January 5, 1926.

29 *The Bronx Home News*, January 10, 1926.

upon it in increasing numbers after 1900. Given its area, the American West remained a difficult object of scientific attention, even into the twentieth century. By comparison, Puerto Rico's size, climate, and relative proximity to the urban northeast made it an attractive venue for the prosecution of tropical research.[30] Indeed, before Britton had secured the use of the British government's Jamaican laboratory at Cinchona (see Chapter 4), Lucien Underwood, the third NYBG botanist to undertake extended studies in Puerto Rico, identified the region as a highly desirable candidate for a tropical laboratory. In 1901, Underwood had been impressed by the pleasant Puerto Rican climate, which, despite the island's tropical latitude, was tempered by trade winds. Puerto Rico, he wrote, provided nearly "ideal conditions" that, when combined with the construction of new government-subsidized roads, made the island "a most delightful resort the year round." However, his party was temporarily disabled by impure, untreated, drinking water. Because these roads were still under construction in 1901, Underwood journeyed to the mountainous interior and the southern coastline by rail and by horseback, marveling as he did at the "luxuriantly tropical" landscape in which fruit trees were cultivated adjacent to the "shacks" of native peasants (dubbed "*jíbaros*"). There tobacco, sugar cane, and coffee flourished. Tramping across the island from north to south in the company of US Department of Agriculture botanists eager to assess the economic potential of local croplands, Underwood offered up the island to Garden members in New York as a "rare prize," ripe for development and modernization. By the end of his expedition, Underwood had concluded that Americans "owe[d] the people [of Puerto Rico] the possibilities of Anglo-Saxon civilization," which the island natives, he asserted, were "seeking with an interest more than commensurate with their appreciation of what it really [meant] to them."[31]

Botanical Adventurers and a Gentle Imperialism

The New York Academy's survey of Puerto Rico linked that colony with New York and introduced the tropical island's elite to a set of northeastern botanical adventurers. Underwood's colorful tales found eager readers among New Yorkers proud of their scientific Garden and curious about the nation's new foreign possessions. In the early 1900s, most of New York City's half-dozen or so daily newspapers reported regularly on these NYBG activities. Articles frequently contained long passages from Britton's reports to the Garden Board of Scientific Directors. Such pieces commonly highlighted the novelty of American tropical possessions ("where the native builds his hut open at all sides to every breeze that blows") and their suitability as venues for the display of American know-how. (Particularly astute newspaper editors might in addition have

30 Regular shipping schedules linked New York and San Juan.

31 L. M. Underwood, *Journal of the New York Botanical Garden* 2 (1901): 166–173.

noted the dependence of the US upon the tropics for 40 percent of total imports.[32]) Although, in these explorers' narratives, "an abundance of food just for the picking" awaited the grasp of the indolent inhabitants of rude huts, natives were also being schooled in the art of US business practices. Puerto Rican weavers of Panama hats, though previously inclined to "count [their] time of little value," were "learning rapidly under American influences" to charge northern visitors higher prices for protection from the tropical sun.[33]

Regular press attention of this sort, providing portraits of heroic scientific voyageurs culling exotic knowledge from the dangerous realms of colonial natural history, was certainly not unique to New York. Adventure tales starring itinerant naturalists had been a commonplace in the early nineteenth century. Field naturalists traditionally embodied a "heroic ideal of manly action," men who enjoyed access to "unbounded spaces" in which to pursue their work.[34] Historian Gillian Beer has studied the workings of nineteenth-century British colonial natural history, but her observations apply to much of the work of Britton's NYBG. Underwood's voyage, in Beer's terms, "express[ed] the will to control, categorize, occupy and bring home the prize of samples and of strategic information." Natural history, Britton recognized, was "closely interlocked" with the national future.[35] These narrative accounts of fieldwork, rich in detail and often, Beer noted, "self-serving exaggeration" flirted with the danger of being perceived as "fantasy" rather than "controlled observation." The Garden's press coverage, however, was not meant to serve as the primary purveyor of scientific knowledge gained by NYBG workers. (Britton had established the roster of journals discussed in Chapter 4 for this purpose.) Instead, the stories were meant to convey to a receptive, urban, lay audience the importance of the Botanical Garden's

32 Statistic from O. P. Austin, *Colonial Administration* (US Bureau of Statistics, 1901), cited in Garcia, "I Am the Other," 46, see note 1.

33 See, e.g., "The Vegetable Products of Porto Rico," *New-York Daily Tribune*, March 29, 1903 and "Exploring Porto Rico: A Scientific Expedition for the New York Botanical Garden," The [New York] *Evening Post*, April 7, 1906.

34 Dorinda Outram, "New Spaces in Natural History," in N. Jardine, J. Secord, and E. Spary, eds., *Cultures of Natural History* (Cambridge, UK: Cambridge University Press, 1996): 249–261. Outram notes, however, that although this heroic imagery supported claims by outdoor-bound natural historians of discovering essential truths about nonhuman nature, challenges were made to this ideal of objectivity beginning in the early nineteenth century. French zoologist Georges Cuvier held that claims to truth regarding natural history were in fact not dependent upon the labors of the intrepid fieldworker, but instead upon the calculated observations and efforts of institution-based scientists in urban laboratories. See Outram, pp. 261–262. It might be argued that by the beginning of the twentieth century, workers at institutions such as the NYBG, through a combination of both field- and museum-based science, were able to reconcile these divergent trends in science, despite the overall historic trend toward experimentalism and professionalism.

35 Gillian Beer, "Traveling the Other Way," in Jardine et al., eds., *Cultures of Natural History*, 322–325, see note 34.

research and its potential usefulness to Americans in the form of drugs and industrial products.

In the fall of 1902, a decade before Britton conceived the Puerto Rico Survey, one such patriotic New Yorker and self-described "botanical enthusiast" (unidentified) had called for both public and private funds to be allocated for the exploration of new American possessions in Hawaii, the Philippines, Puerto Rico, and Guam. The natural history of these new colonies, the writer declared, was ripe for American investigation. The various islands were a "botanist's paradise," in which "lush tropical forests" of the "greatest beauty" could be studied by the nation's scientists in the "cause of general education" and the "useful and refining influence" of botany. Other nations, the writer worried, were beginning to seize the opportunities for study and, potentially, economic exploitation that American hegemony provided. The colonies' previous European masters, the Spanish, "looking with suspicion upon all requests for exploring privileges and scientific research," had traditionally restricted access to Philippine or Puerto Rican forests to "Jesuit fathers," whose inadequate studies displayed a "lack of individual research and close observation," and were rejected by "accredited botanists." The American people, the writer enthused, particularly New Yorkers, were "clamoring to know something more of the flora of [the new colonies]. To cater to this demand," he concluded, "is peculiarly the province of the New York Botanical Garden."[36]

These various public endorsements of a prominent role for the Garden in American empire building were couched in language familiar to *Times* readers concerned with politics and international affairs. Many educated New Yorkers considered the nation's new role in the international colonial sweepstakes to be, on the whole, a proper one. Much initial hesitancy over the prospect of becoming a colonial power was ameliorated by the successful articulation of an ideology of American imperial exceptionalism. This notion was predicated upon the assumption that the US was peculiarly competent to "improve upon the civilizing mission" of European colonial powers.[37] For example, as American troops battled Filipino insurgents loath to trade one Western master for another, US administrators posted to tropical Manila delighted in reporting that, contrary to expectations, Yankee soldiers thrived in the hot climate. In 1900, William Howard Taft worked to establish a civilian government in the Philippine Islands once US forces had secured firm control of the archipelago. His observations of the ability of US troops to work and administer their new political domain were meant to assuage fears of national hypocrisy among Americans wary of assuming the white man's burden.[38]

36 "'Imperialism' for Flowers," *The New York Times*, Sunday, September 28, 1902.

37 See Michael Adas, "'Improving on the Civilizing Mission?': Assumptions of United States Exceptionalism in the Colonization of the Philippines" *Itinerario* 22/4 (1998): 44–66.

38 Ibid., 44–45.

American adventurism of this sort could be made more palatable to a troubled stateside public if it were understood to be part of a new unprecedented project. Although many Americans admired British and Dutch colonial commercial success, urban "patrician progressives" maintained that US occupation was meant only as a prelude to self-determination and, as such, entailed the exportation of democracy to regions in need of American self-governing know-how. [39] Overall, debates about the nature of American influence abroad in this era were centered on the question of access to markets. Although most Americans at the turn of the twentieth century agreed that foreign markets were important, there remained differences over the degree to which access to such markets would necessarily involve direct American control over colonies.[40] To the extent that procolonial policymakers could describe their aims in terms of democratic tutelage of Puerto Ricans, they were successful in making a case for extended US military occupations. Efforts by US authorities to "civilize" inhabitants of newly acquired regions involved public works projects (roads) and scientific surveys of economically important natural resources. Roads were built, soils tested, sewers dug, agricultural experiment stations founded, and plants collected and catalogued in an earnest comprehensive effort (one that harnessed the applied scientific powers of American engineering and social science experts in addition to New York botanists) to "modernize" the inhabitants of US Caribbean insular possessions.[41] Once these efforts were underway and a "new empire" constructed, identified as the watershed in American history by historian Walter LaFeber, the staff of NYBG and NYAS could serve American national aims.[42]

Metropolitans and Others in Puerto Rico: The Historical Context

Lucien Underwood's attitudes about Puerto Ricans were widely shared by the North Americans staffing the island's fledgling scientific institutions. Elizabeth Stearns Barrett, wife of the American head of an insular experiment station, traveled with her husband throughout US possessions in the Caribbean. The Barretts were based in Puerto Rico, where they, like their friends the Brittons, found a comfortable tropical home. Elizabeth

39 Ibid., 46.

40 Walter LaFeber, *The New Empire: An Interpretation of American Expansion*, 1860–1898 (Ithaca, NY: Cornell University Press, 1963).

41 Adas, "Improving on the Civilizing Mission?" 51–54, see note 37. Adas, p. 60, notes that the modernizing efforts in the Philippines, although "consistent with the reformist Republican tradition" dominant in 1890s and 1900s American politics, actually "did little to alleviate" the "worsening situation" of Filipinos outside the elite class of landowners and merchants. Without substantive land redistribution or the construction of agricultural experiment stations responsive to the needs of peasant agriculture, which were impossible reforms within the context of American laissez-faire ideology, rural prosperity in the Philippines would remain elusive.

42 LaFeber, *The New Empire*, 67, see note 40.

Barrett confided to Elizabeth Britton that although she ranked many Caribbean botanic gardens among the best in the colonial world, she nevertheless missed "the comforts of San Juan." She contended that if "the Filipinos and Porto Ricans so anxious to shake US education, health work, etc. [were] obliged to journey this way," their anti-North American attitudes would change. The "rude, coarse," youth Barrett encountered in Dominica, she reported ruefully, spoke only a "dreadful patois French," while "talking Russia in the streets." Elizabeth Britton concurred: "I firmly believe," she wrote Barrett, "that both Porto Rico and the Philippines are better for our control."[43]

These attitudes, however, were not restricted to North American explorers, scientists, and their families. Elite Puerto Ricans collaborated with North American scientists and engineers. These insular elites, who made up the propertied educated classes of both urban San Juan and rural plantations, discovered over the course of the first decades of American occupation that they shared many of the class interests of US colonizers. The Puerto Rico Survey and the metropolitan science it fostered drew the admiration and cooperation of these Puerto Rican elites, for whom education and a cosmopolitan culture served as benchmarks of progress, modernization, and Americanization.[44]

On an island famous for its population density, only a small proportion of Puerto Ricans were actually metropolitans in training. Since the early nineteenth century, when coastal sugar cane production, in the guise of large-scale hacienda-based monoculture, had become the major cash crop of the island, the majority of the rural population had been displaced from subsistence farming. In a reversal of this peasant economy's traditional patchwork of independent landholdings, Spanish-run sugar plantations (dependent upon the wage labor of displaced rural peasants) had effectively consolidated the most fertile of productive agricultural lands, monopolizing what had been smaller ancestral farms. This trend stimulated both a migration of Puerto Rican peasants to the mountainous interior (where they attempted to recreate subsistence farms) and an influx of immigration from several continents that involved a reinvigorated African slave trade (which had been dormant for much of the previous century). As free immigrants from the Americas and Europe traveled to this formerly marginal Spanish colony, race relations among the island's European, Creole, and African populations deteriorated. Whereas previous decades in Puerto Rican history had been characterized by frequent manumissions, miscegenation, and relaxed race relations, the height of the sugar years in the first half of the nineteenth century involved the imposition of a strict race-based discipline in the service of the sugar economy.[45]

43 E. S. Barrett to E. Britton, March 13 and March 21, 1924, Records of the Puerto Rico Survey, 1923–1926, New York Botanical Garden Archives.

44 García, "I Am the Other," 39–64, see note 1.

45 Francisco A. Scarano, *Sugar and Slavery in Puerto Rico: The Plantation Economy of Ponce*, 1800–1850. (Madison: University of Wisconsin Press, 1984): 162–164.

Although the island did not become the plantation society that Cuba did, the rise of the Puerto Rican sugar plantation, with its concomitant increase in immigration and wealth, had marked a turning point in the ability of the island's elite to participate more fully in the international economy. A colonial backwater, a geopolitically strategic but remote military garrison, became by 1850 the world's second-largest producer of cane sugar.[46]

At mid-century, with a Spanish tax upon sugar exports and a steady drop in sugar prices following the introduction of European beet sugar, coffee usurped cane sugar's place as the lynchpin of the Puerto Rican economy.[47] From 1850 until the arrival of US occupation forces in 1898, the island's experience in coffee production was typical of the integration of Latin American regions into the wider Atlantic economy. As historian Laird Bergad observed, "rising demand for primary agricultural products in the North Atlantic world led to capital investment from abroad, the development of an unexploited internal frontier, European immigration, the destruction of peasant autonomy, labor force regimentation, and the concentration . . . of land." After the bust of the sugar economy, interior Puerto Rican areas developed a coffee export economy, further linking the island to the wider Atlantic community.[48]

The decade immediately preceding the arrival of American troops represented the Puerto Rican coffee industry's golden age. The wealth of the island's mercantile elite, built upon a booming export economy, grew through the mid-1890s. This insular elite would be surprised, however, by the collapse of the coffee economy as world prices dropped in 1897 due to a glut of production and then by the devastation of an 1899 hurricane.[49] Although the interior coffee-producing regions, unlike the lowland sugar haciendas, had been characterized by a large number of modest family farms, the coffee boom, like the sugar economy, had *not* produced widespread prosperity. This lack of economic development had been due to the structure of the coffee trade itself rather than the system of production, because immigrant European merchants dominated the credit and capital resources of the coffee regions and extracted the wealth they produced for transfer back to Europe.[50] It was in this context of economic upheaval that American invaders arrived in San Juan in 1898. In the wake of US occupation, many wealthy European immigrants whose commitment to insular development was negligible found themselves displaced in local governing hierarchies by native members of Puerto Rican criollo society. Many native Puerto Ricans were

46 Teresita, Martinez-Vergne, *Capitalism in Colonial Puerto Rico: Central San Vicente in the Late Nineteenth Century* (Gainesville: University Press of Florida, 1992): 2–8.

47 Ibid., 5.

48 Laird Bergad, *Coffee and the Growth of Agrarian Capitalism in Nineteenth-Century Puerto Rico* (Princeton, NJ: Princeton University Press, 1983): xv.

49 Ibid., 145–207.

50 Ibid., xviii, 208–210.

thus initially grateful for the American-induced end to Spanish colonialism.[51] This, then, was the political and economic milieu in which New York's scientists began arriving on the island in the early twentieth century.

Puerto Rico Welcomes "Millionaire Botanist" Nathaniel Britton

In 1906, after several trips to other West Indian locales, including St. Kitts, Cuba, and the Bahamas, Nathaniel and Elizabeth Britton traveled to Puerto Rico for the first time. They returned again in 1913 as an initial part of the Survey. Then, from 1922 until 1933, the 1,500 mile steamship journey from New York to San Juan became an annual Britton pilgrimage, one that combined scientific research with the trappings of an upper-middle-class holiday. By the time the Brittons retired from active fieldwork in 1933, they had collected over 10,000 specimens; roughly 2,200 of them were bryophytes (mainly mosses), Elizabeth's specialty.[52]

Nathaniel Britton had long been convinced of the commercial and political utility of tropical botanical exploration and was eager to use his post at the Garden to play an active part in the expansion of American influence. His earliest forays into West Indian botany had persuaded him that any work that "tended to make the plants of tropical America more accurately known," would have "an important bearing upon the increasing commercial relations of the United States with the countries to the south of our present domain."[53] Britton's assessment of the Puerto Rican scene was optimistic in an idiomatically racist way not unusual for a man of his class and age.[54] He described the *jíbaros* as "active, intelligent, and hardworking" fieldwork assistants, whose "kindliness, aid, and hospitality," and "knowledge of natural objects" were "remarkable" and whose "stock [was] mainly Spanish with some evidences of the original aboriginal admixture." "Negroid types," he assured his readers, were "little in evidence," echoing a similar claim dating from the antebellum era, when Southern adventurers had coveted Cuba as a potential slave state.[55] His prescription

51 Ibid., 207. "Criollo" refers to Puerto Ricans of Spanish ancestry.

52 Inés Sastre-D. J. and Eugenio Santiago-Valentin, "Bryology in Puerto Rico: Knowledge Prior to and after the Scientific Survey of Porto Rico and the Virgin Islands" in Colon, ed. *The Scientific Survey of Puerto Rico*: 116, see note 21.

53 *Journal of the New York Botanical Garden* 2 (1901): 161–165.

54 Contemporary views regarding Puerto Rico as a venue for American "patrician tutelage" are discussed in Garcia, "I Am the Other," 40, see note 1.

55 Typical of Puerto Rican descriptions of the island's predominately white population was the assertion by Angel Paniagua y Oller, Vice President of the Historical Society of Porto Rico, that "the native of Spain, and, more properly of Castille, from all points of view is the entity genuinely representative of the Porto Rican people . . ." For the benefit of American readers, Paniagua y Oller traced the origins of the majority of Puerto Ricans to a portion of the Spanish nobility that had been instrumental in the reconquest of Spain from the Moors. See E. Fernandez Garcia, ed. *El Libro de Puerto Rico, The Book of Porto Rico* (San Juan: El Libro Azul Publishing Co., 1923): 91.

of Americanization for the island clearly followed Underwood's and was informed by his own experience of the tropics. The task of assimilation, Britton asserted, would be made easier by the establishment of federally funded agricultural experiment stations. By promulgating methods of intensive scientific cultivation of coffee and other cash crops, these stations would stimulate insular commercial development.[56]

That the Brittons became regular visitors to Puerto Rico and established an amiable tropical existence in keeping with urban standards of leisure was unsurprising. Elizabeth Britton was no stranger to Spain's former Caribbean colonies. She had spent a portion of her upper-class girlhood on her family's sugar plantation at Matanzas, Cuba, and spoke fluent Spanish.[57] In addition, by the time they began their annual Puerto Rican working holidays in the 1920s, the Brittons had made numerous similar trips to Cuba, where Elizabeth felt particularly at home.[58] This background enabled them to form strong personal ties not only to the American governing class in San Juan, but also to the educated native Puerto Rican elite who began staffing the new universities and the insular government in the twentieth century. (The Brittons' San Juan friends remembered Nathaniel in particular for his slight physique and lugubrious face, his prodigious memory, and his generally ill-humored demeanor. His dour moods were punctuated by occasional fits of loquacity during which he would hold forth on botanical subjects.[59]) Elizabeth Britton's language abilities also enabled her to lecture at the University on botanical subjects, recounting for 500-member audiences recollections of her Cuban childhood.[60] These Caribbean experiences were made even easier by the Brittons' own wealth. Their relatively lavish spending habits (Britton was dubbed the "millionaire botanist" by locals, a literally apt description before the Depression) enabled them to host both North American and Puerto Rican guests and reproduced in a tropical setting the comforts of home.

56 "Porto Rican Coffee Industry Falls Off: Dr. Britton of N. Y. Botanical Garden Urges Support of Insular American Enterprise," *The New York Times*, April 13, 1924.

57 Elizabeth Britton served as translator of Spanish language articles for her husband.

58 *Havana Daily Post*, September 20, 1910.

59 These are the recollections of Carlos Chardon, a Cornell-educated Puerto Rican botanist who became chancellor of the University of Puerto Rico. See Jose Alvarez Conde, *Historia de la Botanica en Cuba*. (La Habana: Publicaciones de la Junta Nacional de Arqueologia y Etnologia, 1958). Many educated Puerto Ricans recalled the Brittons fondly. Journalist Luis Sanchez Morales recounted in 1927, as the Survey progressed, the proceedings of a banquet given in part to honor Nathaniel Britton, who "presided over a group of eleven modest scholars" who have been "making a thorough study of the natural history of our island. Its geology, paleontology, botany, zoology are being set forth ever clearer in the monumental books that year by year they are publishing." See Luis Sanchez Morales, "Britton, Mitchell, and Baxter," *Porto Rico Progress* February 2, 1927.

60 J. F. Kemp and N. L. Britton, Report to Survey Committee, April 15, 1923, Puerto Rico Correspondence, New York Botanical Garden Archives.

The Brittons took their annual lodgings in the Condado-Vanderbilt Hotel, which, although "somewhat too fashionable and too expensive for successful scientific occupation," according to the NYBG director, was nevertheless, "for social and administrative reasons," desirable as a home base for local excursions. From this posh hotel, Britton confided to a fellow New York scientist, he could "extend hospitality to friends to good advantage."[61] While exploring regions beyond the immediate San Juan vicinity, the Brittons found themselves the guests of wealthy North American locals such as F. W. Maxwell, a sugar planter, who provided their party with a four-bedroom, four- bath cottage, complete with bay views and maid service.[62] Often these winter expeditions to the island comprised an itinerary of tours of various private gardens and estates, as well as government agricultural facilities.[63] With their personal background, then, both Nathaniel and Elizabeth Britton can be considered "cultured annexationists" in their attitudes toward Puerto Rico: their experiences, wealth, education, and their shared profession led them to a "genuine scientific interest and sincere curiosity to learn more about the geography, natural resources, and economic potential" of the island and its people.[64]

The island's American-led government and business community eagerly touted Puerto Rico's material connections to New York. At the height of the Survey in the 1920s, insular authorities and entrepreneurs courted US investment, assuring American manufacturers that the island market "afford[ed] a most fertile sales field" for all exports, save, of course, winter garments, which would not be needed in a mild tropical paradise. The business-minded editor of a large volume called "*El Libro de Puerto Rico*" (The Book of Puerto Rico) considered the new US–Puerto Rican relationship to be a partnership dedicated to "consciously helping the development of a new

61 N. L. Britton to Lewis B. Woodruff, Survey entomologist, January 16, 1925, Puerto Rico Correspondence, New York Botanical Garden Archives.

62 F. W. Maxwell to N. L. Britton, March 16, 1923, Puerto Rico Survey Correspondence, New York Botanical Garden Archives. For typical accounts of the various North American island residents, military personnel, and sugar company functionaries, as well as elite Puerto Rican natives, who "cordially welcomed and aided" members of the Survey, see the *Journal of the New York Botanical Garden* 24 (1923): 94–99 and N. L. Britton, "Descriptive Flora of Porto Rico and the Virgin Islands," *Journal of the New York Botanical Garden* 25 (1924): 129–135.

63 See, for example, Britton's accounts of his trips made in 1923 and 1924, cited above and note 54.

64 The term "cultured annexationism" is Gervasio García's; see "I Am the Other," 51, note 1. García distinguishes between those North Americans he dubs "cultured annexationists" with a sincere interest in the island and others who sought solely to emphasize the bizarre and different. In the case of New York scientists, however, both impulses were reconciled. Botanists from the NYBG sought examples of useful and beautiful tropical plants for display to an appreciative urban audience, but they also established friendships and working relationships with Puerto Rican counterparts and, rather than emphasizing the exotic, sought to establish familiar institutions on the island in an earnest effort at mutually beneficial development.

civilization, full of promise, resulting from the intimate contact of the two currents of opinion that exert their influence in the island." His boosterism was tempered by his recognition that Puerto Rico enjoyed its own cultural heritage, one created on an island "known to the world before North America was known to exist." Nevertheless, the book's editors remained optimistic that the US ("that nation so young in years, so advanced in progress") would be able to incorporate its new insular domain in a meaningful way. In a telling use of horticultural metaphor these elite Puerto Ricans asserted to a wider American audience that

> Subject to the same biologic laws that prevail in the grafting of plants to obtain fruit of a higher quality, the typical characteristics of the graft complementing those of the one grafted, the proud old oak of the ancestral Hispanic civilization [had] been grafted with vigorous Anglo Saxon shoots, [and that] new fruits—the youth educated in the modern [US-built] school—already show[ed] the mixture of the exquisite Spanish sentiment with that practical sense and mental equilibrium so genuinely North American.

For this reason, the editors concluded, there was not "another center in a better position than Porto Rico to favorably influence the closer relations of the peoples of the two Americas."[65] These insular leaders, then, understood Britton's natural history Survey as a necessary precursor to Puerto Rican development. Britton, cast in the role of metaphorical and literal horticulturist, would personally help cultivate the island's culture and economy by applying a democratic, rational, New York science to a tropical civilization with its own noble pedigree.

Of course, elite Puerto Rican opinion varied regarding continued US involvement in island politics. Whereas the Puerto Rican Republican Party favored statehood and direct American integration, the island's dominant Union Party, headed by Resident Commissioner (the island's nonvoting delegate to the US Congress) Muñoz Rivera, continued in the 1910s and 1920s to advocate increased home rule within a close overall US–Puerto Rican relationship. In addition, opposition to US military rule had taken violent form in the years leading up to the Foraker Act of 1900, which had granted

65 Fernandez Garcia, ed. *El Libro de Puerto Rico*: 1188, see note 55. This bilingual volume, published with Spanish and English translations of text on facing pages, was compiled as a commercial encyclopedia, designed to woo American investment. It was an elite production, characterized by portraits of American presidents, governors and insular legislators and dedicated to the proposition that Puerto Rico would serve as the "center of a new civilization that carries . . . influence from the North to the South." Subsequent sections of the volume included overviews of: Natural History, History, Religion, Government, Public Service, Public Health, Education, General Agriculture, Special Agriculture and Industries, Commerce, Finance, Communications, Sociological Studies, Arts and Letters, the "Porto Rican Woman," the "Spanish and Foreign Element," Associations for Social Progress, and Professional Associations.

a degree of civilian self-government to the island. This resistance had initially shaken the confidence of the American foreign policy elite in its ability to carry out its exceptional colonial mission. The pre-1900 violence had led McKinley's War Secretary, Elihu Root, to consider Puerto Ricans far from ready for self-government and in need of an extended period of direct US schooling in democracy.[66]

By the late 1910s, however, even some home rule advocates, seeking increased autonomy within the American commonwealth, were inclined to view a sustained study of the island's natural resources as an opportunity to court northern capital and promote local prosperity for a population of roughly 1.3 million newly created American citizens. Many of these new Americans faced unemployment and wretched living conditions, and, by the 1930s native islanders had begun migrating to the northeastern US in great numbers.[67] With even the Unionists cognizant of the benefits the Survey could bring, the majority of the island's power structure was prepared to welcome the scientists. This welcome took official form when the Puerto Rican legislature appropriated a series of grants to help cover the cost of the NYAS-sponsored expeditions.

The Brittons in the Field and among an Insular Elite

The Brittons' first Puerto Rican expedition, in 1906, included NYBG botanist M. A. Howe; entomologist William Wheeler of the American Museum; John Cowell, director of the Buffalo Botanic Garden; and Delia Marble, Elizabeth's assistant. Beginning at the US Naval base at Culebra, the group collected cacti and marine algae and moved on to the main island to explore interior mountainous regions. While climbing some of the higher peaks near Mayagüez in western Puerto Rico, they collaborated with the staff of the local agricultural experiment station. Then, after traveling to Ponce on the southern coast, they returned to San Juan via the newly constructed military road and steamed home to New York with some 1,700 dried herbarium specimens.[68]

The 1913 trip was conducted on a similar basis, and several frequent Britton collaborators once again accompanied them on a series of excursions that also included the Virgin Islands as well as Puerto Rico.[69] This time they brought along

66 García, "I Am the Other," 41, see note 1.

67 The population figure is cited in *Libro de Puerto Rico*, see note 55, and is taken from the 1920 census. See also Dietz, *Economic History of Puerto Rico*, 110–130, note 1, for a description of the economic context of these migrations.

68 *Journal of the New York Botanical Garden* 7 (1906): 116.

69 Britton's interest in the former Danish colonies was similar, if not quite as extensive, as his fascination with Puerto Rico. Once the Virgin Islands had become, through purchase, an official US possession, Britton argued that the name of the islands should be changed. In a letter to the *New York Evening Post*, he suggested "Antilla," arguing that the "name is appropriate, euphonious,

Joseph Rose as part of the ongoing cactus investigations and J. A. Shafer, as well as the intrepid Miss Marble. Britton considered this trip an extension of earlier European and American fieldwork efforts.[70] The banner day of this entire collecting expedition, Britton said, came on March 14, in an "enchanting forest" 2,000 feet above Ponce. There, he recollected, the party was able to gather

> specimens of not fewer than 213 species, a result submitted as a record breaker in botanical field work. Orchids festooned . . . the tree trunks; ferns in great variety and luxuriance were everywhere, from small filmys up to tall arborescent species; mosses and liverworts carpeted the banks and logs, lichens were abundant and fungi numerous, while the kinds of flowering shrubs and climbers were unexpectedly large; that we came out to the car saturated by a mountain shower, and dried up in the subsequent sunshine were only incidents of this joyous experience, and we came into Ponce at dinner time with both the inside and the outside of the motor, full of packages of spoils.

Nathaniel Britton completed this trip convinced that he needed to explore Puerto Rico further. He also called for a new comprehensive descriptive flora of the island, an educational necessity for students enrolled in the newly established College of Agriculture of the University of Puerto Rico.[71]

During the subsequent botanical expeditions of the 1920s, Britton took care to publicize his efforts as part of the Academy of Science's official Survey. He published travelogues for NYBG members, embellishing his narratives with vivid descriptions of plants whose unusual or striking features were likely to stimulate interest and financial support. Within several years of Britton's proposal of the Puerto Rico Survey, the project, in all its disciplinary manifestations, was well underway. In addition to the Brittons, American Museum zoologists and entomologists, Columbia anthropologists

---

and concise, either as colony, territory, or reservation." "The origin of the name," he continued, "is mythical, but, in the form Antilles, it has been used for the whole West Indian archipelago, especially by European authors. Geographical names of three syllables are the most attractive of any, and "Antilla" is quite as pretty as Montana, Alberta, or Sonora. There would then be Gibraltar at the entrance to the Mediterranean and "Antilla" at an entrance to the Caribbean. Whenever any of the other Lesser Antilles come into the possession of the U S, they would naturally be added to form a larger Antilla." See "A Name for the Islands," *New York Evening Post*, April 6, 1917.

70 Columbia geologist James Furman Kemp noted that upon US annexation of Puerto Rico, the US Geological Survey had done "a little work on the island," including a survey on water resources and a rudimentary contour map by an 1881 Columbia School of Mines graduate. This meant that the earliest New York-Puerto Rico survey link antedated even the Puerto Rico Survey. See J. F. Kemp to N. L. Britton, April 7, 1914, Puerto Rico Survey Correspondence, New York Botanical Garden Archives. North American foresters had also begun studying the island soon after annexation, which will be discussed below.

71 *Journal of the New York Botanical Garden* 14 (1913): 99–105.

and geologists, and federal botanists all began arriving in Puerto Rico and the Virgin Islands. During the summer of 1916 alone, 23 different groups explored the main Puerto Rican island and outlying islets for the NYAS.[72]

These efforts were lauded by insular scientists as "veritable monument[s] of American science" that would "raise Porto Rico to a scientific level much higher than that of any other island in the tropics." With Britton's help, Ramon Gandia Cordova, publisher of the *Revista de Obras Publicas de Puerto Rico*, regularly issued Spanish translations of the Puerto Rico Survey's publications.[73] Britton, in a characteristic display of ambition, claimed that the Academy's efforts would make "Puerto Rico the key to the geology and natural history of the West Indies." He assured Puerto Rican Governor Arthur Yager that the Survey would make "important contributions to knowledge and economic development," while effectively contributing "thousands of dollars worth of labor for the Government of Puerto Rico."[74] Yager himself was convinced of the Survey's utility and served as Britton's ally in San Juan. The Governor strongly endorsed the New York Academy's plan before the insular legislature, asserting

> that by continuing this work and intensively prosecuting this survey along the lines already mapped out that results of practical and commercial value, as well as of scientific importance, will be achieved in the discovery of economic minerals which are believed to exist in the Island, as well as the study of insects which may be injurious or beneficial to plant life, and a complete classification of the soils and the various forms of grasses that may be adapted to them.[75]

Yager, intensely interested in exploring the possible existence of oil-bearing shales off of the Puerto Rican coast, granted the Academy's geologists precedence in this search.

With $10,000 in Puerto Rican government money and $2,500 from NYAS President Emerson McMillan, work by various Survey participants continued throughout the 1910s. The American Museum's Henry Crampton and Frank Lutz continued zoological investigations, while Roy Miner gathered thousands of specimens of marine invertebrates in San Juan Harbor, and US National Museum entomologists examined the

72 Baatz, "Imperial Science and Metropolitan Ambition" in Colon, ed. *The Scientific Survey of Puerto Rico and the Virgin Islands*: 9, see note 21.

73 N. L. Britton to Herbert Schwarz, Editor, New York Academy of Sciences, April 29, 1926, Puerto Rico Survey Correspondence, New York Botanical Garden Archives.

74 N. L. Britton, *Journal of the New York Botanical Garden* 20 (1919): 220; Carlos Chardon, "The Flora of Porto Rico," in *Libro de Puerto Rico*, see note 55 and N. L. Britton to Puerto Rico Commissioner of Interior J. Wilson, April 14, 1922 and N. L. Britton to Arthur Yager, December 1913, Puerto Rico Survey Correspondence, New York Botanical Garden Archives.

75 Yager is cited in the report of Henry Crampton to the Puerto Rico Survey Committee of the New York Academy of Sciences, January 28, 1915: "Report on Porto Rico Trip of December 19, 1914–January 25, 1915, New York Botanical Garden Archives.

island's arachnids. Geologist Edwin Hodge continued Survey efforts by examining over two dozen shale sites in western Puerto Rico in 1915, concluding (to Yager's chagrin) that these rock formations contained no extractable petroleum. Later that year, 15,000 entomological specimens were collected, while Britton's Columbia colleague Franz Boas began his anthropological and archaeological investigations. Boas started testing the theory that children of tropical latitudes reached physical maturity earlier than those of temperate zones, before deciding that it lacked merit. He did, however, continue studies on Puerto Rican adults, specifically a regiment of native soldiers commanded by American officers, and concluded that "a remarkably strong environmental influence upon the racial type of man may be observed." Boas's colleague J. Alden Mason investigated the island's folklore, accumulating an enormous database of local legends, songs, riddles, and tales, all of which, he predicted, would have "not only great value for the study of Romance philology and comparative literature, but [would] also furnish reading matter for the rural schools, attractive and interesting to the children, because based on their own historic environment." Overall, Boas held, these anthropological investigations were "of the very greatest importance for practical measures relating to the hygiene of childhood and for the management of schools."[76]

Other Survey explorers unearthed the bones of large extinct mammals and the sites of prehistoric aboriginal villages; Britton and Boas lobbied the insular government to grant these areas protected status as public monuments.[77] Meanwhile, the Brittons and other NYBG botanists continued collecting mosses, ferns, algae, fungi, and flowering plants from the island. Specimens were initially shipped back to New York with the understanding that once scrutinized, determined, and labeled, some would be returned to the island for installation in a Puerto Rican Museum of Natural History.[78] The disposition of these various collections was of particular interest to insular authorities seeking tangible results from their investment of thousands of dollars in the Survey. Puerto Rican officials, of course, were also driven by the same cultural desire as their metropolitan New York counterparts: acquisition and display of natural history specimens would grant the island a degree of prestige and

76 New York Academy of Sciences, Reports to Puerto Rico Governor Arthur Yager, August 20, 1915, Puerto Rico Survey Correspondence, New York Botanical Garden Archives.

77 Boas urged the Puerto Rican governor that "it was of the greatest importance that the [insular] government purchase and conserve [those] most interesting places and preserve [them] as an insular monument. Conservation of historical and archaeological sites and their protection from vandalism, is fortunately, a movement making much headway in all progressive countries and Porto Rico should not lag behind." See F. Boas to A. Yager, July 30, 1915, Puerto Rico Survey Correspondence, New York Botanical Garden Archives.

78 N. L. Britton, "History of the Survey," in *Scientific Survey of Porto Rico and the Virgin Islands* (New York: New York Academy of Science, 1919–1922): 1–10. Also: Edwin Hodge to C. Berkey, November 23, 1915, Puerto Rico Survey Correspondence, New York Botanical Garden Archives.

cosmopolitanism similar to that sought by Britton and his urban patrons for their own city. Columbia geologist Charles P. Berkey was particularly cautious, nevertheless, about relinquishing the products of his geological surveys. Even encouragement from Puerto Rico governor Yager, a Johns Hopkins classmate of Woodrow Wilson and a former college president, did little to quell the Academy scientists' doubts about sharing their bounty. "I have been a little loath to send a very full series [of fossil specimens]," Berkey confided to Britton, "because of a feeling of more or less certainty that they would be lost in Porto Rico if sent piecemeal." In principle however, Berkey acknowledged the Puerto Ricans' claims, and despite his skepticism toward island facilities, he agreed to "send a box of rocks" to San Juan.

Eventually, Berkey contended, the Academy would have to devise an official plan for the disbursing of the various collections. He proposed the division of all geological specimens into several lots. The "best" lot, containing zoological type specimens, would remain in New York, in the possession of the Academy and placed on display in the American Museum.[79] Other sets of specimens would be distributed to various partners in the collecting enterprise as well as to a completed insular museum. When the American Museum balked at this proposed arrangement, instead claiming ownership of a share of the specimen sets, Britton and Berkey assented to their partners' wishes.[80] Minor disputes of this nature, however, were not the norm, and the allied groups of New York scientists generally conducted the Survey in a spirit of cooperation. Britton oversaw all of these efforts with his usual administrative acumen.[81]

When a rare dispute did arise, the New York scientists, enjoying direct and personal connections to the island's social and scientific elite, never felt their hegemony over Puerto Rican natural history seriously threatened. Britton also recruited scholars unaffiliated with Survey host institutions but whose expertise he deemed essential to the Academy's efforts. Stanford University ethnographer Aurelio Espinoza, for example, was invited to collaborate with Boas on what Espinoza considered "the most important collection of folklore made on the American continent [sic]." Such recruitment efforts,

79 Types were the most significant of all such specimens gathered, for it was from them that taxonomists had proposed new species.

80 C. Berkey to N. L. Britton, June 27, 1917; N. L. Britton to C. Berkey, June 30, 1917; C. Berkey to N. L. Britton, May 21, 1918, and N. L. Britton to C. Berkey, May 22, 1918, Puerto Rico Survey Correspondence, New York Botanical Garden Archives.

81 There were differences among participating institutions in degrees of financial commitment to the Survey. Such differences were the results of financial pressures faced by the home institutions of each of the participating NYAS scientists, each of which was responsible for paying the salaries of the Puerto Rico explorers. Whereas Britton personally financed much of the botanical work, the American Museum, for instance, was not always willing or able to match Britton's enthusiasm with funds. In 1923, Frederic Lucas, H. F. Osborne's successor as AMNH director, had had to demur when Britton inquired whether the Museum could increase its support for Puerto Rican archaeological investigations.

besides incorporating new expertise, were shrewd fundraising maneuvers as well, since new American allies might contribute additional funds to the overall Survey.[82] Other investigators, however, sometimes made "surprise" appearances on the island, and their work was viewed as an encroachment upon the Academy's legitimate scientific domain. The late December 1918 arrival in Puerto Rico, for example, of two University of South Carolina seismologists at Governor Yager's invitation evoked a wary response from Columbia's Charles Berkey, who complained

> there is no doubt of course but that [the seismologists] have as good a right to do things in Porto Rico as anybody, and the Governor has a right to call on anyone he please, but it would seem as though he might give a little more consideration to an organization that has spent several years in a study of his island and probably knows more about conditions already than any new set of investigators are likely to find out in some time.

Britton met such infrequent challenges by getting completed work into print quickly, thereby staking scholarly claim to as much subject matter as possible.[83]

In general, the Survey remained the unchallenged leader of early twentieth century Puerto Rican science. Britton's effective strategy of inviting available investigative talent into the ranks of Survey personnel extended, crucially, to a small cadre of insular scientists. During the 1920s, the Puerto Rican scientific infrastructure consisted of several agricultural experiment stations and the newly established University of Puerto Rico, a land grant college. Paralleling the structure of the insular government itself, these establishments were staffed by a combination of expatriate North Americans (who frequently occupied key administrative posts) and island natives, some locally trained and others educated in US universities. Britton knew personally many members of this local scientific elite and drew upon the resources of the University and the agricultural stations to bolster both the Survey and the Botanical Garden. The agricultural experiment stations at Mayagüez on the island's western coast and at Río Piedras outside of San Juan actually dated to the 1880s and were legacies of Spanish colonialism. They had been revived by US occupation authorities in 1901.[84]

Also included in the Survey's local institutional network were the new insular forest service stations and the Puerto Rican Department of Agriculture and Labor, whose commissioner, Carlos Chardon, was a Cornell-educated botanist and a personal

82 F. Boas to N. L. Britton, October 13, 1916, Puerto Rico Survey Correspondence, New York Botanical Garden Archives.

83 C. Berkey to N. L. Britton, December 31, 1918, Puerto Rico Survey Correspondence, New York Botanical Garden Archives.

84 Colon, in Fernandez Garcia, ed. *El Libro de Puerto Rico* 469–473, see note 55.

friend of the Brittons. Nathaniel Britton considered the island's Agriculture Commissioner to be "practically the only native Porto Rican who has ever demonstrated high ability as an original investigator in the natural sciences, and as such he is distinguished in a noteworthy way."[85] Chardon was a key figure in the Survey's success, serving as a liaison to the insular legislature, which appropriated funds to support Britton's efforts in the early 1920s.[86] This sort of assistance was essential if the Survey were to continue beyond its initial period of funding. The American Museum's Ralph Tower remarked, "it is refreshing to know that Dr. Chardon has taken such a live interest in the activities of [the Survey]. It is always well to have a friend at court."[87] When it came time to begin publication of the various investigations, Britton personally financed the botany reports, but Commissioner Chardon proved invaluable as he secured government money to underwrite the geological volumes. To do so, he had promised legislators and Governor Yager that the Academy would provide long-desired botanical and geological specimens for the insular museum (established within a newly dedicated Carnegie library building in San Juan) as well as enough sets of the Academy's published Survey reports to "supply all Porto Rico schools and libraries."[88]

Britton also established solid working relationships with transplanted North Americans, including Mel Cook, chief of the Río Piedras Agricultural Experiment Station, and NYBG alumnus Franklin S. Earle. He also cultivated ties to local American military authorities, including Army surgeon Basil Hicks Dutcher, who helped guide Britton through the Puerto Rican power structure. Dutcher was a noted public health expert and an authority on yellow fever, which he felt had been ignored by insular authorities out of fear of losing tourist business. He served as Britton's informal man in San Juan, and helped sell the Survey idea to Governor Arthur Yager in 1914.[89]

Economic Aims of the Puerto Rico Survey

Critical aims of the Survey were the promotion of Puerto Rican agriculture and forestry and the introduction of scientific farm methods from US experiment stations

85 N. L. Britton to Baily K. Ashford, School of Tropical Medicine, San Juan Puerto Rico, June 8, 1926, Puerto Rico Survey Correspondence, New York Botanical Garden Archives.

86 Carlos Chardon to N. L. Britton, June 5, 1923 and N. L. Britton to C. Chardon, June 15, 1923, Puerto Rico Survey Correspondence, New York Botanical Garden Archives. Chardon was also a shrewd bargainer, offering his mentor Nathaniel Britton an opportunity to open a branch of the NYBG in Puerto Rico in exchange for an infusion of New York funds. Britton declined.

87 R. Tower to N. L. Britton, August 21, 1922, Puerto Rico Survey Correspondence, New York Botanical Garden Archives.

88 C. Chardon to N. L. Britton, October 9, 1923, and N. L. Britton, H. O. Hovey, and H. F. Osborne to A. Yager, December, 1913, Puerto Rico Survey Correspondence, New York Botanical Garden Archives.

89 N. L. Britton to Basil Hicks Dutcher, April 10, 1914; Dutcher to Britton, April 20, 1914, Puerto Rico Survey Correspondence, New York Botanical Garden Archives.

to the island. Early in the Survey, Columbia geologist James Furman Kemp wondered if the island's experiment stations had begun systematic studies of soil samples from throughout the Puerto Rican countryside. Kemp felt that the Survey's geologists should join with experiment station soil scientists to assess the agricultural potential of various insular regions. "Puerto Rico is essentially agricultural," Kemp noted, and he predicted that by establishing the "connection of the [island's] soils with its bedrock," the geological portion of the Survey would best serve the Puerto Rican government in its efforts to catalog and harness natural resources.[90]

This economic focus, harnessed as it was to the "pure" scientific aims of the Survey, made it an even more attractive venture to wealthy New York philanthropists and the colonial Puerto Rican government. "There will be members in the [NYAS] Council," Academy president McMillan warned Britton, "who will not appreciate pure scientific work, and they are likely to raise the question, whether this is of as much value as some other things in which they could invest their money." McMillan's advice, of course, was not lost on the NYBG director. Britton, despite a temperamental inclination to just the sort of investigation McMillan questioned, readily understood the politics of philanthropy. Years of negotiating municipal support for the Garden and securing Carnegie funds for cactus research gave the director-in-chief a fairly sophisticated conception of the vicissitudes of science funding. "As to economics," Britton assured his NYAS patron, "considerable is being worked in from time to time," and he noted that the Survey's geological reports contained information of economic relevance, as did the papers published by NYAS entomologists and ichthyologists.[91] McMillan's own support was contingent upon similar commitments from the insular government. Academy Secretary Henry Crampton also assured the island's governing authorities of the economic relevance of the Survey's zoological investigations. Making recourse to ecology, Crampton noted that "while invertebrates are not always directly useful to mankind, they may serve as food for, or may affect in other ways, the species that are important to human interests." Entomological investigations, he added, were meant as an adjunct to scientific agriculture.[92]

Such arguments were not always persuasive. In 1916, the Survey's funding was interrupted when the insular legislature failed to appropriate what Britton hoped would be an annual $5,000 grant. During the previous two years, the Puerto Rican government had expended nearly $10,000 on the Survey, and Britton had expected

90 J. F. Kemp to N. L. Britton, November 5, 1914, Puerto Rico Survey Correspondence, New York Botanical Garden Archives.

91 E. McMillan to N. L. Britton, February 1, 1915 and N. L. Britton to E. McMillan, April 2, 1915, Puerto Rico Survey Correspondence, New York Botanical Garden Archives.

92 H. Crampton, "Report on Zoology of NYAS Survey," to A. Yager, June 30, 1916, Puerto Rico Survey Correspondence, New York Botanical Garden Archives.

approximately $15,000 more over the ensuing three years. In response, Britton appealed to Governor Yager, noting that the Academy had received $4,500 in private contributions, and NYBG had contributed $6,500 (much of it from Britton's own pocket) to cover all expenses incurred by botanical investigations. Britton estimated the total nongovernmental contribution to the Puerto Rico Survey at $31,000 by 1916. Yager offered his condolences, lamenting the inconstancy of legislatures, noting that "like juries [they] cannot be relied on in advance to pursue any prescribed line of action." The governor's pleas for further appropriation went unheeded by the legislature in the following year. (Though the US government had granted him veto power over legislative acts, he could not compel legislative appropriations.)[93]

Prior to 1917, the Puerto Rican government had been the creation of the 1900 Foraker Act, which had established the first civilian administration since the US occupation in 1898. Under this first "Organic Act," US civilians, appointed by the US War Department, effectively governed Puerto Rican affairs. The island governor, a US citizen, was appointed by the US President and presided over an 11-member council (only five of whom were statutorily required to be island natives) that determined legislation. In addition, the island's attorney general, auditor, education commissioner, and treasurer were all US appointees. The Foraker Act also created a 35-member legislature, which served as a lower house to the governor's council. The governor and the US Congress enjoyed veto power over all Puerto Rican legislation, and Washington exercised ultimate authority in determining which portions of the US Constitution were applicable to the tropical colony. As both a political and an economic plan for colonial administration, the Foraker Act regulated insular economic relations with the Atlantic world. It prohibited the Puerto Rican legislature from entering into commercial treaties with other nations and required the shipment of all insular imports and exports on US ships. Such economic policies, enacted over the opposition of the US Democratic Party (which contended that the act was inconsistent with the exceptional American form of colonial administration), helped resuscitate a monolithic local sugar industry.[94]

93 Minutes of the Porto Rico Committee, New York Academy of Sciences, January 26, 1916; N. L. Britton to A. Yager, June 22, 1916; A. Yager to N. L. Britton, August 16, 1916; N. L. Britton to A. Yager, April 3, 1917, Puerto Rico Survey Correspondence, New York Botanical Garden Archives. By 1923, income from sales and subscriptions to the Survey's publications were under $1,100; total expenses to date on the seven parts issued to that point were $6,400. In this context, Britton's ability to underwrite the cost of publication of the botanical volumes was essential.

94 By 1935, sugar accounted for approximately 60 percent of Puerto Rican exports, most of which was sent to the US. The total acreage under cane cultivation roughly doubled between 1899 and 1930, with an increased tendency toward concentration of land ownership by absentee landlords. See Dietz, *Economic History of Puerto Rico*, 86–98, note 1 and Arturo Carrion, *Puerto Rico: A Political and Cultural History* (New York: W. W. Norton; Nashville: American Association for State and Local History, ca. 1983): 158–173.

By the time insular funds for the Puerto Rico Survey dried up in 1916, debates in the US Congress, influenced by the continuing European war, led to a reassessment of the Foraker Act's framework of colonial governance. In 1916, fearing German occupation of the Danish West Indies, Congress had appropriated funds to purchase the Virgin Islands from Denmark. The deal was consummated, but the Danish government insisted that the islands' residents be granted American citizenship, which Congress felt politically compelled to grant to Puerto Ricans as well. With the March 1917 passage of the Jones (or "Second Organic") Act, islanders became US citizens. All adult Puerto Rican males were also granted the right to elect a newly designed bicameral legislature. The governor and his cabinet, however, remained US appointees, with the same veto powers. No changes were made to the Foraker Act's policies on migration, trade, tariffs, treaties, or the judiciary.[95]

This, then, was the political context of the first years of the Survey: native islanders sought a greater degree of political and economic autonomy despite a restrictive legislative structure. Seeking regular appropriations from the insular treasury, Britton was careful, in published descriptions of Survey investigations, to address the island's perceived economic needs; insular agriculture remained the Survey's foremost economic priority. Nevertheless, not until 1923, with Chardon's help and a personal appeal from James Kemp (who traveled to Puerto Rico to argue the Academy's position), did the Puerto Rican legislature appropriate additional funds for the Survey. A large proportion of this new appropriation financed publication of Survey reports, the first of which had been published in 1919.[96]

Britton's expansive vision of American botany naturally influenced his understanding of the Survey's role in Puerto Rican development. If harnessed scientifically, he assumed, Puerto Rico's agricultural potential and its tropical natural resources would benefit Puerto Ricans as well as their North American mentors. Encouraged by Puerto Rican elites, Britton immersed himself in several aspects of insular development, beginning with agriculture and forestry. As part of his Survey work, he published papers and popular articles, gave talks promoting island forestry, and collaborated with insular agricultural authorities on the introduction of economically valuable plant species.

95 Carrion, *Puerto Rico*, 193–199, see note 94 and Dietz, *Economic History*, 97–106, see note 1.

96 N. L. Britton to J. F. Kemp, July 5, 1923, Puerto Rico Correspondence, New York Botanical Garden Archives. The earliest Puerto Rico Survey report to be published, *Volume One of the Scientific Survey of Porto Rico and the Virgin Islands*, appeared in 1919. Britton accepted sincere plaudits from his colleagues in New York science: the American Museum's Henry Osborne wrote, "My dear Britton, I send you my warm and heartfelt congratulations on the appearance of volume one. This timely survey, which originated in your fertile brain, promises to be one of the most valuable contributions that the New York Academy, in conjunction with other institutions, has ever made to American science." Henry F. Osborne to N. L. Britton, October 14, 1919, Puerto Rico Survey Correspondence, New York Botanical Garden Archives.

Britton and Puerto Rican Development: Forestry

Britton had been aware of the economic possibilities of colonial tropical agriculture at the start of his NYBG career. In 1899, he consulted members of the Belgian consulate in New York regarding the potential introduction of the economically valuable rubber plant into the southern US and American tropical possessions.[97] By the time he began directing the Survey 14 years later, he had established close ties to Puerto Rico's network of agricultural experiment and plant propagation stations, as well as to several large private gardens. Each of these gardens introduced citrus crops from Florida and California to the island. Their staffs also cultivated varieties of cane sugar less susceptible to local pathogens.

American David W. May, an agronomist at the Mayagüez Experiment Station, planted species collected in "various countries under the Stars and Stripes" (and elsewhere) and sent to him by USDA botanists. Among these introductions were African, Javan, and Guatemalan grasses intended as cattle feed, as well as soybeans acclimatized to South America. Britton colleague O. W. Barrett administered a similar institution at Trujillo, where he maintained a catalog of the food crops grown on the island. With NYBG assistance, Barrett advocated the introduction of hundreds of types of fruit trees into the island's cultivated fields. Britton also convinced Virgin Island authorities to introduce valuable mahogany trees into their landscape.[98] By the 1920s, private Puerto Rican gardens and nurseries were, with Britton's encouragement, following Barrett and May's example, cultivating introduced species with economic potential as ornamental or food plants. Many of these species were grown nowhere else in the West Indies.[99]

Britton was similarly adept at using the Survey to help develop an insular forestry policy, a pressing need by the 1910s. With its high population density, Puerto Rico had, by the time of the American invasion in 1898, suffered severe deforestation. US Geological Survey surveyor Robert Hill, writing in 1899, had described the island as "largely deforested from a commercial point of view," and blamed "300 years of exhaustive cultivation of heavy plantation crops" for robbing a beautiful island of its

97 Pierre Mali to N. L. Britton, January 17, 1899, N. L. Britton Administrative Correspondence, New York Botanical Garden Archives.

98 D. W. May, "Plant Introductions," in Fernandez Garcia, ed. *El Libro de Puerto Rico*, 491–493, see note 55; "Brings New Plants from West Indies," *The New York Times*, April 19, 1925; "Island Has Many Food Plants but Not Half Its Quota," *Porto Rico Progress*, January 27, 1926; and J. B. Thompson to N. L. Britton [undated], Puerto Rico Survey Correspondence, New York Botanical Garden Archives.

99 N. L. Britton, "Botany and Horticulture of Porto Rico and the Virgin Islands," *Journal of the New York Botanical Garden* 26 (1925): 97.

once dense forests.[100] Hill had been optimistic, however, that a concerted forestry policy would result in considerable reforestation of denuded mountain slopes within a decade. In 1912, Lewis Murphy, another American forester, found the same conditions that Hill had lamented at the turn of the century. Surveying the Puerto Rican landscape for the insular Department of Agriculture, Murphy examined a variety of physiographic features and other factors affecting forestation, including soils, drainage, climate, land distribution, and population. After six months' fieldwork, he reported to his patrons the troubling fact that island residents annually consumed three times as much timber as Puerto Rican forests produced. Much of this consumption Murphy traced to the habits of charcoal makers, who cut young growth needed to replenish forest lands. "Failure to put an end to the destructive practices that are rapidly reducing the forests or to provide the means of developing and fully utilizing them in a scientific manner has already brought about a shortage in the domestic supply of wood and consequent hardship to the people [of Puerto Rico]," Murphy warned.

In good Progressive-era fashion, Murphy recommended that insular authorities establish a forest service, an innovation that, he predicted, would bring over $350,000 in economic gain. Murphy lamented the "total lack of property survey, lax title registration, and the free and unmolested operation of the prescriptive right" that made it easy for Puerto Ricans to practice swidden agriculture in the interior. This practice of clearing forest land by girdling and burning invariably claimed the "best types" of forest first, Murphy asserted, because such trees produced the richest ash for fertilization and burned more quickly than areas of shrubby undergrowth. He also advocated the elimination through reforestation of existing cattle pastures. Because these small ranches furnished no beef for export, Murphy reasoned, "there seems to be little economic justification for any longer retaining the bulk of the cleared uplands in pasture. Their partial or complete reforestation would add materially to the productive wealth of the island." In sum, Murphy, writing just before the Puerto Rico Survey began in earnest, called for direct governmental action to reform what he considered to be typically "primitive" tropical agricultural practices. Reform would entail a

> serious educational campaign combining, unifying, and extending the work of the public school systems, the agricultural experiment station, and any other agencies working for rural betterment, until there can be instilled into the mind of the 'conuco' farmer a proper regard for the fundamentals of economic agriculture, by which continuous cultivation under a suitable rotation of crops will be substituted for the present nomadic system.

100 Robert T. Hill, *Notes on the Forest Conditions of Porto Rico* (Washington, DC: Government Printing Office, 1899).

Rational, scientific agriculture, as well as forestry, Murphy concluded, would transform the Puerto Rican economy to the benefit of both the colony and the US. Failure to adopt such measures, however, would lead simply to greater privation on the island.[101]

One of Murphy's key sources on Puerto Rican botany had been Nathaniel Britton, whose keen interest in tropical forestry led him to campaign vigorously for the establishment of an insular forest service and a dedicated program of reforestation. Britton's commitment to forestry dated to his early career in the Columbia College botany department. He had been a leading figure, along with Columbia's president Seth Low and Gifford Pinchot, in the creation of a course of study leading to the degree of "forest engineer."[102] In April 1915, the NYBG's director-in-chief prepared to address the National Academy of Sciences, to which he had been elected the year before, on the subject of Puerto Rican forests.[103] Echoing Hill and Murphy's pessimistic assessments, Britton contended that the remaining forests were insufficient for the needs of the island, either for fuel or for the conservation of soil and the retention of moisture. In January 1903, Theodore Roosevelt had in fact established by executive order a National Forest Reserve of 100 square miles in the Luquillo Mountains in the northeastern part of the island; this land would be placed under the administrative control of the US Forest Service in 1917.[104] Yet, Britton asserted, further action was warranted. The Luquillo reserve, he noted, was the last significant forest remaining in all of Puerto Rico, and the deplorable state of the island's forests called for "an effective . . . policy" that would bring "the entire higher portion of the main mountain axis, in which most of the larger streams arise, from the region of Aibonito [in the south-central portion of the island] westward to the region of Maricao into public ownership and control." Much of this mountainous region, he contended, was unsuitable for agriculture or pasture and would be "vastly more valuable to the colony in forests." Britton gave this speech as planned, save for substituting the word "island" for "colony" on the advice of Governor Yager.[105]

101 Lewis S. Murphy, *Forests of Porto Rico Past, Present and Future and Their Physical and Economic Environment* (Washington, DC: Government Printing Office, 1916): 13–14.

102 N. L. Britton Administrative Correspondence, with Columbia University Department of Botany, 1892–1896, folder 30, New York Botanical Garden Archives. In 1901, Britton had also considered intervening to secure for the Garden a parcel of 5,000 acres of New Jersey forest land that Abram Hewitt had offered to the New Jersey state government. He was not seeking to establish another forestry school, of which the northeast had several, but was merely engaging in a brief speculative exercise in institution-building. No such deal was made.

103 On Britton's election to the National Academy, see Daniel T. MacDougal to N. L. Britton, June 7, 1914, Britton–Rose Correspondence, New York Botanical Garden Archives.

104 Murphy, *Forests of Porto Rico*, 55, see note 101.

105 N. L. Britton, address to the National Academy of Sciences, April, 1915, manuscript copy in Puerto Rico Correspondence, New York Botanical Garden Archives.

Britton's forestry plan was an audacious one; the distance between the regions he cited before the Academy of Sciences is equivalent to roughly half the length of the entire island, or approximately 45 miles. Though choosing to elide the issue of Puerto Rico's colonial status in public, Britton nevertheless clearly maintained a scientist's utilitarian attitude toward the tropical possession's timber resources. He suggested that the insular government "condemn and purchase all lands lying above 2,200 feet, conserve all forests, and reforest the denuded slopes and summit with trees valuable for their wood." He also called for demonstration forests to be planted at intervals along main roads through the interior of the island. Long-term bond issues, he advised, would pay for these land acquisitions, and would "place most of the expense in the future," presumably when the island's booming timber and other industries would enable its taxpayers to meet those expenses. Only such a bold action, Britton maintained, would begin to reverse the "calamity" of deforestation that left thousands of acres of Puerto Rican land bare of trees.[106]

As he administered the NYAS Survey from his Bronx office and on his annual Caribbean trips through the 1920s, Britton continued to be a staunch public advocate of a concerted forest policy for Puerto Rico. During his winter visits, he toured the island's agricultural experiment stations, conferring with administrators and writing of his travels in the Garden *Journal* and in San Juan and New York newspapers. By the 1920s, Britton could happily report that the Río Piedras Forest Service Station was supplying nearly one million saplings annually for reforestation purposes on both public and privately held lands. (Extending forest policy and making it relevant to private owners would be key elements in the success of any insular reforestation effort. Fully 94 percent of Puerto Rican land remained in private hands. For lack of funds, Britton's ambitious 1915 plan to make public a huge swath of the island's interior had never been carried out, despite its private endorsement by Governor Yager.[107]) Encouraged by the success of the insular forest service in producing future timber crops of both Old and New World tropical species, Britton called for the expansion of existing government nurseries and the creation of a permanent government-funded arboretum, in order that enough parcels of island land could remain fallow in between growing periods.[108]

Despite these outward signs of success and the outspoken support of a prominent New York scientist (whom San Juan journalists had dubbed the greatest authority on

106 Ibid.; see also "Now My Idea Is This! Daily Talks with Thinking New Yorkers on Subjects They Know Best," *New York Evening Post*, May 21, 1923 and "For Porto Rican Forest Reserve," *Philadelphia Record*, July 6, 1924 for evidence of Britton's continued interest in tropical forestry.

107 A. Yager to N. L. Britton, April 12, 1915, Puerto Rico Survey Correspondence, New York Botanical Garden Archives.

108 N. L. Britton, "The Porto Rico Forest Service Nursery," Porto Rico Progress, March 21, 1925.

the subject), insular forest policies were criticized by some within the US forestry community as mere political window dressing lacking substance or efficacy.[109] Writing in the *Journal of Forestry*, University of Puerto Rico forester W.D. Durland contended that the public accolades afforded the insular forest service's reforestation program were unwarranted, because nothing had been done to convince the private owners of 94 percent of the island's land of the economic necessity of rational forestry policies. "Though eight years have passed since forestry became a part of officialdom," Durland wrote, "it is to be regretted that the insular government is in no better position today than at the beginning to intelligently advise the private owner of forest lands as to what he may best do and expect." No "definite or sincere effort," he admonished, "had ever been made to seek the solution [to deforestation] in spite of the fact that government paraphernalia and financial support is supplied for this purpose."[110]

To some extent, Durland's criticism merely reflected a degree of institutional rivalry. Though Britton maintained close personal and professional ties with a large number of federal as well as insular scientists, it is likely that his famously abrasive personality and the forceful promulgation of his opinions in the interest of the New York Botanical Garden did alienate some members of the scientific community. And in 1925, at the time Durland published his paper, the Garden's director-in-chief had been a vocal critic of the University of Puerto Rico for not devoting additional land for the purposes of an experimental arboretum. Britton himself attributed Durland's rebuke to the University's antipathy to forestry in general and Britton's plans for an insular arboretum in particular.[111] In addition, the very public arrival of the Survey in Puerto Rico in 1914–1915 had met with cautious approbation from some US foresters in Washington. US Forest Service official H. S. Graves, upon reading a 1915 newspaper account of the Survey in which Britton promoted reforestation, suggested privately to Governor Yager that NYAS researchers contact the Forest Service in order to avoid duplicate efforts. In fact Graves pointed to Murphy's 1912 fieldwork (which, however, was not published until 1916) as an example of "extensive" US government work already conducted in the region. Yager duly informed Britton, who, undaunted by a hint of rivalry, continued his very public efforts for the remainder of his career. Such actions were not likely to win Britton unalloyed approval, and Durland's criticism can be understood only in this context.[112]

109 For an example of the esteem in which Britton was held by many influential Puerto Ricans, see "Forest Needs," *Porto Rico Progress*, March 28, 1925.

110 W. D. Durland, "The Status of Forestry in Porto Rico," *Journal of Forestry* 23 (1925): 913–918.

111 N. L. Britton to C. Chardon, December 7, 1925, Puerto Rico Survey Correspondence, New York Botanical Garden Archives.

112 A. Yager to N. L. Britton, July 8, 1915, Puerto Rico Survey Correspondence, New York Botanical Garden Archives. However, Britton also had many allies in the federal scientific agencies, and the Survey's botanical publications were eagerly sought after by federal botanists not directly involved

Britton's involvement in the Puerto Rico Survey and his participation in insular debates over forestry were undertaken with the interests of NYBG in mind. Reforestation, he knew, would not only provide economic benefit to the island, but would, if carried out under the direction of New York's botanists, add to the Garden's prestige as well. To this end, he continued dispatching fieldworkers to Puerto Rico from New York to conduct ecological surveys designed to garner publicity for the Garden's strong stance in favor of evolutionary theory (1925 was the year of the Scopes trial) and to collect specimens of rare native trees for display in the tropical houses of the great Conservatory. Henry Gleason spent three months in the field with Mel Cook and took photographs of the island's vegetation in a concerted effort to document the island's flora "before it has been completely destroyed or more seriously modified by the activities of man."[113] Other members of the Survey team returned to New York bearing specimens of rubber trees introduced in Puerto Rican nurseries, as well as a dozen types of tree ferns, some unique to the island. These specimens were displayed to the Garden's urban audience, where their exoticness brought further publicity for NYBG in the city's press.[114] By the mid-1920s, Britton had reason to be pleased with the progress of the Survey (despite the financial setbacks); with advances in Puerto Rican forestry and horticulture; and, not least, with the subsequent acclaim afforded the Botanical Garden. As he asserted to Garden members in 1926

> altogether the representation of exotic plants in Porto Rico is now more comprehensive than anywhere else in tropical America and it must compare well with establishments in the Old World tropics. The distribution of exotics from these government plantations to gardens and forests, as well as plants brought in by individuals, gives continuously greater diversity and interest to private gardens and grounds and tends to diversify

in aiding the NYAS in field or laboratory work. See, e.g., October 26, 1925 letter from Paul Russell, Asst. Botanist at USDA's Bureau of Plant Industry to Britton asking for a copy of a set of the botany volumes of the Survey, calling these publications "of great value to this office in its foreign work."

113 H. A. Gleason, "Ecological Studies in Porto Rico," *Journal of the New York Botanical Garden* 27 (1926): 104–106; "To Apply Evolution in Plant Life Study: Botanical Garden and Porto Rico Government to Chart 2,000 Species on Island; Seek Ancestors of Flora; Will Use Ecology for First Time in Tropics to Organize the Plant Communities," *The New York Times*, January 4, 1926. The *Bronx Home News* quoted Gleason further: "Ecology is vital to agriculture. It determines today what lands should be used for agriculture, what for grazing and what for forests." Quoted in "Bronx Garden Experts Will Study Tropical Plant Ancestry to Support Evolution Theory," January 10, 1926. See also, "Dykes Urged to Save Shore of Porto Rico: Island Slowly Losing Its Fight with Seas, Says H. A. Gleason after Survey of 10,000-Acre Swamp," *The New York Herald Tribune*, August 15, 1926 and "Pumps Operated by Trade Winds Would Aid in Reclaiming Area," *The Christian Science Monitor*, August 17, 1926.

114 "Succeeds with a Very Rare Fern," *The New York Times*, January 17, 1926.

> agriculture and forestry. Forest protection and reforestation are beginning to show important results. . . .[115]

Within a decade of his speech to the National Academy on Puerto Rican deforestation, it appeared that Britton had begun making good on his promise to the island government that "if the scientific survey should lead to the establishment of a forest policy, this would be worth far more than its total cost to the colony."[116] The work of the various experiment stations in conjunction with the members of the Survey was proceeding at a pace that gratified the NYBG director. The insular forest service planted new saplings at a rate of 2,000 per day, and the Río Piedras station alone cultivated 300 species for reforestation.[117] "That means have come to be taken to repair this damage [deforestation] to the country is a noteworthy contribution to the future welfare of the colony," Britton proclaimed.[118]

Meanwhile, the director-in-chief had been advising officials of the Polytechnic Institute at San German, Puerto Rico, on developing a tropical arboretum. In a letter to J. W. Harris, the school's president, Britton drew upon his own experience in establishing the Botanical Garden, giving tips on the planting and layout scheme of the trees themselves. Based upon his knowledge of New York municipal politics, he urged Harris to maintain "the advantage of permanent corporate control, [as] this is almost essential for the future of such a collection." The number of tropical tree species that would thrive in San German, Britton wrote enthusiastically, "is rather enormous, probably aggregating 5,000 or more" and he advised Harris that approximately 500 acres would be needed to cultivate this many. Although several hundred species could be procured from the island forest service, Britton augmented the number of species available for the tropical arboretum by introducing seeds from the US mainland. To T. B. McClelland, director of the Mayagüez Experiment Station, Britton sent a set of seeds of Florida palms collected by Garden botanist J. K. Small. The resulting seedlings, Britton instructed, were to be introduced at a new San German arboretum.[119]

115 N. L. Britton, "Further Botanical Investigation in Porto Rico" *Journal of the New York Botanical Garden* 27 (1926): 97.

116 N. L. Britton to A. Yager, April 2, 1915, Puerto Rico Correspondence, New York Botanical Garden Archives. A more current assessment of the island's timber resources can be found in Peter Franco, Peter Weaver, and Susan Eggen-McIntosh, *Forest Resources of Puerto Rico* (US Forest Service Southern Research Station, 1990): 69.

117 N. L. Britton, "Further Botanical Studies in Porto Rico" *Journal of the New York Botanical Garden* 28 (1927).

118 N. L. Britton, "Forestry and Agriculture in Porto Rico," *Journal of the New York Botanical Garden* 29 (1928): 101.

119 N. L. Britton to J. W. Harris, President, Polytechnic Institute of Puerto Rico, April 13, 1927 and November 12, 1927. N. L. Britton to T. B. McClelland, Puerto Rico Survey Correspondence, New York Botanical Garden Archives.

Britton also worked to introduce mahogany trees into the island, while lobbying a new governor, H. Towner, for the establishment of a School of Tropical Agriculture. Proposed by the National Research Council and eventually approved by the Puerto Rican government, this plan affiliated the new agricultural school with Cornell, much as the island's school of tropical medicine was linked with Columbia University.[120] This work in tropical horticulture and silviculture, Britton proclaimed, "provided an unsurpassed opportunity for the observation of many kinds of plants hitherto unknown to us or known to us only from greenhouse cultivation." All of these efforts, he knew, resulted in hundreds of seeds and specimens for the Garden's greenhouse collections in New York. As Head Gardener Kenneth R. Boynton reported to Garden members in 1926, "much interesting material is now flourishing in our greenhouses as a result of last year's Porto Rico trip. From the Río Piedras Forest Station we have *Funtumia elastica* (Lagos Rubber)," as well as timber from Africa, Central America, and India.[121]

Britton knew that the introduction of fast-growing exotics was not a complete answer to Puerto Rico's reforestation problem. Even as the Rio Piedras Forest Station distributed hundreds of saplings of introduced species, he advocated rescuing "from total extinction native valuable endemic trees, which have become very rare."[122] Efforts were underway to preserve certain native trees valuable as sources of building timber or as ornamentals.[123] R. A. Harper, Chair of the Garden's Board of Scientific Directors during the 1920s, agreed with Britton's positive assessment of the NYBG's involvement in tropical forestry. In light of contemporary concerns about supplies of northern hardwood timber, Harper encouraged Garden scientists in their efforts in other Caribbean locales, including Cuba, where they had begun a preliminary survey of the eastern forests in conjunction with American sugar, fruit, and railroad companies.[124]

Britton and Puerto Rican Development: Mining

Britton took an active interest in the investigation of mineral resources and potential mining prospects on the island, in addition to forestry and horticultural work. He served on a governmental committee even after retiring from the Garden. Columbia geology professor J. F. Kemp, a Britton intimate, had been the Survey's key geological investigator. Much of Kemp's Puerto Rican work was accomplished between February and April 1923, when he journeyed with the Brittons from New York to San Juan, where they met with insular senators and representatives and began touring copper

120 Britton, "Further Botanical Studies," see note 117 and "Forestry and Agriculture," see note 118.

121 K. R. Boynton, "Some Recent Accessions to the Living Plant Collection," *Journal of the New York Botanical Garden* 27 (1926).

122 *New York Evening Post*, May 21, 1923, see note 106.

123 Britton, "Botany and Horticulture of Porto Rico," see note 99.

124 *Bulletin of the New York Botanical Garden* 13 (1925).

mines with a prominent local mining engineer. Later that spring, the New Yorkers toured a placer gold district, "where, guided by the prehistoric surface pits made by the Arawak Aborigines," they explored small gold-quartz veins. Then, as guests of a local sugar planter, the party examined fossil beds and a promising district in the northeastern part of the island rich with iron ore. With the construction of a relatively short rail line, Britton and Kemp concluded, this site gave "promise of supporting a not inconsiderable mining industry" in the near future.[125]

An enthusiastic economic geologist, Kemp would have continued leading the New Yorkers' efforts at Puerto Rican development; his death at a relatively young age, however, left Britton, as chair of the Survey committee, to continue the campaign as its most respected authority. Though a botanist by profession, Britton had been trained as a geologist at Columbia College's School of Mines and maintained an active interest in both paleobotany and economic geology. By the time of his retirement from the Garden in 1929, his standing within Puerto Rican science and his reputation in government circles was such that he was asked to chair an insular Committee on Mineral Resources. This effort was the last major undertaking of his life.

Organized in early 1932 at the Puerto Rican Bureau of Commerce and Industry located in San Juan's gubernatorial palace, the Committee consisted of native members of the insular legislature; local eminences and bureaucrats (including the island's secretary of the interior); and another Survey member, Smith College geologist Howard Meyerhoff. Funded by the Puerto Rican legislature, the Committee on Mineral Resources was to identify sites worthy of exploitative attention and suitable to sell to "local or foreign capital." Such potentially lucrative regions included deposits of marble, iron, kaolin, and lignite. Committee members advocated legislation promoting private prospecting, while arranging for the University's geologists to assay ore samples free of cost.

Meeting weekly, the Committee, chaired by Britton, established a list of priority metals and minerals that included gold, iron, magnesium, copper, silver, platinum, mercury, zinc, and lead. Britton suggested that the Committee investigate the availability of nonmetallic resources as well. Key materials included clay (for bricks), lime, lignite, petroleum, and guano (as a fertilizer). In 1932, Britton's Committee sent 15 tons of iron ore samples to US Steel and Bethlehem Steel. Sending samples to the states was necessary because assaying equipment was in short supply at the island's scientific institutions. The Committee also consulted regularly with US Bureau of Mines personnel. Among other suggestions, the Bureau noted the possibility of creating mountain reservoirs with hydroelectric power generators, which could be used to run mines in addition to local development. The overall judgment of the Bureau

125 J. F. Kemp to N. L. Britton, April 15, 1923, Puerto Rico Survey Correspondence, New York Botanical Garden Archives.

(and various NYAS geologists) regarding prospects for Puerto Rican mining was highly positive. One engineer "was under the impression that mining development could become one of the important industries in the Island." Working with this Committee, Bureau consultants identified a number of sites worthy of commercial exploitation. In addition, Committee members conferred with University of Puerto Rico officials regarding the establishment of a geology curriculum and solicited the legal opinions of the insular attorney general regarding the validity of old Spanish-era mining laws.[126]

The Committee's key responsibility was to lure American mining companies into the region in order to develop the island's mountainous interior. Upon learning of particularly promising manganese deposits near Adjuntas, the Committee determined to recruit private investors for a commercial mining operation. After using military roads to conduct reconnaissance work on the island during the summer of 1933, Britton's protégé Howard Meyerhoff concluded that the Committee must give investors "every encouragement and help," because commercial mining concerns would have "both the interest and access to the capital for the development of the manganese." Even if the deposits proved unprofitable, Meyerhoff counseled, "considerable employment [would] be available to local residents while that conclusion is being reached, and a moderate tonnage of manganese will be obtained and shipped in the course of the development work."[127]

Meyerhoff's pragmatism extended to his assessment of the Puerto Rican political landscape of the early 1930s. In a candid aside to Britton, the geologist asserted that the island's most recent governor, a Roosevelt appointee, was too self-interested and potentially dishonest to last long in public life. In the interim, Meyerhoff suggested, the Mineral Committee might best exploit this corrupt new administration by playing upon the insular governor's innate greed:

> He [Governor Gore] is interested in every field of development where he thinks he can put his own and his friends' money at a profit; he is not in the political game for any other motive that I could detect. Between the two of us, I think we can play his game to the Committee's and the Island's benefit without much risk to our standing or moral integrity, for I doubt if he will live a long gubernatorial life unless he changes his tactics or Roosevelt is exceptionally loyal. If we are willing to stoop to it, we maybe able to get the Committee on a permanent footing by using his self-interest for what it is worth."[128]

126 Minutes of the Insular Committee on Mineral Resources in Porto Rico, Nathaniel Lord Britton Records, Box 5, New York Botanical Garden Archives.

127 Based upon Meyerhoff's [unpublished, internal memo] "Progress Report on Mineral Investigations, July 3–July 20, 1933," New York Botanical Garden Archives.

128 An article in *The Nation* during the fall of 1933 echoed Meyerhoff's negative assessment of Puerto Rican Governor Gore's administration. In Gore's case, FDR did *not* prove "exceptionally loyal," and he was replaced by Governor Horton (a Meyerhoff acquaintance).

Although he successfully shepherded the Survey's botany volumes through the press (and paid for their publication), Nathaniel Britton did not live to see either the end of this undertaking nor the outcome of efforts by the Committee on Mineral Resources to stimulate insular development. His involvement in each of these endeavors, though incomplete, was representative of the cosmopolitan scientific outlook he and his staff cultivated in order to gain prestige within the international botanical community.

The Scientific Survey of "Porto Rico" in Retrospect

The New York Academy of Science's natural history survey of the US colony of Puerto Rico remains a significant and relevant scientific achievement. Nathaniel Britton's work with the Survey "stands to this day as one of the most complete descriptions of the natural history of any tropical area ever attempted, and for some disciplines it remains the principal reference available for the islands." Indeed, into the latter half of the twentieth century, US government scientists would make use of the various publications that came out of Britton's ambitious vision.[129] In contrast to those European scientific agencies whose presence at the nexus of imperial power bestowed institutional prestige, legitimacy, and crucial financial solvency, most American research institutions with roots in the nineteenth and early twentieth centuries remained dependent for support upon private philanthropy. Public funds, when available to US researchers, were usually to be found at only the state or municipal level.[130] This pattern was clearly apparent in the case of the Survey. Britton's role in securing funds from both private and public (Puerto Rican) sources represented a reprise of his efforts decades earlier when he founded New York City's botanic garden.[131] In doing so, he astutely tied his own institutional and research goals to an economic agenda attractive to authorities interested in fostering insular development. Puerto Rico proved to be a particularly suitable venue for the practice of Britton's metropolitan science.

As objects of this specialized attention, the regional flora, fauna, mineral deposits, and, eventually, the human populations of Puerto Rico fulfilled the institutional and personal desires of New York's itinerant scientists quite well. San Juan and its tropical environs provided a congenial setting for the practice of natural history, as well as the newer sciences of ecology, forestry, and anthropology. In addition, the island

129 Colon, ed. *The Scientific Survey of Puerto Rico*, see note 21. See also an oral history interview of Henry Allen Gleason conducted by William Campbell Steere on March 7, 1973 in New York Botanical Garden Archives. Personal telephone interview with Butch Grossman, retired U S Geological Survey hydrologist and former president of the Puerto Rican Natural History Society, November 21, 2001.

130 Baatz in Colon, ed. *The Scientific Survey of Puerto Rico*, see note 21. Baatz noted that the exception was the US Geological Survey, which enjoyed the regular support of the federal government.

131 Ibid., 4.

allowed North Americans an opportunity to display their institution-building prowess. Puerto Ricans became in a sense the *objects* of such metropolitan expertise, a largesse bestowed with as much attention to the benefits that accrued to the donor as to the recipient. That situation had the potential to relegate native Puerto Ricans to peripheral roles in the practice of insular science.

For the set of New York scientists working under Britton on the Survey, Puerto Rico enjoyed a reputation as a place rich in raw materials for projects that were potentially ambitious in scope, lavish in execution, or both. The Survey spawned several other collaborative efforts to chronicle insular natural history in a variety of formats. After the NYBG botanists' tropical work had begun garnering the acclaim Britton anticipated, the wife of Smithsonian Institution Secretary C. D. Walcott requested his help publishing a lavish volume of watercolor plates (modeled on the nineteenth century publications of Audubon's birds) that depicted the island's wildflowers. The wife of another American scientist also conferred with Britton on the preliminary compilation of a separate monograph of a *Flora Borinqueña*, or Puerto Rican flora. Meanwhile, Britton himself, years into retirement, worked to establish a public museum of insular archaeology. He reported to Puerto Rican Governor Theodore Roosevelt, Jr., the results of his canvass of various private collections of the island's antiquities and passed judgment on the suitability of sets of specimens for museum purposes.[132] All three of these examples are, in fact, typical of the practice of a metropolitan botany and museum science: like the NYAS surveyors themselves, these urban North Americans had traveled to a tropical Caribbean colony and had discovered an abundance of natural historical material (wildflowers, native tropical plants, and even the archeological remains of extinct insular populations) to underwrite their agenda of discovery, collection, and display of tropical resources in cosmopolitan settings.

Despite the paternalistic assumptions behind them, US colonial governing institutions allowed for gradually increasing levels of Puerto Rican self-determination. Consequently, island elites retained some control over their own status as scientific curiosities. As Britton learned to his chagrin in 1916, Puerto Ricans were not automatically prepared to support his particular agenda, especially when they controlled the insular legislature. Only the active intervention of local allies, along with Britton's skill in articulating the Survey's scientific aims in terms of economic utility, ensured its long-term success as a research project.

In a 1988 assessment of the causes of tropical underdevelopment, Daniel Headrick identified the introduction of Western methods of resource extraction as a primary

132 See N. L. Britton, "Notes on Collections of Porto Rico Archaeological Specimens in Porto Rico," manuscript in New York Botanical Garden Archives. Britton's margin notes on this manuscript indicate that the report was "prepared for Gov. Roosevelt," and it was dated April 1930.

culprit. Western colonial administrators and scientific authorities, Headrick noted, were not content merely to tax subjects. Rather they sought to increase production of tropical commodities via industrial methods, which, in turn, involved a massive transfer of technology from the West to the tropics, stimulating further international trade and bringing "greater security, more regular food supplies, and a growing demand for [tropical] labor." Tropical populations grew further, but, Headrick asserts, this technology transfer neither diversified tropical economies nor raised per capita incomes significantly.[133] Among North American metropolises, New York City, by the time of the Scientific Survey of Puerto Rico, had long been a primary importer of tropical goods. By the 1870s, the city dominated the Brazilian coffee trade, and, in an effort to rationalize this market, New York merchants had established the New York Coffee Exchange in 1881.[134]

To recognize the element of self-interest inherent in undertakings such as the Survey and the kindred projects it spawned does not obviate the proper acknowledgement of a degree of genuine "humanitarian and reformist" impulses behind the development of colonial US science. Accounts of metropolitan tropical endeavors must go beyond mere utilitarian interpretations.[135] The Brittons and their New York cohort, to some extent products of that nebulously defined American historical period, the "Progressive Era," were in earnest as they couched their scientific proposals in the language of benevolent economic and institutional development. And, to some extent as well, their prescriptions of "wise use" conservation policies, once implemented, proved genuinely effective in safeguarding Puerto Rican resources from utter depletion. By agitating vigorously for a rational set of forest regulations and the establishment of government-sponsored tree nurseries, Britton set in motion a long process that enabled late-twentieth-century foresters to conclude:

> Puerto Rico's forest resources are recovering from the land use practices prevalent in the nineteenth and first half of the twentieth centuries. [A] 1990 survey revealed an

133 D. R. Headrick, "Economic Botany and Tropical Plantations," in D. R. Headrick, *The Tentacles of Progress: Technology Transfer in the Age of Imperialism, 1850–1940*. (New York: Oxford University Press, 1988): 7.

134 Richard P. Tucker, *Insatiable Appetite: The United States and the Ecological Degradation of the Tropical World*. (Berkeley: University of California Press, 2000): 190–191.

135 Richard H. Grove, *Green Imperialism: Colonial Expansion, Tropical Island Edens and the Origins of Environmentalism, 1600–1860* (Cambridge, UK: Cambridge University Press, 1995): 7–8. Britton's utilitarian concern for the productive forests of Puerto Rico lends credence to Grove's linkage of the rise of a modern conservation movement and the Euro-American encounter with tropical landscapes. Grove contended, "as colonial expansion proceeded, the environmental experiences of Europeans and indigenous peoples living at the colonial periphery played a steadily more dominant and dynamic part in the construction of new European evaluations of nature and in the growing awareness of the destructive impact of European economic activity on the peoples and environments of the newly . . . colonized lands." The problems of environmental degradation and resource

> increase in forest land, better stocking with larger and higher quality timber trees, and a maturing secondary forest. . . . The results of [this] survey portray Puerto Rico as an example of how once heavily degraded tropical forests can recover to a level of productivity and usefulness. Knowledge of these trends can facilitate the development of management techniques to enhance the value of the timber resource and provide other forest-related benefits for the residents of Puerto Rico.[136]

These words echo those spoken and written by the founder of the New York Botanical Garden almost seven decades earlier. It is in this sense of possibilities recognized and courses of action delineated, then, that the Scientific Survey of Puerto Rico deserves recognition as one of the most significant US scientific projects conducted beyond the continental boundaries of the US prior to the Second World War.[137]

Overemphasizing the Survey's success, however, risks obscuring a final, ironic reality, one that reflects a clash of competing visions of agricultural development. The ability of the United States to govern and exploit its tropical colonies has historically been contingent upon the efforts of American researchers in various branches of economically important sciences, including agriculture and forestry.[138] And yet such efforts in turn made possible, despite the best intentions of men like Britton, the Puerto Rican landscape of 1940, in which only 16% of cultivated land was found in farms under 20 acres and 50 US corporations, many based in New York, controlled 250,000 acres of the island's prime, agricultural, real estate. Much of this acreage was devoted to the cultivation of cane sugar, involving a set of agricultural practices with particularly devastating social and ecological consequences: sugar monoculture involved clear-cutting thousands of acres of Puerto Rican forests.[139]

---

depletion in colonial states such as Puerto Rico and the Virgin Islands, Grove argues, pp. 3–7, ultimately stimulated "rigorous analytical thinking" about the processes of ecological change and about ameliorative strategies of land control on the part of colonial policymakers. In the case of Puerto Rico, of course, Britton identified Puerto Ricans themselves as the primary agents of deforestation.

136 Franco, Weaver, and Eggen-McIntosh, *Forest Resources of Puerto Rico*, see note 116.

137 Baatz in Colon, ed. *The Scientific Survey of Puerto Rico*, see note 21. Baatz errs, however, in asserting that the Survey was the "only substantial endeavor by North American science before 1940 to go beyond the boundaries of the US." He overlooks Britton's previous activities in the decades leading up to the Survey, including the Garden's establishment of a tropical botanical laboratory in Jamaica in the early 1900s; the series of ambitious cactus-collecting expeditions conducted throughout the Western Hemisphere; and the work of Garden-affiliated scientists in Puerto Rico prior to 1913. Britton's contributions to Puerto Rican forest policy were recognized in his lifetime, when the forest service, with his permission, designated a mountain peak in the island's Luquillo National Forest as "Mt. Britton."

138 Tucker, "Insatiable Appetite," 4, see note 134.

139 Ibid., 55–56.

Chapter 7
The New York Botanical Garden in the 1920s: Negotiating Priorities

The year 1915 marked the twentieth anniversary of the Botanical Garden's official arrival in Bronx Park. Speeches were made, conference papers delivered, and a dinner held. Membership by 1915 had climbed to 1,000, and an average of 40,000 New Yorkers entered the Garden's gates on a typical summer Sunday or holiday. Despite these modest signs of success, Nathaniel Britton's institution had been faced almost continually with a shortfall in revenue. Its endowment, which was relatively substantial by nineteenth century standards, had failed to produce an income commensurate not only with Britton's nationalistic ambitions but with the basic institutional needs of a twentieth-century public park, museum, and research center. Operating funds from New York City's Department of Public Parks, although crucial, were disbursed by the local political structure. Annual appropriations were subject to the vagaries of the city's economy and contingent upon Britton's ability to negotiate with municipal officials while couching his projects in terms of civic utility.

Within several years, these financial pressures undermined Britton's previously unchallenged administrative control over the Garden, when a long-time member of the Board of Scientific Directors, Columbia University's animal physiologist Frederic Schiller Lee, ascended to the presidency of The New York Botanical Garden (NYBG) Corporation. Convinced of the Garden's inadequacies as a research center, Lee sought to alter its institutional priorities, refurbish its physical plant, and secure additional funds for its endowment. The botanic garden's institutional weaknesses (either perceived or real) and the threat that these shortcomings could pose to its scientific reputation meant that the struggle during the 1920s between Lee and Britton was more than simply a personal feud between cranky, aging colleagues. To the extent that Lee's negative assessment of the quality and relevance of the research program and priorities was accurate, the Garden, hampered by a small income from the city and an insufficient endowment, would see its scientific reputation suffer. This diminished reputation, in turn, would undermine Britton's efforts to secure the private donations upon which the Garden depended as a research institution and as an open, accessible, public space. It was a potentially devastating cycle for an institution funded on a public-private partnership basis. Thus, even as Britton managed the Puerto Rico Survey with typical zeal, he faced pressure from within the Garden's managerial hierarchy. Lee's challenge reflected differences in disciplinary orientations among the NYBG's scientific managers. It also indicated yet again the extent to which the Garden's activities took place within a contested urban landscape.

Lee, a controversial figure in contemporary debates over animal vivisection, presided over the NYBG Board of Managers from 1924 to 1927. Unlike his predecessors, who had deferred to Britton in administrative and scientific matters, Lee used the authority granted him by the NYBG charter in an attempt to redefine the Garden's scientific priorities. He aimed to mold the NYBG research agenda so that it matched his own understanding of contemporary experimental and applied botany. Wisely, Lee also sought new income for the Garden; he hoped to enhance its prestige as an ornamental landscape. To this end, in an ironic twist, he hired the Olmsted Brothers firm, whose advice had been so controversially absent at the Garden's inception during the 1890s.

Was Lee correct in his scathing assessment of the Garden's scientific relevance under Britton's guidance? Or was his judgment clouded by an alternative conception of American botany? Were the priorities of a physiologist and public health researcher simply inappropriate for a botanical garden? Finally, what was the status of the Garden's *garden*? Had either the outdated research priorities Lee ascribed to Britton or the minimal endowment and low annual city appropriations undermined the Garden's beauty? Lee considered Britton's passion for the herbarium somewhat useful, but outdated. Taxonomy, Lee contended, had been replaced by a new experimental focus on the plant, the function of its parts, and the applicability of this research to broader problems of public health. The Olmsted evaluation of the Garden's landscape inaugurated another in a series of fundraising drives to expand the NYBG endowment and to secure its place among the city's leading cultural institutions. This chapter assesses the Lee years, evaluating his arguments against Britton's brand of botany, while placing the debate between these men in the context of the development of a unified science of life, "biology," during the early twentieth century.

Economic and Scientific Concerns

Frederic Schiller Lee joined the Board of Scientific Directors of NYBG in 1903. Prior to assuming the presidency, he had not been highly active in Garden administration. In 1912, however, he had been asked by several colleagues to discuss with Nathaniel Britton the relatively low salaries received by the NYBG curatorial staff. In May of that year, Lee addressed the Board of Managers to assert that low pay ("pitifully inadequate both to the present cost of living and to the high standing of the men involved") hindered the ability of Garden botanists to perform first-rate research.

NYBG's prestige, he admonished, was consequently in jeopardy. The Managers, heeding his warning, approved salary increases, provided the annual city appropriation was adequate in 1913.[1]

Lee's concern with the Garden's prestige was a continuing theme throughout his NYBG career. As a university-based scientist and a wealthy donor to scientific institutions, he recognized the extent to which research was contingent upon support from private American sources, given the relative paucity of governmental support for science at the time. Many donors, Lee knew, wanted to invest in perceived "winners," i.e., high profile, prestigious, scientific projects. In addition to fieldwork by the NYBG's itinerant taxonomists, research was pursued on the third floor of the Garden's Museum building. Much of this work consisted of traditional taxonomy conducted in the NYBG Herbarium, as Britton and his staff classified the plant specimens collected during their surveys of the Western Hemisphere. To encourage the use of the Garden's growing Herbarium, Britton in 1903 had established a small resident-scholarship fund from which he could pay living and travel expenses for out-of-town researchers. These $50 monthly stipends were designed to "meet the needs of professors or instructors in colleges, officers of museums, or other botanists of equivalent training or experience who desire to utilize the facilities of the Garden for continuing investigations already commenced or for carrying out investigations which require facilities additional to those at their command at home." From Britton's point of view, these small outlays were good investments for the Garden. In addition to promoting the use of his own authoritative herbarium, it enabled him to exercise some control over the publication of much of the resulting work conducted in NYBG laboratories.[2]

The Garden, through its affiliation with Columbia University, was also the scene of alternative types of experimental botanical work. R. A. Harper, Chair of the Scientific Directors and Torrey Professor of Botany at Columbia, encouraged a wide array of investigations at NYBG, hoping to develop an even closer relationship between the University and the Garden in the process:

> The committee [of Scientific Directors who considered the possibilities of expanding Garden research work] is entirely agreed as to the desirability of offering at the garden the fullest possible opportunities for research in all lines of botanical work. One of the chief difficulties to be overcome in meeting the needs of graduate students desirous of specializing in botany is in the provision of opportunities for work in related biological and physical sciences. The students who have come to Columbia University for graduate work in botany in recent years have for the most part been

1 Minutes of the Scientific Directors of the New York Botanical Garden, New York Botanical Garden Archives, April 13, 1912; June 8, 1912 [Subsequently cited as "Minutes of the Scientific Directors"].

2 *Journal of the New York Botanical Garden* 4 (February 4, 1903).

> desirous of specializing in the lines of physiology, cytology, and pathology with minor and accessory work in zoology, chemistry, bacteriology, animal physiology, etc. In the opinion of your committee it would be highly advantageous if all lines of biological research work at least could be concentrated at or in the neighborhood of the botanical and zoological gardens.

Harper acknowledged that due to space constraints within the Museum (a building that lacked electric lighting until 1928), there had developed in effect a division of labor among these Columbia students. Traditional taxonomic work, plant geography, plant breeding, and horticultural investigations were pursued at the Garden, whereas studies in physiology, cytology, and pathology were conducted at Columbia. In light of the University's relocation to its uptown campus, Harper and his colleagues hoped that much if not all of this work would soon be concentrated at the Garden.[3]

Although a traditional taxonomist by training and temperament, Britton was also an astute administrator. He welcomed proposals designed to advance the Garden's scientific prestige. Even as he granted stipends to taxonomists working in the herbarium, Britton appointed a Garden pathologist, who specialized in the diseases of ornamental plants. He also encouraged NYBG staffers such as A. B. Stout to continue research of commercial, agricultural, and horticultural import.[4] In fact, even as he built the herbarium and live plant collections to establish NYBG as the undisputed leader in New World taxonomy, Britton hoped to transform the Garden into a "horticultural university." To this end, he proposed a detailed program of research into the biology of soil, which involved hiring a mycologist, bacteriologist, algologist, pathologist, chemist, and plant physiologist. Britton's overall aim was the study of minute organisms (fungi, algae, and bacteria) that inhabited the soil. "If the life history of soil organisms and their functions in nature were better known," he asserted, "it might be possible by selection and breeding to amplify their beneficial action." Conversely, those microorganisms whose presence harmed vegetation would be studied with an eye toward control and eradication. Such an ambitious plan would be expensive: his proposed horticultural university, Britton estimated, would require an additional greenhouse, another wing to the main museum building, and equipment for new laboratories in addition to the salaries of an enlarged research staff. Overall, this new project would

3 R. A. Harper, Report of Scientific Directors to N. L. Britton, director-in-chief, December 6, 1917, New York Botanical Garden Archives. On the progress of installation of electric lighting in the Museum, see Minutes of the New York Botanical Garden Board of Managers, 1928, New York Botanical Garden Archives (subsequently cited as "Minutes of the Board of Managers").

4 Minutes of the Scientific Directors, December 13, 1919 and October 11, 1919. Stout had continued studying chestnut trees after the 1912 northeastern chestnut blight decimated the species; see *The New York Times*, February 12, 1912. As perhaps the most visible botanist in the city, Britton was called upon for advice when Central Park's trees were faced with disease.

cost almost $200,000, much of it to be secured (Britton hoped) by yet another subscription drive.[5]

Despite Britton's skill at securing personal, political, and foundation philanthropy, financial support for a soil biology program never materialized. Andrew Carnegie, whose backing was crucial for the success of many New York institutions, turned Britton down, noting that the various state agricultural experiment stations were already researching soil biology. Without Carnegie's approval and on the advice of E. D. Merrill, who similarly counseled the NYBG director-in-chief to avoid duplicating the efforts of agricultural stations, Britton decided not to continue this particular effort.[6]

In general, funds for the Botanical Garden remained in short supply throughout the 1920s. The annual municipal appropriations from the City of New York never met the operating costs of the Garden, and Britton was forced to allocate scarce funds according to his own institutional agenda. His first research priority, of course, was a traditional taxonomic one. This was the type of botany that had helped him construct the Western Hemisphere's largest public display greenhouse and compile one of the world's most comprehensive herbaria. The overseas expeditions that built up these collections had been financed almost exclusively through annual "special fund drives" to which Britton would contribute a modest but substantial sum toward exploration that would be matched or exceeded by several of the wealthier members of the Garden Corporation. But this traditional taxonomic agenda did not necessarily exclude the possibility of other types of research being conducted within the Garden's third floor laboratories.

Ironically, given the amount of difficulty he caused Britton during his few years in office, Frederic Lee had actually attempted to resign from the Garden's Board of Managers in 1920, citing a pending trip to Europe. Britton demurred, assuring him that an extended absence would not preclude him from remaining on the Board.[7] Britton's willingness to cultivate absentee Managers was a part of a shrewd administrative strategy. He secured the goodwill of prominent New Yorkers, while maintaining a maximum of direct control over the Garden's day-to-day affairs. (This Board of Managers met only several times a year.) Keeping Lee associated with the Garden was good policy in two respects: he was a Columbia professor and, thus, helped to strengthen the traditional affiliation of the two institutions, and he had married into

5 Minutes of the Scientific Directors, April 14, 1917.

6 Minutes of the Board of Managers, June 21, 1917. See also E. D. Merrill, October 27, 1917 to N. L. Britton. Merrill, then director of a nascent California botanical garden and later Britton's successor at NYBG, argued that the latter should concentrate on both taxonomy and the investigation of the pathology of specifically urban plants, leaving rural and crop research to the agricultural stations.

7 F. S. Lee to N. L. Britton, May 14, 1920 and Britton to Lee May 18, 1920, N. L. Britton Correspondence, New York Botanical Garden Archives.

the Billings family, whose railroad wealth had built the city in Montana that bears its name.[8] Britton lived to regret his decision to court Lee. When the Columbia physiologist took over the Garden presidency in 1923, he arrived convinced of the Botanical Garden's inferiority in both research methods and landscape design, and he determined to remedy this perceived deficiency by undermining Britton's authority.

Frederic Schiller Lee and the Garden

Born on June 16, 1859, Frederic Schiller Lee (FIGURE 20) was Nathaniel Britton's junior by only six months. He had been born into a family well-acquainted with the academy: his father, a Universalist minister, was the first president of St. Lawrence University. Both of his brothers also entered academic life: his elder brother as a professor of biology at Bowdoin College and his other brother as a literary scholar and later a college president. Lee took a doctorate in physiology at Johns Hopkins under the direction of Henry Newell Martin, who also served as a mentor to T. H. Morgan, Edmund B. Wilson, Edwin Conklin, William H. Howell, and A. B. Macallum. In 1885, Lee traveled to Germany to study with Carl Ludwig, a prominent physiologist responsible in part for the promulgation of the central physiological idea "that biological phenomena are explainable in accordance with the laws of physics and chemistry rather than by some vital principle."[9] Upon returning to the US, Lee began teaching at Bryn Mawr. By 1891 he had moved to Columbia's College of Physicians and Surgeons as a demonstrator in physiology. Appointed Dalton Professor of Physiology in 1904, Lee succeeded John Curtis as chair of the department in 1909 and served until 1920, when he retired from administrative duties to become a research professor.[10]

Lee became known for his integration of small-scale experimental demonstrations into his lectures. When physiology was elevated to the level of a department within the Faculty of Pure Science, he was given permission to offer courses in general and "special" physiology to Columbia undergraduates. By 1903, Lee was instructing

8 Mrs. (Laura Billings) Lee was the daughter of Frederick Billings, whose wealth was derived from the Northern Pacific Railroad. The definitive biography of Frederick Billings is Robin W. Winks's *Frederick Billings: A Life* (New York: Oxford University Press, 1991).

9 Lee's professed commitment to the primary importance of experimental research, despite the "pragmatic" orientation of much of American medical science, can be traced to his experience as a student in German universities. See Robert G. Frank, Jr., "American Physiologists in German Laboratories, 1865–1914," in Gerald L. Geison, ed., *Physiology in the American Context* (Bethesda, MD: American Physiological Society, 1987): 39.

10 J. Gardner Hopkins, "A Tribute to the Work of a Leader in Modern Physiology: Frederic Schiller Lee," *Columbia University Quarterly* February, 1941: 48. See also Alejandra Laszlo, "Physiology of the Future: Institutional Styles at Columbia and Harvard," in Geison, ed., *Physiology in the American Context*, 76–77, note 9.

FIGURE 20
Frederic Schiller Lee, 1940.

medical students in laboratory methods. That year, he also helped establish the Society for Experimental Biology and Medicine. Lee's own research methods had been honed at the Marine Biological Laboratory at Woods Hole, Massachusetts, where in the early 1890s he had investigated the function of certain organs in fish. His later career involved investigations into the physiology of muscular fatigue, a topic of prime importance to conceptions of industrial efficiency and public health in the first decades of the twentieth century.[11] The bureaucratic coup, temporary as it proved to be, on behalf of Columbia physiology was the result of Lee's steady lobbying of university president Seth Low in the early 1890s. Lee had long contended that the practice of physiology needed to be divorced from the practical purpose of instructing medical students in human anatomical functions: "Shall our task as a department be teaching medical students the known facts and principles of human physiology for practical purposes," he asked Low, "or shall it be the advancement of science?" Without awaiting Low's answer, Lee asserted "that the advancement of science by physiological research may best be promoted by the department being divorced from medicine altogether. [Physiology] is a biological science as much as zoology or botany, it is a pure natural science as much as chemistry or physics." "In our present location at 59th Street," Lee continued, "the atmosphere is, as it should be, almost wholly medical, and this must of necessity retard and prevent the complete growth of a biological physiology."

11 Ibid. (Hopkins) and F. S. Lee to Seth Low, December 8, 1893, F. S. Lee Correspondence, Central Files, Columbia University, Columbiana Collection: 12/1893-6/1910 [subsequently cited as F. S. Lee Correspondence, Columbiana Collection]. Lee had been granted research space at Woods Hole through the largesse of Low, at the time President of Columbia College. For a discussion of experimental research on fatigue and industrial efficiency, see Richard Gillespie, "Industrial Fatigue and the Discipline of Physiology," in Geison, ed., *Physiology in the American Context*, 240–241, note 9.

In this way, Lee implored Low to include his department in the ranks of the "pure sciences," which he felt would garner serious students interested in research. He cited the case of his alma mater, Johns Hopkins, to show the desirable results of including physiological research in the pure science facilities along with zoology, chemistry, and physics and concluded his appeal by contrasting Hopkins's admirable situation with that of Harvard, which had kept its physiologists as part of the medical faculty and consequently had "produced few working physiologists."[12]

Lee's assessment of the state of physiological research at institutions that had divorced these studies from clinical settings (or medical school curricula) was a product of late-nineteenth-century efforts to reform medical education. Even as medical courses in physiology became standardized, friction at various medical colleges often resulted from the clash of disciplinary goals with the needs of clinical education and practice. Lee, though never a stellar researcher himself, was nevertheless a university-based physiologist, determined to carve out a space within Columbia University for the pursuit of physiological knowledge for its own sake. And Low, as Columbia's president, was particularly receptive to such arguments. Defining a working relationship between the core units of Columbia College and the professional schools had been his priority upon taking office, and he had managed to integrate the College of Physicians and Surgeons into Columbia by 1891. Lee's lobbying efforts had also been presaged by those of John G. Curtis, head of Columbia's physiology department.[13] Historian Philip Pauly noted that there was in the early twentieth century an inverse relationship between a discipline's involvement in practical medical education and its success as a branch of "pure" (experimental) research.[14] Lee contended that physiology, like chemistry, should have two branches: an applied branch that would continue its work allied to the medical school and a pure branch in the department of biology.

12 F. S. Lee to S. Low November 15, 1894, F. S. Lee Correspondence, Columbiana Collection: 12/1893–6/1910. For a fuller portrait of Lee's views on physiology, see his article "The Scope of Modern Physiology," *The American Naturalist* 28 (Philadelphia: Edwards and Docker, 1894): 380–388, 473–482, in which he argues for an experimental method, the formulation and testing of hypotheses, and the central place of physiology within the realm of pure biological sciences.

13 Alejandra Laszlo, "Physiology of the Future: Institutional Styles at Columbia and Harvard," in Geison, ed., *Physiology in the American Context*, 67–78, see note 9.

14 Philip Pauly, "The Appearance of Academic Biology in Late Nineteenth-Century America," *Journal of the History of Biology* 17 (1984): 369–397. Pauly's study of the development of academic biology, in which he identified relevant institutional idiosyncrasies at several American universities that accounted for the historical development of biology as an academic discipline independent of the study of medicine, is particularly instructive as an assessment of the validity of Frederic Lee's criticisms of NYBG of the 1920s. Within these academic settings, natural scientists struggling to create and defend an autonomous intellectual niche sought to define departmental boundaries with recourse to any and all social or institutional resources they could muster. Lee had considerable resources at his disposal.

Only through this partial separation from the College of Physicians and Surgeons would physiology prosper as a life science. Lee's arguments were undeniably bolstered by his ability to draw upon his wife's fortune to purchase new laboratory equipment and underwrite the salary of an assistant for the new department.[15]

With this institutional background and inclinations, Frederic Lee, Columbia physiology professor and wealthy patron of science, accepted a seat on the Board of Managers of NYBG, and, in 1923, became the President. Given his professed interests, Lee was not inclined to be a vociferous advocate of continued taxonomic research at the Garden and, upon assuming the presidency, stated his position explicitly: "At present the scientific staff of the Garden includes thirteen members, of whom ten are systematic botanists." He called for the addition to the Garden's staff of a physiologist, chemist, and pathologist, each with an assistant. "We need in particular a chemical staff, Lee asserted. "The chemistry of life processes has proved one of the most fruitful methods of attack on a great series of biological problems. Physiological research is based in a large degree upon chemistry."[16]

Until Lee took the office, the Garden presidency had been something of a ceremonial post, whose occupant aided in fundraising, but who normally deferred to Nathaniel Britton's wishes with regard to making policy. Lee, however, set out to wield his personal wealth to achieve institutional goals of his own devising. His primary aims from 1923 until his resignation in 1927 were the reshaping of the Botanical Garden's research agenda and a substantial revitalization of its physical design laid out by Britton's basic plan of 1896. Such ambitions could be achieved, Lee recognized, only if adequate funds were available.

As a professed devotee of experimental methods, Lee began his presidency by commissioning a study designed to evaluate the Botanical Garden's research programs and to devise a comprehensive plan for future investigations.[17] Early in his tenure, Lee told newspaper reporters that up until that point, the Garden's scientists had focused almost entirely on systematic botany and added that there were "questions in other fields [that were] timely and intimately connected with the welfare of mankind. We should undertake their solution promptly as an additional contribution to human knowledge." To conduct this review of Garden research practices, Lee turned to the Board of Scientific Directors. Members of the Board of Managers encouraged Lee's efforts, although one, Kenneth Mackenzie, was quick to recognize that deficiencies in the Garden's efforts were likely the result of the "utterly inadequate resources" with

15 Laszlo, "Physiology of the Future," 72, see note 13; Pauly, "The Appearance of Academic Biology," 393–395, see note 14.

16 *The New York Times*, February 8, 1925.

17 Minutes of the Scientific Directors, Oct 11, 1924, New York Botanical Garden Archives.

which NYBG scientists were forced to work.[18] Mackenzie also noted the vast amount of knowledge that could be sought within any study of "the Vegetable Kingdom" and argued that it would be impossible for any single institution to excel in all areas of botanical research. Despite real handicaps, the Garden in its brief history, Mackenzie asserted, had managed to produce a substantial quantity of research in systematic botany and horticulture.[19]

In their evaluation of the Garden's priorities, the Scientific Directors underscored the centrality of research as a "duty of the garden" in a time when "science permeates and is indispensable to all human activities." Because this obligation to engage in scientific study was written into the Garden's founding charter, the Directors noted, it was a given that the NYBG staff would undertake regular investigations. What the Directors hoped to determine, however, was "along what lines [the Garden's] research [should] be pursued."[20] The traditional tasks of systematic botany, they reported, were intrinsically valuable for their ability to provide information on the geographic distribution of plants, as well as for the light they shed on the evolutionary affinities of plant groups. Taxonomy had also yielded knowledge of great value to medicine and to industry. Echoing Mackenzie, the Scientific Directors contended that such work needed to be prosecuted continually, because the percentage of the earth's landmass that had been sufficiently explored remained small. Overall, the Directors took pains to defend taxonomic botany as "the traditional, time-honored form of [botanical gardens'] scientific research" and to assert the NYBG's leading position in this regard among American institutions. They seconded Mackenzie's call for additional trained assistants to aid senior Garden taxonomists.[21] The Scientific Directors' report moved from a consideration of the Garden's taxonomic prowess to its work in paleobotany and economic botany, each of which was judged adequate.[22] Homage was paid to each of these traditional areas of research in turn, with specific attention given to Britton's own work on *The Cactaceae*, the Flora of Northern South America, and systematics in the various West Indian islands.[23]

18 The Board of Scientific Directors was responsible for setting scientific policy at the Garden, subject to the overall approval of the Board of Managers. The Managers included the city's philanthropists; the Scientific Directors' Board was comprised of academic botanists and other university professionals.

19 Kenneth Mackenzie to F. S. Lee, December 11, 1924, New York Botanical Garden Archives.

20 "The New York Botanical Garden: A Report by the Scientific Directors on the Scientific Research of the Garden," December 1924, Records of the Scientific Directors, New York Botanical Garden Archives: 2 [subsequently cited as "Report of the Scientific Directors"].

21 The original report, as submitted to a committee of three that consisted of Britton, Lee, and R. Harper, had also included the statement, "a larger staff of taxonomic specialists does not seem to be demanded," but Britton struck it from the text.

22 Report of the Scientific Directors.

23 Ibid., 3–4.

Although generally laudatory of the Garden's research practices in traditional fields, the report subsequently evaluated the NYBG's efforts in other areas and found them lacking. Research in various fields of "experimental botany," including dendrology, studies of root growth, breeding and selection, ecology, soil analysis and composition, and plant pathology, although needed, was hindered by a lack of resources. The report's authors recognized the value of similar work conducted under state auspices at experiment stations and agricultural colleges. Yet, they contended, such work was designed "to meet immediate exigencies in agriculture from year to year." Echoing Lee's argument for a separate sphere for experimental research in physiology, the NYBG's Scientific Directors contended that such services, often in an effort to control plant diseases, were analogous to the work of "practicing physicians," rather than that of "research pathologists." This use of metaphor was a telling one, as the Directors sought to establish for their own staff (which was *both* privately endowed and publicly funded) a key role as long-term researchers. Consequently, they argued that necessary investigations into "fundamental problems" in pathology and other research of economic import demanded the efforts of workers in institutions free from "changeable political control." Endowed organizations like NYBG, the Directors maintained, enjoyed a greater degree of stability and the "greater freedom [that] is practically indispensable" for the solving of crucial research problems.

Although the federal and state governments had expended great quantities of research dollars seeking to control the diseases of both crop and ornamental plants, the causes of most of these potentially devastating botanical ailments remained, the Directors reported, undetermined. "The breeding of types resistant to the rusts, smuts, and mildews of our great agricultural/crop plants," the report read, "is only beginning." In this context, they concluded, forestry work, pathology, horticultural development and breeding, plant nutrition, and the effects of industrial pollution upon plant growth were all especially suitable problems for NYBG researchers to tackle in the coming years. Such work would also have serious and beneficial implications for public health: by making the Malthusian argument that "before many decades the human race will be facing the problem of providing sufficient food for its members," the Directors attempted to carve out a further space for experimental work designed to safeguard the national food supply. "Botany," Frederic Lee declared, was "one of the livest of sciences [sic], touch[ing] many of the most acute problems of the world's needs."[24]

Lee's direct influence on the report's recommendations is apparent in its further insistence upon an analogical relationship between the diseases of plants and those of animals and humans. Such a comparison was a particular favorite of Lee's. The potential to substitute research on plants for experiments on animals was undeniably

24 Ibid. and *The New York World*, January 11, 1925 "Plants May Save Animals from the Vivisector's Knife."

attractive to a man who had very publicly defended vivisection in the popular press.[25] "There are," the Directors reported

> reasons for believing that many of the fundamental phenomena of disease processes applicable to all living beings may be studied in plants as readily as, if not more readily than, in human beings. With New York gradually becoming the leading medical center of this country and with the opportunity for medical contributions from the side of botany, it would seem most appropriate for the Botanical Garden to offer the facilities of its unusually large collections for fundamental researches into the problems of disease.[26]

Such an effort to associate botanical with medical research was essentially a pragmatic attempt to raise the institution's scientific profile and garner prestige by forging an even stronger link to Columbia University. The city was becoming a center for medical research and education, Lee noted, and he asserted that "the Garden is in duty bound to offer all its facilities for these important researches into the problems of disease."[27] Given Lee's 1894 objections to too close a relationship to medical education, this would seem an ironic strategy. Yet by the 1920s, Lee's efforts to establish a physiology free from disciplinary bondage to medical education at Columbia had failed, and a closer affiliation of the Garden to the school's scientific departments, he felt, would serve only to increase the quantity and quality of research conducted at NYBG by attracting more of the University's graduate students to the Bronx.[28] In addition, drawing analogies between human and plant diseases was becoming increasingly fashionable among botanists. In 1923, Britton colleague Melvin T. Cook had published an article in the *Journal of the New York Botanical Garden* on battling "plant cancers," tumor-like growths known to be caused by bacterial infection that infested apple trees and other horticulturally important species. Research into this particular pathology, Cook reported, had been proceeding at the Garden that year, and he professed the hope that this work might eventually prove useful in clinical medicine's fight against human cancers.[29]

25 See, for example, an open letter to *The New York Times*, November 21, 1912, in which Lee defended the creation of "spinal animals," i.e., decapitated animals whose bodies were "kept alive" for the purposes of physiological research.

26 Report of the Scientific Directors.

27 *The New York Times*, February 8, 1925

28 Despite the NYBG's original institutional relationship with Columbia University's Department of Botany, other suitable avenues for graduate research began luring the University's advanced students in the 1920s. Among these alternate research institutions were the Carnegie Institution's laboratory at Cold Spring Harbor, Long Island and in 1924, the Boyce-Thompson Institute, which will be discussed below. See E. Matzke, "History of Botany at Columbia University," 1953: 33–34, manuscript at New York Botanical Garden Library.

29 Mel T. Cook, *Journal of the New York Botanical Garden* 24 (1923): 70. Knowledge of plant pathology had increased markedly since 1880 with the demonstration of the link between bacteria and plant

The 1924 research report did not effectively draw a distinction between research on plant pathology and investigations, using plants as analogues, into human disease. This was a function of Lee's personal outlook, in addition to its growing frequency among American physiologists. (In his own limited experimental work conducted while at Columbia, Lee had been known to "give" sets of plants diabetes and then proceed to "cure" them with injections of insulin.[30]) Yet the report's main recommendations for reforming the NYBG research agenda dealt largely with the actual problems faced by the Garden's plants themselves, as well as the pathologies of plants found in other urban parks. In addition to complementing the work of agricultural stations, the Garden, the Directors concluded, needed a competent staff of botanical pathologists simply in order "to protect itself." This knowledge would also prove necessary for solving the riddles of urban forestry. "In this particular," the report's authors contended optimistically,

> the Garden is confronted with a necessity which is no less an opportunity.... The solutions of the problems with which we ourselves are confronted... would constitute a contribution of the highest value to those who are concerned with the care of decoratively planted estates—among whom many of our members are included—and to those who are concerned with the maintenance of public parks and planting for city decoration. They would thus be a direct return for both our municipal and privately contributed support.[31]

In this way, the Scientific Directors sought to associate new Garden research priorities with worthy, public-spirited projects (and those with private applications) that would bestow the benefits of botanical science upon the city. In a sense, Lee's insistence upon the transformation of the Botanical Garden into an institutional affiliate of medical and scientific education within the city represented an attempt to repeat and to *complete* his performance at the College of Physicians and Surgeons two decades earlier. Although he had assisted John Curtis in securing a place for physiology within the Faculty of Pure Science at the University, this victory had been somewhat hollow. Within several years (by 1908), Curtis had retired, and a new medical dean had begun to tip the balance of control of the medical school once again toward the clinicians.[32]

---

disease, as well as the later discovery of the link between insects and the spread of such microorganisms. By the time Cook wrote his report on the Garden's research, plant quarantines and plant breeding were common methods of checking the spread of disease, and the ecological relationships among host, microorganism, and insects had long been recognized and were being studied intensely. See Kenneth B. Baker, "A Plant Pathogen Views History," in *History of Botany: Two Papers Presented at a Symposium Held at the William Andrews Clark Memorial Library, December 7, 1963*, (Los Angeles: The Clark Memorial Library, 1965): 59–60.

30 *Philadelphia Public Ledger*, May 25, 1926.

31 Report of the Scientific Directors, 11, 17–18.

32 Laszlo, "Physiology of the Future," 76–78, see note 13.

Given the short-lived nature of physiology's victory at Columbia and the continuing tensions at Physicians and Surgeons between clinicians and researchers, Lee's desire to establish the Garden as an integral part of New York's scientific research infrastructure was to be expected. If he succeeded in wresting control of NYBG from those he considered outdated taxonomists, he might finally be able to establish his own research ideals within a major institutional setting.[33]

Taxonomic Botany and the Development of Biology

Lee's vociferous animosity toward taxonomic botany as practiced at NYBG might seem somewhat ironic, given the fact that his own experimental research protocols with "diabetic" plants would today be considered sheer quackery. To assess his critique of Britton's conception of the Garden, it is important to consider the intellectual context in which both men operated by the 1920s. Nathaniel Britton's career coincided with a period of pronounced intellectual ferment in the study of living organisms. The era between 1889 and the early 1920s was, according to one historian of science, a "rocky road [leading to] organized biology in America." The task of unifying a varied set of methods for examining life into one overarching discipline was considered a formidable one into the 1920s. As late as 1929, the year of Britton's retirement, biology was criticized as an immature discipline lacking in the secure foundations and methodologies of the experimental or mathematical sciences. Biology appeared hampered by residual "metaphysical" elements. Its widely differing methodologies still seemed to some observers in the late 1920s to be doomed to continued balkanization.[34] Within this context, it was inevitable that Lee, a self-professed "physiologist," considered Britton's traditional natural history ripe for reform.

Botanical systematists had defined several key sets of taxonomic tasks. First among these tasks was the identification of a given plant specimen. In this process, the researcher made explicit reference to previously published plant descriptions in an attempt to determine whether or not the newly collected specimen was similar to, or completely different from, known plants. The next steps, naming the plant (if necessary) and placing it within a classification scheme designed either to aid in identification or to delineate phylogenetic (evolutionary) relationships, were understood to be inextricably linked. The earliest "natural systems" of taxonomy, developed in botany by Alphonse de Candolle, George Bentham, and Joseph Dalton Hooker

33 Lee's failure at Columbia had an effect on that department's productivity within the wider community of academic physiologists. Columbia physiologists were relatively underrepresented in the production of publishable research in the first four decades of the twentieth century. See Gerald L. Geison, "International Relations and Domestic Elites in American Physiology, 1900–1940," in Geison, ed., *Physiology in the American Context*, 134–135, see note 9.

34 Vassiliki Betty Smocovitis, *Unifying Biology: The Evolutionary Synthesis and Evolutionary Biology* (Princeton, NJ: Princeton University Press, 1996): 98–102.

(among others) in the nineteenth century in response to the artificiality of the Linnaean sexual system, represented pre-Darwinian efforts to portray a Divine order in nature. Whereas modern systematists employ a set of taxonomic tools that include knowledge of plant genes and structures derived from biochemistry, morphology, anatomy, cytology, genetics, paleobotany, phytogeography, ecology, and geomorphology, taxonomists of the nineteenth century generally relied upon morphology (the observable external characteristics) to categorize plants. After the 1859 publication of Darwin's *Origin of Species*, taxonomists began positing ancestral affinities among plant groups by developing "phylogenetic" classification systems. In 1868, German botanist Julius von Sachs was among the first to propose modification of Candollean natural taxonomy to account for Darwin's theory of evolution through natural selection.[35]

This increase, post-*Origin*, in the formulation of evolutionary theories was coincident with an overall specialization in what today would be known as the "life sciences." Zoology and botany split into subfields such as embryology, cytology, genetics, and ecology. At the same time, gaps undeniably emerged in methodologies: experimental biologists began asking questions and using methods markedly different from those botanists and zoologists working with whole organisms. By the turn of the twentieth century, just as the Botanical Garden buildings were being erected, the rediscovery of the mid-nineteenth-century work of Gregor Mendel had launched the modern study of genetics, with a concomitant increase in understanding of the mechanisms of inheritance in living organisms.[36]

Although knowledge of this new genetics was crucial for a full understanding of processes of evolution, it was slow to spread through the various emerging disciplines that made up the diversity of interests that comprised "biology" at that time. Much of the knowledge of evolution in the early twentieth century was spread through the work of early geneticists, including Columbia University's T. H. Morgan and Dutch scientist Hugo de Vries. These men worked, however, without a conception of a species as a biological population and downplayed the importance of natural selection in evolution. For them, mutation pressures, rather than Darwinian natural selection, drove evolution. By the 1920s, the various intellectual exercises that comprised the study of living organisms remained essentially isolated, because practitioners from one discipline rarely read the work of those in others. The resulting confusion, according to zoologist and historian of biology Ernst Mayr, was "an almost complete alienation between experimental geneticists and naturalists."[37]

35 George H. M. Lawrence, *Taxonomy of Vascular Plants* (New York: Macmillan Company, 1952): 3–6.

36 Ernst Mayr, "Some Thoughts on the History of the Evolutionary Synthesis," in Ernst Mayr and William B. Provine, eds., *The Evolutionary Synthesis: Perspectives on the Unification of Biology* (Cambridge, MA: Harvard University Press, 1980): 6–7.

37 Ibid.

Mayr, the mid-twentieth-century architect of what came to be known as the "evolutionary synthesis" (an effort to unite these disparate sectors of biological work under one intellectual framework grounded in a Darwinian understanding of evolution)[38] looked back on early-twentieth-century biology as something of a contest between two opposing intellectual camps. He identified a "biology of proximate causes," characterized by scientists studying the observable traits of individual organisms (their "phenotype") in an effort to examine the interplay between genetics and environment. In contrast, he said, evolutionists (who might be termed natural historians) were interested in the historic origin of the genotype (the pattern of inheritance of a group of genetic markers), that is, the ways in which a particular species came to possess a certain genetic makeup. This latter group was interested in what Mayr dubbed "ultimate causes." Prior to Darwin, such questions about "ultimate" causes, e.g., the reasons behind a particular physical adaptation, were answered by recourse to the supernatural (a deity). Darwin's theory of natural selection, Mayr wrote, "provided the first rational approach to the study of ultimate causes." But well into the twentieth century, as late as the 1930s, taxonomists of these differing camps misunderstood one another, and this confusion resulted in distinct taxonomic arrangements based upon very different criteria. Botanists and zoologists interested in proximate causation contended that certain bodily characteristics were the results of physical laws playing out within individual bodies, rather than the results of evolutionary history. Therefore, they contended that experimentation on individual organisms was the best way to investigate the causes of various observable physical traits. Experimentalists claimed that any other efforts at tracing evolutionary lineage beside their own was speculative; unscientific; and, thus, an insufficient basis for the development of an overall evolutionary theory. Because of this schism, certain taxonomic schemes might result in the grouping of organisms with similar physical traits (due to the workings of similar internal mechanisms), but lacking a common evolutionary ancestor.[39]

In the early 1900s, as Britton set up the Garden's research facilities, many geneticists, including NYBG laboratory director Daniel MacDougal, were enthusiastic about the power of de Vries's mutation theory to account for the development of new species. They viewed this theory, initially, as incompatible with a Darwinian understanding of evolution that posited the emergence of new species as a result of natural selection acting upon small anatomical differences among individual members of a given population of plants or animals. Prior to the 1910s, many plant and animal

38 Ernst Mayr, *Systematics and the Origin of Species from the Viewpoint of a Zoologist* (New York: Columbia University Press, 1942). Julian Huxley was responsible for coining the phrase "evolutionary synthesis" in his *Evolution: The Modern Synthesis* (London: Allen and Unwin, 1942).

39 Ibid. (Mayr), 9–11.

geneticists considered such a process to be ineffective in altering the genetic composition of a population; only large mutations, they felt, could accomplish this. Only after much experimental work in genetics did this predominant view change, as geneticists discovered that, in fact, natural selection operating on small "continuous variations" could change populations' genetic makeup sufficiently to create novel species. As a consequence, by 1918, most geneticists had come to consider Mendelism and Darwinism complementary features of the evolutionary process.[40]

Although a great deal of knowledge of genetics was contributed by the study of plants in the first decades of the twentieth century, botanical science did not make a major contribution to the development of the evolutionary synthesis (and, thus, to the unification of biology) during the 1930s. Into the late 1940s, botany's various subdisciplines (taxonomy, morphology, genetics, ecology, and paleobotany) consisted of "a much more heterogeneous assemblage of practices" and data. This was partly a function of the more complex genetics of plants, which hindered the development of a unified understanding of species among botanists. Because of the unique behaviors of plants during reproduction (resulting in a profusion of varieties of plants, with various mixtures of genetic material), older ideas (those derided as "speculative" by other sciences) such as Lamarckian beliefs in the inheritability of acquired characteristics took longer to dispel among botanists.[41]

Thus, by the time Frederic Lee assumed the Garden presidency, Darwinism (the idea that natural selection drives evolution) was on its way to forming the basis for a "modernized evolutionary theory" that incorporated the latest advances in genetics, mathematical models of population genetics, as well as field studies by systematist-naturalists.[42] Although in the 1890s, he had called for a modernized botanical taxonomy that accounted for an up-to-date understanding of evolution, Britton personally played no active role in the evolutionary synthesis itself; neither did Frederic Lee.[43] A product of an earlier era, Britton retired from the Garden at the end of the 1920s, before the most important botanical work that contributed to twentieth-century evolutionary theory took place.[44]

40 William B. Provine, "Adaptation and Mechanisms of Evolution after Darwin: A Study in Persistent Controversies," in David Kohn, ed., *The Darwinian Heritage* (Princeton, NJ: Princeton University Press, 1985): 840–841.

41 Smocovitis, *Unifying Biology*, 136–37, see note 34.

42 G. Ledyard Stebbins, Jr., *Variation and Evolution in Plants* (New York: Columbia University Press, 1950). See also Stebbins's own retrospective of his role in the synthesis, "Botany and the Synthetic Theory of Evolution," in Mayr and Provine, *The Evolutionary Synthesis*, 137–49, note 36 and Smocovitis, *Unifying Biology*, 114–19, note 34.

43 Smocovitis, *Unifying Biology*, 20–21, see note 34.

44 See Sharon Kingsland, *The Evolution of American Ecology, 1890–2000* (Baltimore, MD: Johns Hopkins University Press, 2005): 69. Keith Vernon characterized such late-nineteenth-century talk by classic herbarium taxonomists as "little more than lip service paid to evolutionary considerations." Until

Not all historians of science have been entirely convinced of a vast methodological divide developing within the life sciences during the late nineteenth and early twentieth centuries. In the early 1980s, some writers questioned the historiographic notion of mutually exclusive realms of botanical inquiry, separated into methodological camps devoted either to a descriptive naturalist taxonomy or an "experimentalist" biology. Such a portrait of a radically split biology unified only in the mid-twentieth century by the evolutionary synthesis, they claimed, was an oversimplification.[45] Rather, as historian Joel Hagen asserted, a strict dichotomy between naturalists and experimentalists never existed, although he conceded the existence of substantive methodological differences among various sets of botanical investigators.

Nevertheless, Hagen pointed out, collaboration among field and herbarium taxonomists, cytologists, ecologists, and geneticists *did* occur, and such cooperative efforts helped to place botanical taxonomy, in its efforts to elucidate evolutionary relationships among groups of plants, on a firmer empirical footing. Some of the pioneers of what Hagen dubbed "experimental taxonomy" were well-known to Britton, including ecologist Frederic Clements, whose work had been supported in part by the Carnegie Institution of Washington.[46] Britton was also a close collaborator with "new botany" pioneer Charles Bessey in the nomenclature debates, as well as in the promotion of professionalization in botany. Such colleagues, along with NYBG staff ecologist Henry Gleason (who conducted the ecological portion of the Puerto Rico Survey), certainly informed Britton of the growing variety of botanical research techniques.[47]

Britton, who had published his first new species in 1884 and who by the mid-1920s was approaching 70, did not personally adopt newer methods of "experimental"

"the 1930s and 1940s," Vernon acknowledged, "it is unclear how many taxonomists made more than rather vague commitments to evolutionary issues." See Vernon's "Desperately Seeking Status: Evolutionary Systematics and the Taxonomists' Search for Respectability 1940–60," *British Journal for the History of Science* 26 (1993): 207–227.

45 Stebbins, "Botany and the Synthetic Theory of Evolution," in Mayr and Provine, eds., *Evolutionary Synthesis*, 149, see note 36.

46 Joel Hagen, "Experimentalists and Naturalists in Twentieth-Century Botany: Experimental Taxonomy, 1920–1950," *Journal of the History of Biology* 17 (1984): 250. Hagen's call for a less dichotomous view of late nineteenth and early twentieth century life-science methodologies is echoed in Ronald Rainger, Keith Benson, and Jane Maienschein, eds., *The American Development of Biology* (New Brunswick, NJ: Rutgers University Press, 1991). For an influential statement of the schismatic version of early twentieth-century American life sciences, see Garland Allen, "Naturalists and Experimentalists: The Genotype and the Phenotype." *Studies in the History of Biology* 3 (1979): 179–209 and "The Transformation of a Science: TH Morgan and the Emergence of a New American Biology" in Alexandra Oleson and John Voss, eds., *The Organization of Knowledge in Modern America, 1860–1920* (Baltimore, MD: Johns Hopkins University Press, 1979).

47 Hagen, "Experimentalists and Naturalists," 250–251, see note 46.

taxonomy. Up until his death in 1934, his publishing career consisted primarily of a series of ambitious floristic treatments of various Western Hemisphere regions, with additional work (as in the case of *The Cactaceae*) on monographic portraits of particular plant families. He nevertheless encouraged the adoption of new research methods in NYBG laboratories when such methods proved promising. When assessing the state of the research agenda of the Botanical Garden in the 1920s, it is important to recognize that Britton's traditional methods, derided by some, remained vital parts of the taxonomic endeavor, even as taxonomy incorporated the new tools of genetic and ecological research. "This fusion" of research methods, Hagen contended, "constituted neither a repudiation of descriptive botany nor a complete revision of taxonomic theory or practice."[48]

Despite its persistence at places like NYBG, classic herbarium taxonomy certainly suffered a considerable loss of prestige during the first half of the twentieth century.[49] From the vantage point of the 1940s, American vertebrate paleontologist/geologist G. G. Simpson recalled that

> A generation ago taxonomy had fallen into disrepute. It was believed to be a blind alley, a pursuit for hack workers only or a phase happily outgrown. 'Taxonomist' was a term of reproach... taxonomy seemed... little more than sorting dead specimens according to a few rather superficial morphological characters, printing out labels and then stowing the specimens away in drawers.[50]

Herbarium botany during this period was not emphasized in university departments, and contemporary taxonomists found few employment opportunities outside of botanical gardens. Even scientists, like Simpson and Mayr in the 1940s, who sought

48 Ibid., 255.

49 Henry Gleason, "The Scientific Work of Nathaniel Lord Britton," *Proceedings of the American Philosophical Society* 104 (1960): 206, 226 and E. D. Merrill, "Nathaniel Lord Britton." *National Academy of Sciences Biographical Memoirs* 19 (1938): 147–159. Similarly, historian Lynn Nyhart argues that a viable taxonomic botany persisted throughout the twentieth century as natural historical methods were appropriated by younger experimentalists who incorporated them into the various emerging subdisciplines of biology. See Nyhart's "Natural History and the New Biology," in N. Jardine, ed., *Cultures of Natural History* (Cambridge, UK: Cambridge University Press, 1996): 430. Although it was true, as Keith Vernon reported (in Desperately Seeking Status, note 44), that by the 1920s many university departments shunned the sort of taxonomic work Britton espoused, nevertheless, as Steven Conn notes, institutions such as the Garden have again become increasingly relevant since the rise in the 1960s of ecological thinking precisely because they retained connections to natural historical knowledge and methods. See Steven Conn, *Museums and American Intellectual Life: 1876–1926* (Chicago: University of Chicago Press, 1998): 72 and Paul L. Farber, *Finding Order in Nature: The Naturalist Tradition from Linnaeus to E. O. Wilson* (Baltimore. MD: Johns Hopkins University Press, 2000).

50 Simpson is cited in Vernon, "Desperately Seeking Status," 214, see note 44.

the rehabilitation of taxonomy (in an effort to attract research funding) insisted upon the creation of a "new" systematics. This new endeavor, they claimed, was no longer the province of men arguing about nomenclature and the possession of type specimens. Rather the field and the laboratory both belonged to taxonomists studying populations in the wild, conducting breeding experiments, and analyzing ecological and physiological data. Such efforts would enable "new systematists" to posit classifications that raised questions about evolutionary relationships among the organisms studied.[51]

Britton's staff clearly recognized herbarium taxonomy's loss of prestige. In response to the 1924 Scientific Directors' report, Curator of Economic Botany H. H. Rusby dissented, if mildly,

> I cannot help feeling also that the way in which you have treated the subject of Taxonomy may tend toward depreciation of it. It is true that it has received an undue ratio of support as compared with the other things that you mention, but the amount of support has not been too much, by any means. It is a very great achievement to have done what we have in this direction, and this had been done under great handicaps, all things considered. If it were necessary to choose between giving further support to Taxonomy and to secure the things that you ask for, I should unhesitatingly declare in favor of the latter, but this is not the case. We should ask for both, at a time when we are likely to get both needs supplied by so doing.[52]

Rusby, a Britton contemporary and a classic herbarium botanist, clearly perceived an inherent criticism of NYBG botany at a time when taxonomy's reputation suffered. Given this intellectual climate, the potential for acrimonious debate among American scientists in general, and between Frederic Lee and Nathaniel Britton in particular, remained high. This was the context in which Lee assumed the NYBG presidency, and it was with this backdrop that he challenged Britton directly for control of the Garden.

Frederic Lee rose to prominence at the Garden in the 1920s and enjoyed the status conferred upon him by a Columbia professorship. These attainments were in fact largely the result of his fortuitous marriage into the vast Billings railroad fortune.[53]

51 Ibid., 215. "Traditional" taxonomists, Vernon noted, pp. 224–26, resisted the steady erosion of their prestige, contending that the skills of collecting, describing, naming, and categorizing organisms were indeed scientific and intrinsically important, even without any subsequent attempt to use this new information to posit evolutionary relationships.

52 H. H. Rusby to R. A. Harper, December 5, 1924, Minutes of the Scientific Directors, New York Botanical Garden Archives.

53 Lee's position at Columbia was undoubtedly strengthened by his (and his wife's) willingness to contribute generously to the University's Department of Physiology for equipment and to meet other expenses. See, e.g., May 18, 1912 Lee to Nicholas Murray Butler, F. S. Lee Correspondence, Columbiana Collection. In 1914, the Lees had contributed $20,000 to the department so that it might maintain its own library of physiological journals; ibid., September 12, 1914.

Historian Richard Gillespie recognized that Lee came of age professionally in an era when one of the most pressing questions facing experimental physiologists was the investigation of the role of muscle fatigue in industrial efficiency. At the turn of the twentieth century, such work, Gillespie wrote, allowed successful physiologists to claim a scientific role as experts on an issue of great social importance:

> Fatigue, a scientific concept whose range of meanings was restricted to academic physiological knowledge, suddenly resonated with new meanings, as it became associated with overwork, ill health, racial degeneration, and most importantly, inefficiency. This transformation cannot be attributed to a new scientific discovery opening up the possibility for the application of physiology to a major social issue; rather it was the reverse. Social and political changes created an environment in which new meanings were constructed for the concept of fatigue. Industrial fatigue emerged out of the interactions of several factors: the development of scientific management in industry; increasing concern in an imperialist age with national efficiency; and a new confidence—voiced by the state, social reformers, and scientists alike—that social problems and conflict could be resolved by the application of scientific knowledge.[54]

Lee had tried with only limited success to carve out a niche for himself as a researcher in industrial efficiency, which would have given him this very sort of public authority on a set of social issues of prime importance.[55] Having had his ambitions essentially thwarted by the failure of biological physiology to take root at Columbia apart from the medical school, and the decline of the concept of "industrial fatigue" as a socially important field of research, he turned his attention in the 1920s toward the Botanical Garden, an alternate venue in which he might finally achieve scientific prominence and wield power in the public arena.[56] Because the work of other physiologists was cited by attorneys arguing for limited working hours for women and children before the Supreme Court, Lee would easily have recognized an opportunity for the Garden to utilize scientific expertise to achieve similar public prominence and authority. By making botany a critical arm of physiological research on pressing questions of both medical science and urban horticulture, NYBG, he hoped, would only grow in stature. He then would be in a position to claim credit for these developments.[57] The

54 Richard Gillespie, "Industrial Fatigue," 240–241, see note 11.

55 Geison's study, in "International Relations and Domestic Elites in American Physiology" (note 33), does not characterize Lee as one of the leading researchers of American physiology during the years in which the discipline became unambiguously dominated by US researchers. As a wealthy and socially prominent man with a Columbia professorship, Lee did enjoy a certain amount of press coverage for his work, part of which was attributable to the very public stand he took in favor of vivisection. See, e.g., *New York Times*, November 21, 1912

56 Gillespie, "Industrial Fatigue," in Geison, ed., *Physiology in the American Context:* 253–254, see note 11.

57 Ibid., 241 for a discussion of the role a physiological study in changing American labor law. See also

Scientific Directors, recognizing the opportunity to raise the profile of the Garden while tapping potential sources of both research funds and graduate students, endorsed Lee's vision of institutional growth and, in turn, expanded this mandate to include the additional development of NYBG into a leading center of horticultural (applied) expertise.

Yet in fact, such a strategy was not the radical departure from institutional precedent that Lee portrayed it to be. This pragmatic vision of Garden development echoed Britton's own 28-year habit (seen perhaps most clearly in his orchestration of the Puerto Rico Survey) of associating the Garden's institutional aims with the wider public interest, from tropical colonial development to the cultivation of hardy varieties of horticultural mainstays. In this regard, then, the 1924 report did not represent much of a break from past Garden practice, although, under the influence of its new president, the Board of Scientific Directors seemed for the first time to grant equal prominence in the NYBG research agenda to pursuits other than floristics and monographic portraits of plant families. Despite his relative openness to such alternatives (as noted above, he had encouraged those botanists on his staff with interests and abilities outside the realm of taxonomy to pursue such research, if not at the Garden, then in conjunction with other institutions), Britton probably considered Lee's influence to be little more than meddling. And yet, even as Lee tested the veteran director-in-chief's patience with his attempt to reorder the institutional research agenda, he initiated a similar investigation into the physical layout and construction of the Botanical Garden grounds.

Reforming an Ornamental Landscape: A Bronx Model

As a member of New York's plutocracy, Frederic Lee belonged to a class intimately familiar with the decades-old seasonal pattern of retreating from urban centers during hot weather to landscaped suburban or rural estates. In summer, Lee and his wife, Laura, traveled from their well-appointed Upper East Side Manhattan townhouse to the Billings family's estate in Woodstock, Vermont, where Lee's father-in-law had purchased the former home of conservationist and diplomat George Perkins Marsh. On the Marsh land, Billings had practiced a progressive experimental forestry, planting rows of Norway spruce and other economically important species, and setting an example that was not lost on his son-in-law.[58]

The New York World, January 11, 1925, "Plants May Save Animals from the Vivisector's Knife," in which Lee was quoted as saying that plant research in medicine would never abolish animal research, but that it could help advance science and that NYBG was planning to advance botanical medical research.

58 *The New York Record and Guide* of September 17, 1904 described the Lee's newly built home at 125 and 127 E. 65 St.: "[T]he front will be made of brick, with steel frame, marble trimmings, concrete

At NYBG in the 1920s, Lee sought to build the institution's reputation in urban forestry (a goal he shared with Britton) by encouraging experimental research on the effects of air pollution on street and park trees.[59] Lee also worked simultaneously to secure the appointment of Samuel Trelease, son of the former director of the Missouri Botanical Garden and a professor of botany and plant physiology at the University of Louisville, to a position in Columbia's Department of Botany. Trelease's appointment was meant to coincide with the development of a coordinated and "comprehensive program of research in experimental botany" conducted by Columbia and the Garden.[60] This program would, Lee assured *The New York Times*, "represent a sweeping departure from [the research] methods which [had] prevailed at the Garden for the previous 30 years," and would provide new scientific tools for the preservation of the beauty of urban parks. Though similar work was then underway at state agricultural experiment stations, these latter institutions, the *Times* opined, were "too often . . . subject to influences—politics and fertilizer companies—that [would] be absent from our Botanical Garden." The *Times* concluded that this new line of research and its concomitant emphasis upon landscape beauty maintained through unbiased, apolitical, scientific expertise was worthy of high public expectations.[61]

Faith in the importance of open space and landscaped public parks for public health and morals had been a staple of elite politics, of course, since the dawn of the US urban parks movement in the nineteenth century. Lee believed that reform of the NYBG landscape in accordance with the tenets of landscape art, when coupled with a revised research agenda, would enable the Garden to exert even more influence in the public life of the city. NYBG's ability to do so, however, was contingent upon the physical beauty of its Bronx setting, an aspect of the Garden that by the 1920s was open to criticism.

---

and expanded metal floors, metal and glass skylights, one elevator, marble cornices, copper bay windows, steam heat, electric light at cost of $60,000." This building would be 35 feet wide, with a depth of 65 feet. On the Woodstock home and Billings's interest in estate horticulture and experimental farming and husbandry, see Winks, *Frederick Billings*, 297–300, note 8.

59 *Philadelphia Public Ledger*, May 25, 1926: "Botanical Garden Scientists See Chance to Add Five Years to Growth of Poplars: Experts have Obtained 5000 Seedlings by Crossing Trees Which Promise to Be Speedy Growers –Now Investigating Whether Automobile Gas Fumes Are Killing Trees," and *The New York Herald Tribune*, undated: "Including the Study of Air Pollution in American Cities, Effects of Automobile Exhaust Gases on Vegetation and Losses to Society through Plant Disease," in F. S. Lee Correspondence, Columbiana Collection.

60 *The Evening World*, January 6, 1925; *The New York Sun*, January 6, 1925.

61 *The New York Times*, editorial, January 8, 1925. See also *The New York Times*, January 20, 1927, "Declares Science Can Save Our Trees." Here Lee is quoted as asserting that only dedicated research at NYBG could determine the cause of many tree deaths in Central Park. He used the occasion as a further call for increased philanthropic donations.

Turning to Landscape Authorities

Lee's efforts to improve the Botanical Garden's physical setting began in earnest in 1923 when a donation by two Garden Managers enabled him to hire the Olmsted Brothers. Lee commissioned this firm of landscape architects headed by Frederick Law Olmsted, Jr. to evaluate NYBG and prepare a landscape plan to attract the attention of the "general public," especially "that part of the public particularly interested in gardening . . . and in landscape beauty." Completed in 1924, Olmsted's report included extensive and expensive recommendations for reforming the NYBG grounds.[62]

In 1923, the Olmsted Brothers firm enjoyed the loftiest of contemporary reputations as landscape architects, having designed Fort Tryon Park, the Cloisters, and Forest Hills Gardens in New York and portions of the White House grounds, Lafayette Park, and the landscaping of the Washington Monument, the Jefferson Memorial, and the National Cathedral in Washington. The Olmsted firm, of course, was well-known to Britton and other Garden veterans from the assessment of the original Garden design that J. C. Olmsted had made with C. S. Sargent in 1897. As part of the Parks Department's "expert commission," Olmsted, Sargent, Charles Eliot, and Samuel Parsons had been responsible for banishing Britton's house from the grounds of the planned Garden and relocating the Conservatory from the site initially chosen by Britton's planning committee.[63]

When, over a quarter of a century later, an Olmsted was again given the opportunity to evaluate the Garden's layout and assess the work of Britton, he readily identified the lack of sufficient maintenance funds as the primary problem confronting the Garden. Regular upkeep of the grounds and plantings was crucial, he noted, because it was from the "cumulative growth" of such plants that the Garden derived its essential worth as a public landscape. Besides climate, soil, and the "habits of the people who resort to the grounds," Olmsted's report identified several factors to consider when seeking to improve the Garden. Among these was the ability of the landscape reformers to execute several types of landscape "treatments," including use of existing native woodlands, ground covers of shrubs, broad meadowlands or lawns, and "intricate combinations of turf with plantations." Each prospective treatment, Olmsted estimated, would require a different level of maintenance in terms of

62 Olmsted Brothers, "Report on the New York Botanical Garden" (New York: 1924): 5 [subsequently cited as "Olmsted Report"].

63 See Chapter 3. Britton had returned the favor in 1897 when he had commiserated with the president of the Buffalo, New York Parks Board in an exchange of letters in which each man expressed "much dissatisfaction with the Messrs. Olmsted." See Andrew Langdon to N. L. Britton, May 29, 1897 and N. L. Britton to Langdon, June 5, 1897, N. L. Britton Correspondence, New York Botanical Garden Archives.

man-hours; forested acreage would presumably require less direct upkeep than closely trimmed lawns and flower beds.[64]

The report included estimates of labor needs, calculated in numbers of men per acre needed for maintenance of each of the general categories of land enumerated. In making this assessment, Olmsted found the Garden's budget for such necessary landscape upkeep for *any* kinds of "treatments" to be insufficient. The Garden, he concluded, had "bit off more than it could chew under present conditions of cost and of funds available." In light of the friction between the Boston landscape experts (including Sargent) and Britton at the Garden's inception in the mid-1890s, the harsh conclusions and condescending tone of the Olmsted report are perhaps unsurprising. In a historical irony that would not have been lost on Nathaniel Britton, Boston landscape authorities were finally having their say in the Bronx. "We have observed in the Garden some such infections, notably of scales, which no well-conducted commercial establishment would have permitted to go as far as they have gone," Olmsted noted[65]). "Taking the Botanical Garden as it is, and assuming the proper upkeep of the sort of thing that appears to have been intended and attempted," Olmsted continued, "we believe the maintenance force necessary for adequate care of what now exists should be about 110 men, which is about 2½ times as many as are now employed." In dollars, Olmsted estimated that NYBG would have to increase its annual maintenance budget from $70,000 to a minimum of $200,000. These estimates were based in part on the firm's assessment of the state of the market for skilled horticultural labor. There were no indications, Olmsted warned, that an influx of competent gardeners into the American labor market would lower the costs of maintenance. Competition from private estates for skilled horticulturists would only drive such costs upward, and whereas private gardens could afford to cut back, the Garden, as a public institution committed to providing landscaped urban space as its "most essential function," could not without "manifestly failing as a garden."[66]

After drawing this gloomy portrait of the state of the Garden's garden, Olmsted counseled action in the form of triage: given that costs were unlikely to decrease appreciably and income remained insufficient for the time being, NYBG should maintain areas of "first-class" upkeep as examples of what might be done with adequate funding and as a stimulus to the morale of the horticultural labor force. That force, in turn, would need an able superintendent to organize the work and report to the director-in-chief. Included among the superintendent's tasks would be the improvement of the "general scenic qualities" of the Garden's systematic collection of shrubs and herbaceous plants. Olmsted recognized the need for such synoptic collections of

64 Olmsted Brothers Report: 6–7, see note 62.

65 Ibid., 10.

66 Ibid, 8–10.

plants "assigned to certain localities for reasons quite other than aesthetic" and conceded that these features of NYBG were "sophisticated" and "pleasant." Nevertheless, he asserted the need for adjustments in these areas, including the addition of "'background' and 'filler' plants not part of the exhibit proper, but serving functions not unlike those of frames and cases and backgrounds and partitions in an indoor museum." Olmsted did not stop there, however. His critique of the Garden extended to the plan of pathways throughout the plantations, which, he asserted, was at times "extremely unsatisfactory, confusing, arduous and uninviting to follow, failing to lead to the best points of view and [not] conforming pleasantly to the natural topography. . . ." Certain areas of the Garden, he predicted, would require radical alterations, including changes in grading and the layout and connections of internal roadways. Such alterations would be crucial, the report contended, in order for NYBG to achieve its full potential as a landscape. He also counseled the closing of the Garden's borders at night and the posting of uniformed guards throughout the premises during the hours NYBG was open to the public.[67]

In addition to these suggested changes to existing Garden features, Olmsted and his associates drew up a list of potential landscape features of an entirely new character. The first of these was the addition of a rhododendron and azalea glade to be planted in the southeastern corner of the Botanical Garden near the Rose Garden that had been recently installed adjacent to the public picnic grounds.[68] These plantings "would be such as to produce in a large general way and at all times the esthetic qualities of *beauty and picturesqueness*. The effect of hybrid rhododendrons or azaleas in variety at times of bloom could be gorgeously magnificent," and, Olmsted promised, "the beholder might be fairly carried away in his admiration." Planted as an understory to these main attractions would be heaths, heathers, ferns, and other smaller plants to provide scenic ground cover and an artistic contrast with the rocky ledges of that portion of the Garden. Olmsted even proposed planting a small cranberry bog for scenic effect and a simultaneous lesson in economic botany. The rhododendron glade, Olmsted acknowledged, would require a great deal of money and maintenance, and he counseled patience until sufficient annual funds could be secured for its upkeep.[69]

67 Ibid., 12–17.

68 Britton had solicited a design for a rose garden from Beatrix Farrand in 1915. During 1916, she had provided NYBG with a plan for a two-acre, 90-bed garden with a central arbor under a domed roof and the entire plot enclosed by lattice fencing meant to accommodate climbing varieties. Farrand's full version of this rose garden, however, was never executed. What Olmsted encountered was a scaled-down, less expensive garden based on her more opulent design. Only in 1988, after the discovery of Farrand's original plans, was her vision implemented as today's Peggy Rockefeller Rose Garden. See Jane Brown, *Beatrix: The Gardening Life of Beatrix Jones Farrand, 1872–1959* (New York: Viking, 1995): 108–109.

69 Olmsted Report: 17–18, 44–46, see note 62.

Other suggested additions to the Garden as a whole included the construction of model urban and suburban gardens along the southern border where passing motorists would appreciate them and the development of a more "formal" garden, a "frankly man-made thing," in which humanity's "skillful artistry and complete command" over the surroundings would be in evidence. Finally, Olmsted also lamented the lack of long vistas of "sufficient length to give the sense of spaciousness and the mystery of distance" and argued that the gorge of the Bronx River with its "picturesque precipitous slopes" should be restored to as much of a "natural" state as possible.[70]

Given his concern with creating a newly harmonious set of artful landscapes within the Botanical Garden, it is unsurprising that Olmsted identified automobile traffic (which by the 1920s traversed the NYBG's interior roads with increasing frequency) as a particular problem to be solved. That the Garden's interior roads had been open to public vehicular traffic and beyond the direct control of its administrators represented a situation "quite unprecedented... in any of the important botanical gardens of the world," Olmsted noted. The Royal Botanic Gardens at Kew admitted no motor vehicles and contained no through routes; the Arnold Arboretum, though open to the public, also banned automotive traffic. In sharp contrast, NYBG was situated, thanks to the development of surrounding roads and the growth of the Bronx's population in the first decades of the twentieth century, at or near the nexus of several important highways, including the Bronx River Parkway, the Mosholu Parkway, the Boston Post Road, and the Grand Concourse. Traffic within the Garden had become a problem, as motorists utilized the park's interior roads in an effort to avoid local congestion on these exterior arteries.[71]

Britton had earlier placed observers at two main Garden entrances on Sundays in September 1924 in the hope of arriving at an accurate estimate of the number of motor vehicles utilizing the NYBG's interior roadways. Yet the results of these informal surveys led him to question Lee and Olmsted's desire to restructure the various Garden drives. He asserted to Lee that "anything like real crowding on the driveways is limited to a few hours a week during the months from spring to autumn, about seven months, and this mainly on Sundays and holidays." In total, he estimated, there were only "about 180 hours a year on which there is any real crowding. Nothing serious happens even when the heaviest pressure is brought upon the roads, and this pressure is hardly comparable at any time to that on hundreds of city streets and avenues." He was

70 Ibid., 19–22.

71 Construction on the Bronx River Parkway, begun in 1907, had been interrupted by the First World War and was not completed until 1922. See *The New York Times*, September 17, 1922, "City Opens Bronx Parkway to Traffic."

hopeful, furthermore, that once completed, the new road connecting the Bronx River Parkway to the Mosholu Parkway would divert such traffic from the Garden itself.[72]

Olmsted hoped that the city would complete several connections among these major thoroughfares that would actually bypass the Garden's grounds. To the extent that any road continued to pass through the Garden at all, Olmsted insisted, heavily planted embankments should be used to screen passing railroad cars as well as bordering buildings from the view of Garden goers, while providing passing motorists with a scenic overlook from which to comprehend much of the landscape at a glance. A through road along the westernmost boundary of NYBG land, Olmsted concluded, would provide the best means of reconciling local traffic patterns with the aesthetic and horticultural needs of the Botanical Garden.[73] Though desirous of minimizing congestion and the use of the Garden as a traffic route, Britton nevertheless questioned Olmsted's assertions. Worried about the diversion of funds from other uses, Britton doubted the necessity of substantially rebuilding the NYBG's entire roadway system, which, when designed (for carriages) in 1895, had determined much of the Garden's physical character. Beside the financial burden such a radical alteration entailed, Olmsted's critique of the internal Garden circulatory system represented for Britton yet another in a series of rebukes for the design choices he had endorsed at the Garden's inception in the 1890s.[74]

Finally, Olmsted turned his attention to the immediate vicinity of the Garden's Museum, where he identified several key problems. Foremost among these was "the appearance of the museum in relation to its surroundings." In his "frank opinion," the Museum building's condition and its relationship to its physical setting were "aesthetically very bad." As architecture, Olmsted commented, "apart from relation to surroundings and without allowance for any limiting conditions which may have necessitated the present design, the exterior of the Museum building could not, we believe, be regarded by any competent critic as an artistic success."[75] He speculated hopefully that the entire building might in the future be enclosed by additions that

72 N. L. Britton to F. S. Lee, September 4, 1924, N. L. Britton Correspondence, New York Botanical Garden Archives.

73 Olmsted Report: 25–26, see note 62.

74 N. L. Britton to F. S. Lee, September 4, 1924, N. L. Britton Correspondence, New York Botanical Garden Archives.

75 Olmsted's full critique of Gibson's work on the Museum building included a damning assessment of the upper story: "The story above the cornice is particularly unfortunate in its effect on the silhouette and general proportions of the building. And for a building situated as this is, in a large open landscape, the high-shouldered effect thus produced is particularly unhappy." He concluded somewhat drastically (and optimistically, considering the NYBG's small endowment), that "it would seem worthwhile seriously to consider the total elimination or radical change of the present top story at this time."

would alter its appearance entirely. Museum architect Gibson's incompetence in neoclassical style had become, two decades later, fair game for the Garden's critics, adding to Frederic Lee's determination to remake the institution and wrest control from Britton. Olmsted continued his withering critique by commenting upon the disproportionate relationship between the four rows of maturing tulip trees (*Liriodendron tulipifera*) that led up to the Museum's main entrance and the axis of the Museum's wide façade itself. Though intended to impart a formal symmetry to the Museum approach, these rows of trees, planted 20 years earlier, had merely outlined asymmetrical swaths of landscape, and, Olmsted wrote, formed too narrow a setting for the Museum. Overall, the building and its surroundings, he concluded, were the results of "an unfortunate compromise" in design choices.

In place of the current treatment, Olmsted proposed an alternative involving the development of a series of exhibition gardens laid out individually, more "formal" in character, and set off from one another by plantings of both deciduous and coniferous trees. These new plantings would be designed to greet visitors who entered this most heavily trafficked area of the Garden. Pedestrians arriving via public transport tended to be "lazy about walking far," he maintained, and, thus, needed the inducement of interesting and "markedly attractive prospects" that would lead them along further into the Garden and induce crowds to disperse eagerly into its interior. Such alterations, while necessary, he warned, would be "radical, extensive, and costly."[76]

In sum, Olmsted Jr. presented his 1924 report as a starting point, a summary of possibilities for reforming the Garden's physical landscape. The Garden grounds had been, in his opinion, rather haphazardly constructed. Though at times detailed in its criticisms and proposed remedies, the report on the whole was designed not as a specific course of action but as a "useful basis for a program to be adopted by the Managers." Britton own response was, predictably, ambivalent. Although he recognized the need for additional endowment funds and for horticultural improvements, he nevertheless questioned several of Olmsted's specific suggestions, including the installation of terraces in the iris garden, the creation of an east-west vista, the rearrangement of the driveway system, and the development of more formal garden sites. In addition, Britton remained committed to building further wings on the Museum to provide new space for research, something, he noted, that was not addressed by the Olmsted Report.[77] In spite of the director-in-chief's concerns, the report was printed and distributed among the Garden's management; at a meeting in November 1924, it was "approved in principle as a guide for future treatment of the grounds in the Garden." Only Olmsted's general suggestions concerning the modification of the

76 Olmsted Report: 32–37, see note 62.

77 N. L. Britton to the Executive Committee of the New York Botanical Garden, November 5, 1924, New York Botanical Garden Archives.

Garden's roadway system were "reserved for future consideration." (Such changes would involve some of the most radical and costly alterations to the NYBG landscape.) The next steps, of course, would entail commissioning thoroughly detailed plans and maps and, ultimately, securing funding adequate to pay for all proposed projects.

Finding Funds

Olmsted's report, at times sharply critical of Britton's treatment of the Garden design, was generally confident in the promise of the Bronx Park landscape itself. As an optimistic document, it provided Frederic Lee with an opportunity to initiate an ambitious campaign to augment the NYBG's endowment. Since the Garden's creation, its operating funds had been derived from an annual appropriation from the City of New York, income from the endowment, membership dues, and money Britton secured from individual or institutional donors. From the start of his career as Director-in-Chief, Britton had been aware of the precariousness of his institution's finances. The very first volume of the Garden's *Journal*, in fact, had included a brief statement of the "needs of the Garden," in which Britton, citing the 1900 endowment of just over $280,000, called for additional members and increased donations from public spirited New Yorkers.[78] Henry Rusby, Britton's curator of economic botany, noted that in the first years of the Garden's existence, "every dollar . . . was made to do the work of four."[79] Britton's 1900 appeal was followed by subsequent systematic efforts to secure funds dedicated specifically for exploration and horticulture. A $250,000 expedition endowment, Britton calculated, ideally would yield $12,000 annually, while a horticultural fund of $100,000 would yield $5,000 annually. Britton targeted his appeal to benefactors interested in underwriting the costs of specific portions of the Garden enterprise: other desired income-yielding "funds" that he had enumerated in 1909 had included an herbarium fund ($50,000); lecture fund ($25,000); library fund ($50,000); laboratory fund ($20,000); horticultural and botanical prize funds ($20,000); and several research funds (ranging from $5,000 to $50,000). Although these sums were never secured, the numbers provide a glimpse into Britton's institutional priorities during the heyday of his career at the Botanical Garden.[80]

78 *Journal of the New York Botanical Garden* 1 (August, 1900).

79 Cited in E. D. Merrill, "Nathaniel Lord Britton." *National Academy of Sciences Biographical Memoirs* 19 (1938): 147–159.

80 *Journal of the New York Botanical Garden* 10 (1909): 218–219. The city's appropriation for the Garden's budget was sharply reduced in 1910, prompting Britton to complain to the Board of Estimate that the income from endowment and municipal appropriations was insufficient to meet the needs of an institution open daily to the public and under increasing use as an educational resource by the city's growing population. Britton was a regular and vigorous advocate and frequent correspondent with the city institutions that oversaw appropriations for the Garden. See Minutes of the New York Botanical Garden Executive Committee, February 15, 1910, New York Botanical Garden Archives.

Finally, over a decade before Olmsted's critical assessment, Britton had attempted to increase the Garden's endowment to $1 million. Convening an endowment committee, he had printed a circular similar to that drawn up by Lee in 1924 in which the achievements of the Garden in science and in park-building were touted to a presumably grateful public. In 1914, NYBG operated on an annual income from membership dues, sales of its publications, and endowment income of $30,000, $25,000 of which was needed merely to supplement the City of New York's appropriation for annual maintenance. At that time, he estimated a minimum need of approximately $20,000 more annually. In an effort to reach the $1 million mark in endowment funds, Britton, a year before the outbreak of the First World War, had successfully solicited over $117,000 from a roster of 89 donors, including Andrew Carnegie, J. P. Morgan, William Vanderbilt, and Frederic Lee.[81]

Later efforts met with only partial success, as when Britton secured a pair of $50,000 donations in 1917 from the brothers Murray and Daniel Guggenheim to construct a second range of glass horticultural houses in the northeastern quadrant of the Garden, but had never freed the director-in-chief from constant concern over the NYBG's financial health.[82] Britton's courtship of the Guggenheims had been carefully planned and involved the advice of J. P. Morgan, Jr. and other members of the Garden's Board of Managers. In January 1917, Daniel Guggenheim, a Garden Manager, had chaired a new Garden Endowment Committee and inaugurated its campaign for funds to erect new structures by pledging the $50,000 that his brother matched within weeks. The Guggenheim glasshouses, completed and opened to the public by 1919, were designed with space for public lectures and other areas reserved for orchid culture. Though simpler in style than Lord and Burnham's Conservatory, its roof echoed the main glasshouse's design.[83] Later that spring, Britton, at Guggenheim's urging, began consulting Garden Vice-President Francis Stetson to determine the best ways in which they might approach other prominent wealthy New Yorkers.[84]

81 "Circular of the Endowment Fund Committee," *Journal of the New York Botanical Garden* 15 (1914): 23–28.

82 *The New York Times*, July 22, 1917. In his 1916 report to the Board of Managers, Britton had noted that "owing to the necessity of using all available funds for supplementing the city allowance for maintenance, major botanical exploration has been essentially limited to expeditions made possible by special gifts." See the *Bulletin of the New York Botanical Garden* 8: 307.

83 *Journal of the New York Botanical Garden* 20 (December 1919). The Guggenheim glasshouse no longer exists.

84 N. L. Britton to D. Guggenheim, January 20, 1917, Endowment Committee letters, New York Botanical Garden Archives. F. L. Stetson had been personal counsel to J. P. Morgan and a presidential advisor in the Cleveland administration. See the *Dictionary of American Biography* (New York: Scribner's, 1946–58).

Britton had also commissioned his long-standing Garden architect John Brinley (coauthor of the original 1895 Garden plan) to provide watercolor renderings of various desired structures, each to be annotated with a corresponding cost estimate. Once completed, these drawings were circulated among the Garden's officers, members of the NYBG Corporation, and the Women's Auxiliary and then exhibited at a Manhattan art gallery in 1918. (A subsequent exhibit of Brinley's artwork had been held at Columbia University's Avery Library. The New York Public Library had declined Britton's request, noting that it displayed only books and manuscripts.) A particularly prominent target of Britton's fundraising had been Andrew Carnegie, still a member of the Garden's management board, but no longer active as an officer. (His wife, however, remained a regular visitor and contributor to the Garden's horticultural school for children). Although Britton requested Guggenheim's assistance in approaching Carnegie, Guggenheim declined, citing his business connection to the aging philanthropist. During the war years, Carnegie never bestowed a major gift upon the Garden.[85] Nevertheless, Britton had continued throughout 1917 pursuing major philanthropic gifts. He and members of the Endowment Committee debated possible strategies: whereas some considered waiting for wartime prices of raw materials to fall before beginning construction on the Guggenheim glasshouses, others counseled immediate action as a spur to other donations. Then-President W. Gilman Thompson urged Britton to "conspicuously label all major gifts," including even individual cherry trees, as a way to personalize philanthropic urges. When gifts were "impersonal," Thompson contended, "it is easily assumed that the Garden can take care of itself." It was his hope that such methods would "corral more gifts." With the Guggenheim's donations fresh in the Garden's treasury, it was hard not to succumb to such optimism.[86] Even Frederic Lee, then a Scientific Director, had praised Britton, noting that he was "impressed with [Britton's] skill in devising so many and varied forms of bait."[87] Britton had also attempted to hook smaller donors, reserving the development of the land surrounding the Lorillard Mansion for less prominent members whom he hoped would pledge smaller amounts in great numbers.[88]

85 N. L. Britton to Daniel Guggenheim, March 24, 1917 and Guggenheim to Britton, March 26, 1917; Endowment Committee letters, New York Botanical Garden Archives.

86 Optimism was not entirely without precedent in American science. As Margaret Rossiter notes, the donation of a "named gift" to the scientific departments of Harvard University had been, by the mid-nineteenth century, a "very suitable" path for wealthy, social-climbing Bostonians to associate themselves with that city's Brahmin elite. See "Philanthropy, Structure, and Personality or, The Interplay of Outside Money and Inside Influence," in Clark A. Elliott and Margaret Rossiter, eds., *Science at Harvard University* (Bethlehem, PA: Lehigh University Press, 1992): 15. Similar motivations likely led the Guggenheims to donate the glasshouses at NYBG in the 1910s.

87 W. G. Thompson to N. L. Britton, May 8, 1917; F. S. Lee to N. L. Britton, May 9, 1917, Endowment Committee letters, New York Botanical Garden Archives.

88 N. L. Britton to J. R. Brinley, July 31, 1917, Endowment Committee letters, New York Botanical Garden Archives.

The city's press had been a regular ally of Britton's fundraising efforts through the 1920s, printing letters praising the Garden's attractiveness and editorializing on its usefulness as a source of scientific information.[89] *The New York Times* in particular championed the Botanical Garden, noting that its land was by 1923 valued at $15 million and calling upon the city to increase its annual appropriations for its upkeep. "New York can afford to give what [the Garden] asks," the *Times* editorialized; "it cannot afford to refuse."[90] Despite these grand intentions, by the time Lee became president, the Garden's management had been able to construct only the new greenhouse with the Guggenheim donations, along with some fencing and two small shelter kiosks, one of which Britton paid for himself. Bequests and other gifts to the Garden from the estates of various members and benefactors between 1900 and the war had brought just under $310,000 to the NYBG coffers. By the 1920s, Britton still needed $600,000 for proposed construction projects and approximately $900,000 in endowment funds. The dedicated endowment campaign of the war years had fallen considerably short of Britton's ambitious goal.[91]

When he succeeded William G. Thompson as Garden President, Frederic Lee was not content to rely upon the city's press or Nathaniel Britton to attract favorable attention. In 1924, he published a pamphlet entitled, "The New York Botanical Garden: Its Status and Needs." Written as an overview for potential donors, this brief circular outlined the history, major functions, and the priorities of the Botanical Garden and was designed to stimulate giving to the new $7 million endowment campaign Lee inaugurated that year. The Garden, the report asserted, "has made itself one of the City's most important educational institutions" and had grown to prominence on the strength of its taxonomic collections. In addition, it possessed a varied and beautiful landscape and sponsored both scholarly and popular publications on botany. In an odd bit of anthropomorphic prose, the pamphlet's author (in all likelihood Frederic Lee himself) concluded that "after thirty years of service," however, the Garden had "reached a stage where it feels that it ought to contemplate its own needs." To grow and continue to serve the city and nation as both a park and a scientific research institution, the Garden needed additional funds to underwrite the costs of new

89 See, e.g., *The New York World*, January 29, 1923. In the fall of 1918, for instance, the City of New York Board of Estimate and Apportionment reduced the appropriations granted the Garden and the Zoological Society by $20,000 and $32,000, respectively. (The zoo retained a higher overall municipal appropriation of $175,000.) See Minutes of the Executive Committee of NYBG, October 3, 1918, New York Botanical Garden Archives.

90 *The New York Times*, January 15, 1924, "More Money Sought for Botanical Garden" and a *New York Times* editorial, January 16, 1924. See also the *New York World* editorial, January 29, 1923, "More Funds for the Botanical Garden."

91 "Items obtained or designed by the endowment committee, 1917–1919," and "Summary of Dr. Britton's Statement," Records of the Endowment Committee, 1917–1922, New York Botanical Garden Archives.

buildings, equipment, additional staff, and refurbished grounds along the lines suggested by the Olmsted Brothers firm. "Our large population lives for the most part in daily contact with ugly surroundings," Lee wrote, echoing a Progressive refrain with roots in the nineteenth century parks movement and "City Beautiful" ideology. "In our schemes for social welfare," he continued, "we have devoted too little attention to beauty. To make the Botanical Garden beautiful will help supply a need of the City that is quite as important in advancing our civilization as the City's more material needs." The pamphlet ended with an explicit entreaty for cash: instructions were given for making donations.[92] The *New York Daily News* hailed Lee's new endowment campaign as a blueprint for "making the Garden a model for the world. All that is required," the *News* editors noted, "is the money."[93]

By the middle of 1924, then, despite Britton's misgivings, Lee considered the two major reports he had commissioned since taking office, the research assessment compiled by the Scientific Directors and the Olmsted Brothers' report, to be dual master plans with which he might transform the Garden's scientific priorities and its landscape. Lee attempted do so, while adhering to the explicit mandate of the Garden's corporate charter, which charged it with a responsibility to educate the lay public and to foster botanical and horticultural science.[94] Endowment needs continued to be pressing: the Garden's total income in 1924, from city appropriations and various forms of privately derived income (including membership dues), was $332,000. Of this total, over $324,000 had been spent, almost two-thirds on salaries and labor alone and only $4,300 on "exploration, investigation and lectures." (The second-largest expense had been on light and heat for buildings.)[95] These troubles had been exacerbated by post-war inflation.

As part of his effort to keep NYBG in the news as its endowment campaign continued in 1925, Lee inaugurated a series of competitions for small-garden designs. These contests, in which a winning design would be built on a small plot of NYBG

92 "The New York Botanical Garden: Its Status and Needs," 1–8, manuscript in New York Botanical Garden Archives. Here Lee was not employing the term "beautiful" as it had been used by Andrew Jackson Downing in the 1850s, but instead utilized its colloquial meaning to underscore the potential social uplift provided by a landscaped antidote to New York's industrialized cityscape.

93 *New York Daily News*, December 8, 1924.

94 *Bulletin of the New York Botanical Garden* 9 (1924): 5. Perhaps unsurprisingly, given Lee's professed attitude toward taxonomy, the effort in the 1920s to secure a reputation for the Garden in various forms of research, i.e., to create the "horticultural university" within a beautifully landscaped (and artfully "natural") setting, did not entail a revamping of the museum's various systematic exhibits. Lee had even discussed abolishing the indoor museum altogether with a member of the Board of Managers. With laboratory space at a premium within the building, such a change would not have seemed inappropriate, perhaps, to Frederic Lee. These exhibits, however, were not ultimately dismantled until after the Second World War. Kenneth Mackenzie to F. S. Lee, May 29, 1923, N. L. Britton Administrative Correspondence, New York Botanical Garden Archives.

95 *Bulletin of the New York Botanical Garden* 1924: 58–59

land reserved for demonstration gardens, constituted a savvy attempt to capitalize upon the popularity of home gardening among the growing American suburban middle class. "It is surprising," Lee asserted in the spring of 1926, "the interest that has arisen during the past years in amateur gardening. We are only too glad to have persons write in to us—I don't care from where, what part of the United States or world—and ask us for advice."[96] Competitors were encouraged to provide for a seasonal succession of blooms in their designs and could include small structures, but none that required professional expertise to construct or maintain. Such small gardens, Lee asserted, were rapidly "developing as an important social factor, providing new opportunities for a wise use of the increasing leisure from shorter working days." Garden Board members agreed, noting that "where not many years ago a few dollars were spent on shrubbery, one now finds the average man making a substantial expenditure for ornamental evergreens." Along with Lee, the Board of Managers hoped that the Garden could position itself at "the center of all this vast horticultural interest."[97] Juries consisting of representatives from the Garden, the Federated Garden Clubs of New York State, and the Garden Club of America awarded cash prizes of $100 in addition to constructing the winning design. Members of the Women's Auxiliary to the Board of Managers that had been established in 1914 by Managers' wives were consulted on how the Garden could further its connection to local Garden clubs.[98] Members suggested that closer links to these organizations be established and female members be invited to the Garden for meetings with the Women's Auxiliary, given lectures on rare and ornamental plants, and guided through the conservatories.[99] Garden staff also forged a closer relationship to the still-active American Institute of the City of New York, at whose 94th annual floral exhibition NYBG botanist Marshall Howe served as a judge.[100] By 1926, the annual garden design competition was drawing 160 entries nationwide.[101]

96 "Plans Garden Competition," *The New York Times*, December 14, 1925; "Nationwide Competition Announced to Popularize the Small Garden," *Yonkers Herald*, December 15, 1925.

97 Kenneth K. Mackenzie to F. S. Lee, May 29, 1923, New York Botanical Garden Archives. Mackenzie, however, was highly critical of the taxonomic displays in the museum. Although he professed great interest in the "purely scientific work of the Garden," he suggested that the indoor museum itself was expendable, because its educational value, he felt, was questionable and the space might be better devoted to other uses.

98 On the founding of the Women's Auxiliary, see *Bulletin of the New York Botanical Garden* VIII (1914): 48 and "Women to Boost Botanical Garden: Auxiliary Visits Bronx Park," *The New York Tribune*, May 8, 1914.

99 Martha Prentice Strong to F. S. Lee, February 16, 1923, N. L. Britton Correspondence with F. S. Lee, New York Botanical Garden Archives.

100 *Yonkers Herald*, December 15, 1925, see note 96. For Howe's role with the American Institute, see *The New York Sun*, August 24, 1925.

101 Minutes of the NYBG Board of Managers, 1926. Britton remained somewhat skeptical of the need for model gardens of the sort constructed from the winning plans. Such efforts, he asserted, were

A similar effort to carve out a profitable place for the Garden in social reform had taken place several years earlier when Britton briefly secured federal funds to train over 80 convalescent World War I veterans in horticulture. The Federal Board of Vocational Education had approved the Garden's curriculum of laboratory work, museum studies, field work, greenhouse instruction, garden maintenance, lawn and tree upkeep, and various horticultural specialties such as fruit culture and water-garden construction. Britton had also promised federal authorities the possibility of providing advanced instruction in plant breeding, selection, and pathology to suitable candidates. Begun in 1919, the program had cost the US government $15 per student per month in tuition and fees. The program lasted little more than a year, however. While a "few" of the sick and maimed veterans had displayed an aptitude and interest in horticulture, Britton noted, "the great majority did not, and these will probably be more useful in other occupations." Though some remained as apprentices at the Garden, the program ended in 1920, when NYBG's Assistant Director Gleason complained that the federal Board for Vocational Education had begun interfering too closely in the design of the Garden's curriculum.[102] "All students," Britton concluded optimistically, "were greatly benefited in health while at the Garden, and this was, perhaps, the most important result to be obtained."[103]

The most successful effort of Lee's mid-1920s endowment campaign was a gift from John D. Rockefeller, Jr. of $500,000 between 1925 and 1927. Britton and Lee had begun courting Rockefeller in 1924. Lee had written his philanthropic staff, and Britton had escorted these representatives through the Garden grounds.[104] Rockefeller senior had been an original contributor to the Garden, at the urging of J. P. Morgan, Sr. donating $25,000 for the original endowment fund and an additional $1,000 to begin building up a mycological (fungi) herbarium. The half-million-dollar gift (twice what he donated simultaneously to the Brooklyn Botanic Garden) represented a break from the philanthropic precedent set by the elder Rockefeller, who had turned down repeated requests for funds in the 1900s and 1910s.[105]

The Rockefeller gift, Britton and Lee hoped, would stimulate subsequent giving to the overall endowment campaign. Of the $7 million sought for an expanded Garden

---

"of doubtful practicality and unessential to a botanical garden." N. L. Britton to the Executive Committee of NYBG, November 5, 1924, New York Botanical Garden Archives.

102 Minutes of the NYBG Board of Managers, November 18, 1920, New York Botanical Garden Archives.

103 See *Journal of the New York Botanical Garden* 20 (March 1919): 51–53 and "Instruction of Convalescent Soldiers in Practical Gardening," *Bulletin of the New York Botanical Garden* 19: 220.

104 N. L. Britton to F. S. Lee, August 21, 1924; August 29, 1924; F. S. Lee to N. L. Britton, September 2, 1924, N. L. Britton Correspondence, New York Botanical Garden Archives.

105 Rockefeller Family—Record Group 2, OMR Series: Cultural Interests III 2E 94 866—"Cultural Interests: New York Botanical Garden, 1895–1958," Rockefeller Archives, Tarrytown, New York.

endowment, $4 million, Lee felt, could be secured from private sources, with the balance coming from increased municipal appropriations and later private gifts. Of that initial $4 million, Lee and Britton had earmarked over $800,000 for improving the grounds (along the lines proposed by Olmsted); $2 million for additional needed maintenance; $885,000 for the endowment; and $300,000 for salaries and pensions for the scientific staff. After close scrutiny of the Botanical Garden's financial records and its charter, Rockefeller and his staff of philanthropic professionals were impressed with the degree of coordination between the Garden and Columbia's academic departments. They gave the large sum in the hope of advancing Britton and Lee's explicit research goals. Rockefeller acknowledged the dearth of funds NYBG had faced throughout its history, and his staff, echoing Lee, noted that this handicap had meant that the institution, while providing the city with "good" and "useful" scientific and educational work, nevertheless lacked "vigor." In order to excel, Rockefeller recognized, additional private support would be necessary. His investigation into the Garden's campaign led his assistants to conclude that with its "splendid properties," its "close relations to Columbia, [and] its traditional association with the City of New York," the "opportunities of the [institution], both scientifically and educationally, are very great." In making the donation, Rockefeller acknowledged the Garden's continuing scientific potential. In thanking Rockefeller, Lee asserted that his largesse would enable NYBG to take "the premier place among botanical gardens and to make it the model of what such institutions should be in the future."[106]

A Potential Institutional Rival

On September 21, 1924, as Lee courted a second generation of Rockefeller funds, *The New York Times* announced the dedication of a new institute devoted to plant study that would do for botany what Rockefeller philanthropy had made possible for medical research. Situated in Yonkers, a city just north of the Bronx-Westchester border, the Boyce Thompson Institute for Plant Research had been the pet project of William Boyce Thompson, whose landscaped estate and well-tended gardens were located across the road from the new Institute's property. Thompson had been recruiting its staff of over two dozen plant researchers (including nine women) throughout the previous year. A native Montanan, Thompson had made a fortune in Western mining and had moved east where he indulged a love of horticulture by maintaining a manicured estate in the leafy northern suburbs of New York City.[107]

106 Ibid.

107 "Plant Life to Yield Many of Its Secrets: Laboratory Founded by Col. Thompson to Be a Rockefeller Institute of Botany," *The New York Times*, September 21, 1924. For a fuller account of the Institute's founding and original staff, see *Contributions from Boyce Thompson Institute for Plant*

In 1919, Thompson founded a "farm and research corporation," through which his employees asked various botanists what they considered to be the most pressing issues facing plant researchers.[108] According to the founding premise of Thompson's institute, these needs involved basic pathological research. When interviewed, Thompson and employees told the *Times* that although federal dollars were being spent on research into agricultural diseases, the nation nevertheless continued to lack well-equipped institutions dedicated to research and the development of new principles of plant production. Intended to sponsor "fundamental" research into questions of agricultural import, the Boyce Thompson Institute was conceived essentially as a private agricultural experiment station. Intent upon dedicating a monument to his deceased parents, Thompson had financed a reconnaissance of European botanical laboratories. The results of this survey served as the model for the Institute's research facilities. Designed to accommodate 30 investigators, the new laboratory's mandate was to increase the yield of crop plants. (A latter-day Malthusian, Thompson fretted that by 1960, when the US population was projected to reach 200 million, the nation would cease to be agriculturally self-supporting.) Thompson lured plant physiologist William Crocker from the University of Chicago to head his new institution and recruited eminent botanist (and Britton friend) John Coulter to serve on his Board of Directors.[109] Research by the Institute's staff was conducted in a spacious set of laboratories and greenhouses, each appointed with climate control, electricity, gas, and various safety features. Pre- and postdoctoral fellowships were awarded to young investigators in residence. Studies in the first several years of the Institute's existence included "Hastening the sprouting of dormant potato tubers," "Germination of seeds under water," "Attempts to cultivate an organism from tomato mosaic," and "A physiological study of the effect of light of various ranges of wavelength on the growth of plants."[110]

The timing of Thompson's philanthropic largesse was ironic in light of the campaign underway several miles south of Yonkers to expand the endowment and

---

Research, Inc. 1 (1925–1929): iii–54. Though he was invited (as "T. L. Britton"), it is unclear whether or not Britton attended these ceremonies.

108 Thompson's people seem to have interviewed primarily Midwestern botanists and the staff of some agricultural experiment stations. There is no indication in NYBG records of any direct communication between Thompson and Nathaniel Britton.

109 *The New York Times* September 21, 1924, see note 107.

110 It is interesting to note that of the 13 staff members at Boyce Thompson holding doctorates, four of these were women, two of whom headed particular research "departments." Overall, there were nine women on the Institute's staff of 1924. For a description of the Boyce Thompson laboratories and greenhouses, see *Contributions from Boyce Thompson Institute for Plant Research, Inc.* 1 (1925–1929): 13–18.

research capacity of NYBG. The photographic, laboratory, and greenhouse equipment Thompson furnished his researchers was expensive. These funds, along with those set aside for the new Institute's endowment, might have gone to NYBG.[111] At the 1924 dedication ceremonies, University of Wisconsin plant pathologist Lewis Jones, a member of Thompson's advisory board, had asserted that the new institution had not been created to supplant, but rather to supplement, the work of various national and state research agencies. Whatever "fundamental lead opened by anyone in such other institutions," Jones asserted grandly, "here may be the opportunity and facilities for its ultimate development." In plant physiology in particular, Jones claimed, both universities and botanical gardens, which lacked either "the assurance of financial support for new lines of research or adequately equipped laboratories and greenhouses," had fallen short of the necessary amount of research. Other attendees of the dedication, including NYBG Board member R. A. Harper, had nevertheless echoed praise for such new research opportunities, despite the implicit criticism of other botanical institutions.[112] William Crocker later decided not to "duplicate research that was already being adequately covered in the northeastern United States," especially "systematics and genetics" and consequently identified plant physiology, pathology, and biochemistry as the primary activities of the new institution.[113]

As Thompson's vision was being realized in Yonkers, J. F. Kemp, a Garden Scientific Director and close collaborator of Britton's on the Puerto Rico Survey, noted to Britton that Fred Pope, Thompson's aide in organizing the new institute, had been a Columbia student. Pope had informed Kemp in October 1923 that Boyce Thompson's $750,000 laboratory facilities were within a month of completion and that the new staff planned a thorough agenda of physiological investigations into plant pathology. Pope also confided to Kemp, who duly informed Britton, that Thompson apparently planned a considerable expansion of the Institute's grounds to create an arboretum. Despite this apparent loss to the Garden of a considerable amount of local philanthropy, Kemp maintained a fairly sanguine view of events. He counseled Britton, "we can have no regret that all this has escaped the Garden

111 Thompson's penchant for somewhat repetitive philanthropy became apparent several years later when, despite the continuing work of the Carnegie Institution's Desert Laboratory, he also endowed a "Boyce Thompson Southwestern Arboretum" in Arizona, where he hoped to find "practical" uses for the region's native succulents. See "Boyce Thompson Southwestern Arboretum, Incorporated: Purpose, History, Dedication, Superior, Arizona," July, 1930 (copy in New York Botanical Garden Library).

112 Lewis R. Jones, "America's Need for Plant Research," *Contributions from Boyce Thompson Institute for Plant Research, Inc.* 1 (1925–1929): 44.

113 William Crocker, *Growth of Plants: Twenty Years of Research at Boyce Thompson Institute* (New York: Reinhold Publishing Corp., 1948).

because Colonel Thompson's unchangeable ambition was from the start to establish a monument to his own name similar to the Rockefeller Institute on the East side...."[114] Though Britton agreed with Kemp's assessment of the situation, clearly the founding merely several miles away of a well-equipped set of laboratories and greenhouses for the study of plant physiology and pathology must have rankled the Garden's director-in-chief and its ambitious new President.[115] The millions of Thompson's mining fortune expended on such equipment, with the right sort of philanthropic recruitment displayed successfully with the Rockefeller donation, might have placed the Garden on the most secure financial footing of its almost three-decade history. Such a philanthropic coup would perhaps have enabled Lee to achieve his personal goal establishing plant physiology outside the confines of a medical school (while continuing to underwrite Britton's systematic researches). Thompson's personal plans in any case came to fruition just as Lee articulated his vision for the Garden. (Whether any of Thompson's correspondents and consultants in the academic botanical community argued or even tried to argue successfully for an alternative gift to New York's already-existing botanical research institution remains uncertain.)

A further irony was the method of justification used by Thompson's advisors: their contentions that only an endowed institution could investigate fundamental research questions, especially those of Malthusian import, clearly anticipated those of Frederic Lee in 1924. Given Pope's remarks to his former Columbia professor, J. F. Kemp, it is likely that the Garden simply began its similar campaign too late to harness Thompson's generosity. In addition, Thompson would likely have been determined to establish his own institution, under his own control (and entirely devoid of connections to city agencies) regardless of Lee's timing. Thompson's determination to build a personal "monument" was certainly a disappointment for the Garden, based on sheer dollar figures alone: in 1923, Kemp estimated that the Boyce Thompson Institute enjoyed an annual income from its endowment alone of $250,000. By comparison, the Garden's total income for 1924, including revenue from its endowment and public and private contributions, was $332,000. Andrew Denny Rodgers asserts that Thompson, in fact, had consulted with NYBG in 1919, where he had been encouraged to subsidize a great systematic effort for the Bromeliaceae family along the lines of Britton and Rose's cactus project. Thompson apparently refused, determined as he was to erect his own institution in Yonkers. Rodgers, however, does not give a source for this assertion and there is no record in Britton's papers indicating such negotiations.[116] It is clear,

114 J. F. Kemp to N. L. Britton, October 20, 1923, Records of the Puerto Rico Survey, New York Botanical Garden Archives.

115 N. L. Britton to J. F. Kemp, October 25, 1923, Records of the Puerto Rico Survey, New York Botanical Garden Archives.

116 J. F. Kemp to N. L. Britton, October 20, 1923. See A. R. Rodgers, *John Merle Coulter: Missionary in Science* (Princeton, NJ: Princeton University Press, 1944): 287–288.

however, that Lee's ambitious endowment scheme failed, although the Garden and the new Institute would eventually collaborate, along with the Horticultural Society of New York and the New York State College of Agriculture, to provide fellowship money for the investigation of the diseases of lilies.[117]

Ambitious Fundraising Goals Fall Short

On June 16, 1927, after four years in the Garden's presidency, Frederic Lee submitted a letter of resignation to the NYBG Board of Managers, citing "serious encroachment on [his] professional duties." Although this retirement was to become effective in November of that year, Lee informed Britton in October that he wished to "sever his connection with the Garden wholly" and tendered his resignation from the Boards of Scientific Directors and the NYBG Corporation as well. Britton responded to Lee's terse notes in kind, noting that although there had not been a quorum at the most recent meetings of the Scientific Directors, he nevertheless thought it likely that Lee's resignation could be officially recognized by month's end.[118]

Though Lee's missives to Britton contain no clues regarding the extent of his dissatisfaction with his relations with the Garden, Lee was a good degree more candid with influential members of the Board of Managers. His mention of encroachment upon his "professional duties" by his Garden responsibilities was merely a cover for the discontent he felt after several years of only limited fund-raising success. Since assuming the presidency in 1923, Lee had consulted with J. P. Morgan, Jr. and had confided his hopes that the Garden could "hold in its sphere, the place of leadership which is held in theirs by corresponding institutions," including the Metropolitan Museum of Art, the American Museum of Natural History, and the Zoological Society. With enough funds and the cooperation of "men of substantial wealth in New York," Lee had predicted confidently in the spring of 1923, "there is no question . . . that, sooner or later," the Garden would attain the prominence of the city's other cultural institutions.[119] Morgan had responded politely, but had attempted to discourage Lee from initiating his planned endowment campaign that year, noting that other similar fund-raising efforts for "very large amounts" were about to be launched in New York, and he "seriously doubted" Lee's chances for success at that time. Morgan counseled

117 Minutes of the Scientific Directors of the New York Botanical Garden, October 8, 1927, New York Botanical Garden Archives. Boyce Thompson Institute director William Crocker would also travel to Puerto Rico in 1927 to consult with insular Governor Towner regarding the possibility of establishing a tropical agricultural institute. N. L. Britton to H. Towner, July 16, 1927, Puerto Rico Survey Records, New York Botanical Garden Archives.

118 F. S. Lee to NYBG Board of Managers, June 16, 1927; F. S. Lee to N. L. Britton, October 6, 1927; and N. L. Britton to F. S. Lee, October 10, 1927, New York Botanical Garden Archives.

119 F. S. Lee to J. P. Morgan, Jr., April 7, 1923, J. P. Morgan, Jr. Papers, The Pierpont Morgan Library Archives, Box 7, Folder 370.

patience and careful planning in addition to the choice of more propitious timing before undertaking the effort to add millions to the Botanical Garden's endowment. Lee heeded Morgan's advice by commissioning the Olmsted study of the Garden landscape and the Scientific Directors' assessment of the NYBG research program before opening the campaign in earnest.[120]

Lee's disappointment in 1927 was due to the relatively modest results of his four-year drive to augment the Garden's endowment by several million dollars. The Rockefeller gift of $500,000, although significant, was matched in that time by roughly an equal amount, far short of the $7 million goal Lee had set in 1923. Despite a concerted effort in which both Lee and Britton contributed funds and were joined by other donors including members of the Guggenheim, Morgan, Carnegie, and Roosevelt families, at the time of Lee's resignation, only $1,019,000 had been added to the Garden's endowment.[121] In a parting statement to the Board of Managers, Lee laid blame for the failure to reach his goal on Britton (although he did not name the director-in-chief explicitly). Though NYBG, Lee wrote, possessed land "of great possibilities," a "diligent and devoted staff," and other "commendable features," it nevertheless confined its "activities . . . within its original framework," and thus "its expected leadership among botanical gardens has failed of realization." Since the establishment of the Botanical Garden in the 1890s, Lee asserted, American landscape architecture, botanical science, and public education methods, all aspects of the Garden's public mandate, had "progressed remarkably." Such progress, however, was in evidence at NYBG only sporadically, a shortcoming underscored, Lee noted, by the Olmsted Brothers' report of 1924. In addition, the NYBG's institutional peers, the American Museum and the Metropolitan Museum of Art, Lee contended, had "constantly changed [their educational methods] to meet the needs and stimulate the interest of the public." Finally, and especially damning, in Lee's view, was the Garden's failure to address "the great problems of plant science." "Thirty years of trial of a policy," Lee concluded, "would seem sufficient to test its adequacy," and the Garden as it was run during Britton's tenure, he clearly implied, had failed to live up to its promised reputation.[122]

Although the overall endowment campaign had not met established goals, Lee attempted to sum up his brief presidency in the most flattering light possible. "During the last five years," he wrote,

120 J. P. Morgan, Jr. to F. S. Lee, April 11, 1923, J.P. Morgan, Jr. Papers, The Pierpont Morgan Library Archives, Box 7 Folder 370.

121 "$1,019,000 added to Bronx Garden Endowment Fund," *The New York Herald Tribune*, February 17 1927.

122 F. S. Lee to NYBG Board of Managers, November 12, 1927, J.P. Morgan, Jr. Papers, The Pierpont Morgan Library Archives, Box 7 Folder 370.

> an intensive effort has been made to impress upon those who are directly responsible for the administration of the Garden's affairs the need of a large vision and a broader policy. More than one million dollars have been added to the endowment and a beginning toward an improved situation has been made; but the indifference of many persons and the needless obstacles raised by a few, whose eagerness to assist should have been manifest, have delayed the inevitable outcome.

Lee ended his presidency with the admonishment: "if the New York Botanical Garden is to live, it must evolve. To evolve, it must abandon its self-complacency; its past narrow fundamentalist policy must give place to a liberal modernism. . . . The general plans for such an institution are already before you in various documents which have been presented to the Board during the past five years. In retiring, it is my earnest wish that these plans be put into early execution."[123]

The Scientific Directors, upon Lee's resignation, lauded his effort at reforming the Garden's research agenda along with its physical landscape as "a period of rapid advance." In "working out a broad general program for the further development of the Garden," the Directors asserted,

> both as a great municipal center for the educational and decorative display of plant life in all its phases, and as a center for basic research in plant science, especially as related to practical problems of decorative planting under city conditions, Dr. Lee's leadership must always be recognized as of fundamental significance for the development of applied plant science in its relations with a great metropolis.[124]

The Women's Advisory Council echoed the Directors' sentiments, praising Lee's "enlightened policy of enlarging the usefulness of the Garden to met the needs of the time [and] of enhancing its natural beauty by means of landscape gardening for aesthetic effect."[125]

Although by the late 1920s Nathaniel Britton had outlasted his rival for control over the Botanical Garden, his active career at the head of NYBG was nearing an end. Meanwhile, Elizabeth Britton, NYBG's "Honorary Curator of Bryophytes," had enjoyed an active and serious career as a taxonomist. In 1929, two years after Lee's bitter resignation, Britton and his wife retired from their positions at the Garden to devote themselves to full-time research and frequent travel to the Caribbean, where Britton continued administration of the New York Academy of Science's Natural

123 Ibid.

124 Resolution of the Scientific Directors of the New York Botanical Garden, November 17, 1927. New York Botanical Garden Archives.

125 Resolution of the Women's Advisory Council of the New York Botanical Garden, December 2, 1927, New York Botanical Garden Archives.

History Survey of Puerto Rico almost until his death, several months after that of his wife, Elizabeth, in 1934.

Lee's NYBG Career Ends with Britton Victorious

In considering the history of institution-building, historian E. C. Spary noted that as "functioning sites for the construction of scientific agreement" and investigation, institutions such as the Botanical Garden were established only after their "social walls" had been constructed: i.e., the Garden's "role, purpose, value, and legitimacy must have been agreed upon by the different sides" involved in its creation. She continues,

> the form and legitimation of that agreement, which is what constitutes and enables the institution as both social and material reality, may pass unquestioned for long periods subject only to an ongoing process of maintenance and reinforcement. But in times of social change . . . such forms and legitimations are overtly defended, challenged, or rejected."[126]

By the 1920s, NYBG had been in existence, behind the "social walls" built for it in the 1890s, for almost three decades. Those particular institutional parameters had been defined to a large extent by the vision of one botanist, Nathaniel Britton, himself the product of a particular late-nineteenth-century conception of botanical science and urban institution-building. Skillfully mastering the idiosyncratic art of administering a public institution (of private pedigree) within New York City, Britton retained control over the Garden virtually unchallenged for most of his tenure as director-in-chief. He created an institution dedicated to taxonomic research within a setting designed to resemble his memories of the Royal Botanic Gardens at Kew, England, in the 1880s.

By the 1920s, however, when Frederic Lee, philanthropist and socially prominent public-health researcher, began actively questioning Britton's vision of the Garden's research and design priorities, much had changed. The various life sciences by then known as "biology," and American botanical science in particular, were being pursued by a wide array of practitioners, stationed at an increasingly disparate set of institutions: botanical gardens, agricultural experiment stations, universities, and government and private research laboratories. Lee himself represented a direction in biological research, plant physiology, through which he had hoped to secure personal prominence and to wield the political power available to the "expert" in the public arena. When his efforts to establish this vision at Columbia apart from the medical school failed, he attempted during his Garden presidency to redirect the NYBG's research agenda away from taxonomy into areas (such as plant pathology) he deemed both more profitable and more prestigious. None of Lee's new research priorities meant that he

126 E. C. Spary, *Utopia's Garden: French Natural History from Old Regime to Revolution.* (Chicago: University of Chicago Press, 2000): 7–8.

would neglect the NYBG's physical *garden*, however. In fact, he hoped to upgrade the Garden landscape in such a way as to conform to a decades-old elite vision of beauty, which he hoped would also attract additional philanthropic support, while fulfilling the garden's statutory role as a specifically uplifting public space.

In his stated goals as a scientific reformer, his elite understanding of landscape design, and his personal power derived from both family wealth and a position within a university and institutional hierarchy, Lee represented nothing less than an aging "Progressive," however nebulously defined such a characterization must be.[127] As a public figure with an academic reputation enhanced by personal wealth, Lee recalls the bombastic scientist cleverly lampooned in Sinclair Lewis's *Arrowsmith*. (That Lewis might have derived inspiration for his caricatures from men of Lee's position and temperament is suggested also by the sort of interpersonal relations Lee maintained with students and junior colleagues at Columbia, for example.)[128] Lee's interest in the Olmsted firm's assessment of the Botanical Garden in 1924 stemmed from a long-standing environmentalist faith in the reform efficacy of urban parkland, one that had animated patrician liberals since the days of the senior Olmsted. Botanical gardens, like other public city parks, could help reduce working class drinking, assimilate immigrants, and control crime, while simultaneously signifying the high moral and intellectual status of a city branded overly "commercial" since its founding in the seventeenth century.[129] Unlike other city parks, however, the Garden, through its taxonomic displays and plant labels would be able to add a very specific element of public education (what Progressive theorist Herbert Croly might have dubbed "stirring and responsible" opportunities for learning) to its goal of providing landscaped beauty, while also devising pragmatic scientific solutions to problems of urban horticulture.[130]

Despite the changing scientific climate that appeared to render Britton's taxonomic passions dated and the social context that enabled Lee to chip away at the Garden's "social walls," Lee's personal challenge to Britton's leadership of the

127 Indeed, his lectures as a public health authority during the early decades of the twentieth century had been attended by no less an icon of Progressive Era reform than Jane Addams. See "'Tired Feeling Urged as Limit of Workday: Dr. Lee Declares Efficiency Fails after Fatigue Point," *The New York Evening World*, September 25, 1919.

128 Sinclair Lewis, Arrowsmith (New York: Harcourt, Brace, and Co., 1925). See also Charles Rosenberg, "Martin Arrowsmith, the Scientist as Hero," in Harold Bloom, ed. Sinclair Lewis's Arrowsmith (New York: Chelsea House Publishers, 1988): 17-27.

129 Robert R. Weyeneth, "Moral Spaces: Reforming the Landscape of Leisure in Urban America, 1850–1920," PhD Dissertation, University of California–Berkeley, 1984: 1 and Mona Domosh, *Invented Cities: The Creation of Landscape in Nineteenth-Century New York and Boston*. (New Haven, CT: Yale University Press, 1995): 143.

130 Weyeneth 3–4, see note 129; Paul Boyer, *Urban Masses and Moral Order in America, 1820–1920*, (Cambridge: Cambridge University Press, 1978): 175, 224; Herbert Croly, *The Promise of American Life* (New York: Bobbs-Merrill, 1965): 406.

Botanical Garden fell short. His scathing critique of the Garden's brand of botany was misguided, based upon his own personal predilections and professional ambitions. Lee gleefully blasted the "mere curio collections," of the Garden, "where one gets the thrill of seeing a coffee tree from South America or a tea shrub from China 'in the flesh' as it were," and its "stuffy, dreary museums where dead and dried plants by the thousands are stored for the delectation of the botanically erudite and the boredom of the man in the crowd."[131] Such criticisms of traditional, "object-based epistemology" exhibits were not far off the mark.[132] Row upon row of glass display cases most likely did bore many visitors. Dried plants affixed to sheets of paper or arrayed (pickled) in jars lacked the intrinsic eye-catching excitement of their mammalian counterparts housed in dramatic dioramas at the American Museum. By the mid-twentieth century, these museum displays, in fact, had been dismantled, giving way to the Garden's increasing need for research and office space.

Lee's presidency was a crucial time for NYBG. Given his professional training, his sweeping contempt for Brittonian botany was predictable. His purported embrace of the "scientific" experimentalism of physiology meant that he would never endorse what he considered little more than vestigial nineteenth-century antiquarianism. Upon Lee's assumption of the presidency, the Garden's institutional mission, as established at the outset by Britton, was challenged from within. With Lee's resignation, Britton's vision for the Garden emerged victorious, and NYBG was ultimately strengthened as a haven for systematic botanical research.

Lee's attack on Britton's stewardship of the Garden extended beyond the director-in-chief's research priorities to include a critique of the original layout of the grounds. However, Lee's arguments again were not entirely convincing. Although the Garden, in fact, had been plagued by a persistent lack of funds for regular maintenance, leading to undeniable shortcomings in the physical aspect, Lee's recourse to the Olmsted Brothers' firm ironically represented an outdated response to the problems of an urban landscape in the mid-1920s. Frederic Law Olmsted, Jr., while carrying on his father's efforts (and his prestige) since the late 1890s, nevertheless suffered from an increasing lack of relevance. As Peter Walker and Melanie Simo have noted, by the 1920s

> landscape architecture as an art of design suffered from a curious compromise. Formal, or clearly man-made, geometrically ordered design was taught along with naturalistic design. Both modes . . . were encompassed by the prevailing beaux-arts system of design education . . . but naturalistic design was viewed as a higher art than clearly man-made design.

131 *The New York World*, January 11, 1925, see note 57.

132 Conn, *Museums and American Intellectual Life: 1876–1926*, see note 49.

This attitude, Walker and Simo postulate, helped consign the profession "to the backwaters of an era increasingly marked by the qualities of modernity: rational, scientific, industrial, technological, efficient, [and] orderly."[133] Olmsted, Sr., a "Wordsworthian Romantic," with direct links to American utopianist traditions, had intended his designs to exercise harmonizing, tranquilizing, and refining effects upon the populace of the growing cities and to engender communal sentiments that would ameliorate class divisions.[134]

Well before the 1920s, however, when Lee and Olmsted began arguing for a return of the original Central Park ideal within the confines of the Botanical Garden, environmentalist Progressives had already turned from these traditional ideas to embrace alternatives such as the active playground and the district park. Sylvan scenes were no longer enough; playground settings for constructive, socializing, and regimented games would have to be managed by trained professionals.[135] Not even the inherent educational promise of the Botanical Garden's outdoor "museum" (or its indoor exhibits) would meet these public needs as they were enumerated by early-twentieth-century reformers. And yet, even these reform ideas were on the wane themselves by the 1920s.

Potentially "immoral," unregulated, public spaces (including amusement parks and saloons) had decades earlier begun fulfilling necessary social functions, and the idea of creating such mandatory "moral" public spaces had abated, falling victim to the growing power of mass culture.[136] Ironically, then, even as Lee attempted to portray himself as a scientific modernizer (one committed as a promoter of public health to the ideals of order and efficiency cited by Walker and Simo[137]), one-half of his Garden reform plan made recourse to an increasingly tradition-bound form of landscape art.

The history of the Botanical Garden's troubles in the 1920s cannot be considered outside of the wider context of reform, as well as the intellectual shaping of biological science, in which Lee and Britton struggled to implement contrasting visions of the Garden's future. *Any* of the plans proposed at that time were, as always, contingent upon the ability of the Garden to attract the necessary philanthropic support to implement them. Such funds, however, were not forthcoming in sufficient quantities to underwrite Lee's scientific agenda. In short, Lee failed to remake the Garden in his

133 Peter Walker and Melanie Simo, *Invisible Gardens: The Search for Modernism in the American Landscape* (Cambridge, MA: MIT Press, 1994).

134 Olmsted, by way of example, had been a friend of Socialist editor Parke Godwin, who in turn had been among the original Corporation members of NYBG in the 1890s. Boyer, *Urban Masses*, 238, see note 130.

135 Ibid. and Weyeneth, "Moral Spaces," 162–164, see note 129.

136 Ibid. (Wyeneth).

137 Walker and Simo, *Invisible Gardens*, see note 133.

own putative image. His ambitions had been based upon an essentially inaccurate view of the Garden's necessary role in botanical science, an insufficient understanding of the local opportunities for financial support, and an unrealistic and somewhat dated set of ambitions for the NYBG landscape within the city's public sphere. Given the highly competitive philanthropic arena in which the Garden contended, it is not surprising that the funds to do so were never secured, and that Lee's troubled NYBG career ended in acrimony.

An Epilogue:
The New York Botanical Garden, 1929–2002

In the summer of 1929, Nathaniel Lord Britton resigned as director-in-chief of The New York Botanical Garden (NYBG). During the Depression, E. D. Merrill, Britton's immediate successor, oversaw the Garden through several years of disastrously declining municipal appropriations, although the Works Progress Administration did provide NYBG with labor, including archival help to preserve photographs of the Garden's first decade. By the end of the Second World War, the Garden had an endowment of just under $3 million, a sum insufficient to support it as a highly active research institution. A 1945 fund-raising campaign with a target of over $6.5 million fell far short, netting less than $1.5 million from nonmunicipal sources. By 1949, the Garden did operate on a balanced budget, with an income and expenditures of slightly more than $730,000. Of that figure, $387,000 was received from municipal appropriations; $142,000 from investment income; $79,000 from foundations; $68,000 from membership dues and individual gifts; and $55,000 from sales of publications. But $440,000 of the Garden's budget was spent on maintenance alone, with only $222,000 available for both research and education. By 1950, total expenditures surpassed $1.3 million, with $940,000 received from the City of New York.[1]

The failed 1945 capital campaign necessitated a substantial revision of ambitious plans. Plans for a new administration building, whose projected costs rapidly climbed to $2 million, were scrapped in favor of renovation of the existing structure. All available funds were used simply for necessary repairs to the Garden's basic infrastructure, with nothing remaining to build a proposed separate laboratory building. As NYBG approached its fiftieth anniversary, professional financial analysts at the Rockefeller Brothers Foundation assessed its fiscal health as "sound but tight." Budgets were generally balanced, and the endowment principal, by tradition, was off-limits to meet day-to-day administrative expenses. With the failure of the postwar campaign, however,

1 "New York Botanical Garden—General Financial Situation," Memorandum of Robert C. Bates to Dana S. Creel, June 26, 1952, Rockefeller Foundation. Rockefeller Archive Center, Tarrytown, NY: Rockefeller Family, Record Group 2, OMR, Cultural Interests Series, Box III, Folder 2E.

Director William Robbins considered tapping a portion of the endowment to erect a new laboratory building.[2]

When Robbins solicited a donation from the Rockefeller family in the early 1950s, this money was earmarked for the construction of the proposed laboratory. Rockefeller Foundation executives, however, identified the Garden's overwhelming dependence upon city appropriations, which varied from year to year depending upon "political pressures," as a key weakness. Rockefeller fund analysts noted the NYBG's tradition of soliciting "special gifts" for specific purposes, a habit initiated by Nathaniel Britton in the 1890s as he financed individual expeditions and built the Library and Herbarium. Into the 1950s, such individual, idiosyncratic, and restricted donations supported the majority of the Garden's research agenda and bolstered efforts at beautifying the grounds with specific plantings. However, this tradition was insufficient, and the Garden was further handicapped by public presumption that an institution receiving city money was automatically solvent. Everyone concerned recognized that efforts to increase the endowment were necessary to enable NYBG to continue a serious research program. The Rockefeller Brothers Foundation pledged continued help in the hopes of setting "an example of particular significance in helping insure the institution's future." Such assistance, however, remained small-scale, dedicated to individual projects on a case-by-case basis, and individual members of the Rockefeller family provided no further significant support after the $500,000 granted the Garden by J. D. Rockefeller, Jr. in 1925. By 1953, the Garden had run its first budget deficit.[3]

By the mid-twentieth century, certain improvements to the NYBG grounds had been made with city help, and public eating facilities were opened for the first time in one of the last remaining structures from the Lorillard era, a snuff mill. Nevertheless, the Museum building, though structurally sound due to renovations paid for by city appropriations, was being used in ways Britton and architect Gibson had never intended: corners and hallways had been broken up by temporary partitions in an effort to expand working and administrative space. Research work, then concentrated in nutrition studies, bacteriology, plant pathology, and microbiology, would be the prime beneficiary of any increase in investment income from a larger endowment. The Garden, one member of the Board of Managers noted in 1953, "needs money for everything."[4]

2 Ibid. and Memo, "NYBG–General Financial Situation," R. C. Bates to Rockefeller Brothers Fund, April 8, 1953, Rockefeller Archive Center.

3 Memo, "NYBG–General Financial Situation," 1953, Rockefeller Archive Center.

4 Ibid. Such research was one of the reasons that a major funder of 1950s Garden research was the National Foundation for Infantile Paralysis. William Robbins identified research into polio vaccines, along with efforts to isolate compounds with contraceptive properties, as particular medical research topics for study by Garden scientists.

A decade later, Garden science, according to William Steere, Robbins's successor as director-in-chief, could find regular infusions of government money for research purposes from ecological studies of the Gaspé Peninsula to floristic studies of northern Alaska, Mexico, Central and South America, and the Western US. At the height of the Cold War, the Navy sponsored NYBG trips north of the Arctic Circle, as did the National Science Foundation. The Guggenheim Foundation, the American Philosophical Society, outdoor equipment manufacturers, and pharmaceutical companies such as Eli Lily also underwrote the cost of Garden field research. Shipping companies donated the cost of moving supplies, allowing its botanists to add between 30,000 and 40,000 specimens to the NYBG Herbarium annually. In the 1960s, the National Institutes of Health and the Navy also helped secure the Garden's earliest IBM computing equipment. Significantly, the separate laboratory building had finally been erected at the foot of the hill upon which Britton had installed the Museum. Yet, even as expeditions and laboratory work found this regular, nonmunicipal support, problems with "day-to-day housekeeping" costs, Steere noted, continued to plague NYBG. Municipal support was still not sufficient to "wash the windows, guard the collections, and keep people from picking the flowers."[5]

Taxonomic and herbarium work continued to comprise a large proportion of the NYBG's mission during the last decades of the twentieth century aided by a "Facilities Grant" from the National Science Foundation. By the 1990s, the herbarium had surpassed the five-million-specimen mark, but an estimated 95 percent of the earth's flora had yet to be studied, and approximately 100,000 plant species faced extinction. Indeed in the late twentieth century, the Garden was one of many botanical institutions that began highlighting the conservationist elements of its scientific mission. As Britton's work with the New York Academy of Sciences in Puerto Rican forestry indicates (see Chapter 6), such efforts were not necessarily new to the Garden's agenda Yet the Missouri Botanical Garden (which today defines its mission as an effort "to discover and share knowledge about plants and their environment in order to preserve and enrich life") and its New York counterpart began in the 1980s to use their traditional focus on taxonomic research not only for monographic, floristic, and economic purposes, but to implement species conservation and reintroduction programs. These efforts, in turn, utilized the horticultural expertise of the gardens to propagate large numbers of individual members of threatened species. The US National Academy of Sciences called in the 1980s for fast taxonomic surveys of world floras in order to

5 Gregory T. Hellman, "Onward and Upward with the Sciences: The Tail of Taxonomy" *The New Yorker* (May 30, 1964): 41–77. Computing technology at the time was such that the taxonomist who planned to use the equipment to facilitate his research remarked in 1964 that "there isn't a computer with enough space in its core memory to hold all the information I have to put into it."

document disappearing species; botanists at botanical gardens responded eagerly (although they recognized that naming a plant and the ability to propagate a few individuals is not equivalent to the ability to salvage an entire ecosystem). Just as the NYBG's neighbor, the Bronx Zoo, styled itself the "Wildlife Conservation Society," American botanical gardens, including NYBG, now underscore the utility of their research work in preserving genetic and ecosystemic diversity for public relations purposes.[6]

By the 1990s, as NYBG marked its first century and positioned itself as a guardian of the earth's floral heritage, it had established an off-site arboretum, maintained on a 2,000-acre estate donated by a Hudson River Valley benefactor. Over half a million visitors passed through the gates of the now fully enclosed Garden annually, arriving to stroll through rose gardens, seasonal displays of various species, and the still-preserved hemlock forest that had attracted Britton's attention in the 1890s. In addition, New Yorkers and foreign visitors alike were drawn to more recent crowd-pleasing innovations, not uncommon to late-twentieth-century public institutions: outdoor sculpture (on loan from the collection of the Museum of Modern Art, undergoing renovation in 2002); interactive children's gardens, replete with eye-catching plantings as well as electronic kiosks displaying educational videotapes that highlight the Garden's role in conservation; and demonstration gardens designed to offer inspiration and practical advice to the home ornamental and vegetable gardener. Some areas of NYBG are even available to couples eager to rent a parcel of beautifully landscaped open space for weddings. And yet, the traditional Garden attractions, the exotic, humid or arid interiors of the vast glass Conservatory, are still drawing visitors (at a separate fee from the entry price) into simulated versions of faraway ecosystems. If the static taxonomic displays housed within row after row of wooden display cases are long gone (the casualty of the need for work space), visitors are still drawn by the chance to see, within a short subway or train ride from midtown Manhattan, the spoils of plant expeditions to distant lands throughout the old and new worlds.

On May 1, 2002, a brilliant spring day in the New York metropolitan area, NYBG dedicated its new Library quarters before an assembled crowd of city and borough dignitaries, international scientists, and interested friends. After just over a century, the Library was actually returning to its original home atop the Museum building erected at the turn of the twentieth century. The area under the building's dome, long inaccessible to the public, was once again an integral part of the Library

6 Brendan Gill, "The New York Botanical Garden at 100" *Architectural Digest* (November 1991): 116 and Peter Davis, *Museums and the Natural Environment: The Role of Natural History Museums in Biological Conservation* (London: Leicester University Press, 1996): 230–244.

space. The Library itself had been modernized, with internet terminals; an online catalog; and a permanently reserved, climate-controlled, exhibition space in which the Garden's first-rate collection of rare botanical volumes, maps, and prints could be displayed in rotation to the public. Along with the Library, the Herbarium, now numbering over 7.2 million specimens, had moved into new headquarters in a state-of-the-art International Plant Science Center.

Gathered on the steps of R. W. Gibson's Museum building that day, New York City and State politicians, including the City Council speaker and former New York Senator Daniel P. Moynihan, addressed the assembled crowd in words that echoed those used a century earlier. Speakers hailed the Garden as a refreshing green space within the urban precincts of the Bronx. They also highlighted NYBG's role as a public educator, citing the various programs in horticulture, botanical art, and children's programs run by the NYBG staff. Finally, given the setting, it was only natural that each person at the podium that day paid homage to the scientific research and graduate-level education pursued by the Garden's botanical staff. [7]

The roster of corporations and foundations that had helped fund the construction of these new scientific spaces included many of the names often associated in the twentieth century with cultural philanthropy in New York City, including the Andrew Mellon Foundation and J. P. Morgan and Company. In addition, since the Garden's research agenda had grown to include molecular biological research, as well as continued conservation work on the flora of tropical regions, the US Department of Agriculture, the National Science Foundation, and Pfizer had all contributed funds to create a modern botanical-research institution. Landscaped public space, popular and professional education, and scientific research, all efforts begun in the 1890s, continued to be the institutional missions most readily recognized by the Garden's public, private, and political supporters. Broadly construed, then, NYBG's purpose had remained the same. Of course, during the 70 years since Nathaniel Britton's retirement, there had been shifts in emphasis by subsequent Garden administrations to match the changing contexts of botanical science, city politics, and the institutional landscape of city philanthropy.

The Garden as an Institution

Institutional histories are often indistinguishable from biographies of their key figures.[8] Histories of Harvard's Museum of Comparative Zoology, the Smithsonian Institution, the US Geological Survey, and the US Naval Observatory all attest to this

7 The same themes had been sounded by the city's "master builder," Robert Moses, who, in 1953, had lauded the Garden as a "picturesque" rival to Central Park, worthy of support as a center of scientific inquiry. See *The New York Times*, May 5, 1953.

8 Sally Gregory Kohlstedt, "Institutional History," *Osiris* 2nd Series 1 (1985): 17–18.

tendency. The success and priorities of each institution were intimately tied to the crucial actors involved in its creation. Studies of individual scientists, in tandem with investigations into the history of their associated institutions, provide "an intellectual coherence, [with] a specific origin and identity," often preserved within a single, "major repository."[9] Such has been the case with this partial history of NYBG. The career of Nathaniel Britton, a remarkable scientific and philanthropic entrepreneur in an era of remarkable institution-building, provides a neat parallel and an organizational structure to a historical inquiry into the origins of New York City's largest botanical research institution and public garden.

And yet, as historian Sally Kohlstedt notes, compilers of institutional histories must proceed with caution. Authors who rely too heavily upon the rhetoric and biographical detail of participants in order to discern the process of institutional development risk creating a flawed historical narrative. The stated intentions of institutional founders are not always the most reliable guides to the historic development of specific institutions. Although Britton sought to create a Garden with undeniable echoes of established British institutional botany, his contemporary public was not easily convinced of the necessity of NYBG's role in Bronx life. The intentions of the public are parts of any rich institutional history,[10] as illustrated in Chapter 4 for that very real Garden public.

The Kew model nevertheless remains instructive, at the very least as a convenient literary foil. NYBG got started later than its English model, at a time when Kew and other British gardens entered a new century in which the British empire declined. Britton's New York counterpart, in contrast, allied itself with a rising and expansive US empire. Thus, as in the case of Puerto Rico, Britton could organize a natural history survey by making an appeal to both local (New York and Puerto Rican) and national interests.[11]

A crucial distinction between Kew and NYBG needs mention, however. In the case of New York's botanical garden, imperial concerns, i.e., the conquest of exotic alien nature in the cause of colonial domination and economic exploitation, were decidedly less important than the role played by the city's plutocracy. The ideal of urban display, the provision of landscaped public space, a conservatory stocked with the exotic spoils of well-chronicled expeditions, and publicly endowed scientific research, were all more significant in NYBG's story than American empire. The most basic of utilitarian roles in scientific botany were often fulfilled by federal bureau-

9 Ibid.

10 Ibid., 23.

11 Of course, even though the overall fortunes of the British Empire declined through two World Wars (and Britain lost many colonial botanic stations), Kew remains one of the world's most important botanical gardens.

cracies, including the US Department of Agriculture, thus freeing Britton to some extent to pursue the requisite project of this mandate given him by the city's elite. The Garden's initial supporters, including, crucially, Gilded Age financiers, acting largely out of civic motives, gave Britton the wherewithal to build his "American Kew." They sought to create a public urban landscape commensurate with their collective vision of the city's role in the national and international political, scientific, social, and economic spheres. In addition, another local institution, Columbia University, recognized the extent to which a scientific institution with a public endowment and dedicated to the maintenance of basic scientific resources would benefit its own development. Columbia thus gladly played a critical role in creating the Botanical Garden.

Nevertheless, like Kew, NYBG became a critical haven for research in systematic botany. From his education in the 1880s through his retirement from the Garden in 1929, Nathaniel Britton remained a classic herbarium botanist. Although the last decade of his working life played out within the context of a great deal of intellectual tumult in the life sciences, he had no role in the forging of the evolutionary synthesis. His importance to the history of science is as an institution-builder. In the twenty-first century, the Garden's botanists, through taxonomic research (conducted by varied methods, including molecular biology), continue to identify and catalog thousands of species of plants. These botanists and horticulturists outside the confines of university biology departments serve as scientific service workers, identifying poisonous plants, studying plant diseases, cataloguing endangered species, and exploring the potential pharmaceutical uses for the various species of plants collected on Garden-sponsored expeditions. Although Britton's traditional brand of taxonomic research temporarily lost prestige during his career, nevertheless, he succeeded in creating a permanent and ultimately prosperous space for the pursuit of classic herbarium botany. In this sense, then, Elizabeth Britton's goal of an "American Kew" was fulfilled finally in the decades after her death. During the post-World War II era, NYBG's many and varied collecting expeditions enabled the US to fill the space left open by a diminished British botanical empire, and these economic botanical concerns (seeking reliable sources of cinchona, for example) helped bring the Garden to maturity.

As the Garden grew into its public role, it faced challenges to its institutional mission, usually in the form of financial pressures. In sum, the Garden is the product of the desires of a wealthy elite to showcase New York's urban grandeur united with the desire of a late-nineteenth-century New York botanist to continue the scientific institution-building of a century's worth of botanical predecessors. Following in the organizational footsteps of nineteenth-century learned and horticultural societies, NYBG became the institutional locus for the practice of a middle- and upper-class botanical, horticultural, and ornamental science and was situated physically within surroundings designed to recall both the Old World setting of its British model and the Olmstedian mandate of urban spaces such as Central Park. That space, however, remained contested by an urban public not entirely willing to cede control over a local working landscape. The Garden also served as the venue for the playing out of an important and little-studied American scientific career: that of Nathaniel Britton. Convinced of the necessity for establishing American scientific independence from European systems and repositories of knowledge, Britton used the Garden as a vehicle to carve out a specifically "American" space for the practice of New World botany. From his insistence upon US (NYBG) primacy in the study of characteristically "American" plants such as the cactus to his strategic willingness to marshal NYBG's institutional resources in the effort to expand US influence in the Caribbean, Britton furthered both his own career and that of the institution he helped to establish. In doing so, he managed to become, despite contemporary criticism and intermittent public funding, one of the preeminent institutional entrepreneurs in New York City's history.

Index

The letters f or n following a page number refer to a figure or note on that page. The number following the n is the note number.

Memoirs of The New York Botanical Garden

(ISSN 0077-8931)

The Memoirs are published at irregular intervals in volumes of various sizes and are designed to include results of original botanical research by members of The New York Botanical Garden's staff, or by botanists who have collaborated in one or more of The New York Botanical Garden's research programs. Ordinarily only manuscripts of 100 or more typewritten pages will be considered for publication.

Manuscripts should be submitted to the Editor. For further information regarding editorial policy and instructions for the preparation of manuscripts, contact the Editor.

Orders for published and forthcoming issues and volumes should be placed with:
The New York Botanical Garden Press
200th Street & Kazimiroff Boulevard
Bronx, New York 10458-5126 USA
Telephone: 718.817.8721 • Fax: 718.817.8842
E-mail: nybgpress@nybg.org
Web site: http://www.nybgpress.org

Britton's Botanical Empire

The New York Botanical Garden and American Botany, 1888–1929

PETER MICKULAS

Memoirs of The New York Botanical Garden, Volume 94

Copy Editor

Eileen K. Schofield

Book and Jacket Design

Gloria Shen

Typeset in Dante and Scala Sans

Manufactured by Thomson-Shore, Inc. and printed on 60 lb. Text Nature's Natural paper